I0064598

# ANATOMY OF THE RAT

### Eunice Chace Greene

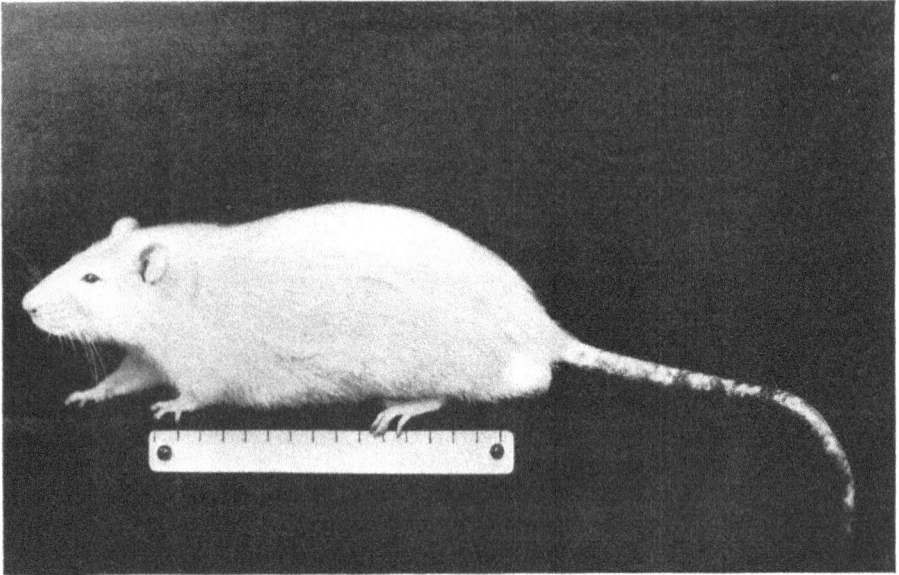

Mus norvegicus albinus.
Extracted Albino Rat (Seventh generation inbred)
Dr. King's Colony of the Wistar Institute.
Male.   Age 521 days.   Weight 551 grams.
Reduced approximately ⅔. Scale in Cm.

# TRANSACTIONS

OF THE

# AMERICAN PHILOSOPHICAL SOCIETY

HELD AT PHILADELPHIA

FOR PROMOTING USEFUL KNOWLEDGE

---

NEW SERIES—VOLUME XXVII

---

## ANATOMY OF THE RAT

Eunice Chace Greene

---

PHILADELPHIA

THE AMERICAN PHILOSOPHICAL SOCIETY

104 South Fifth Street

1935

Copyright 1935
American Philosophical Society
Philadelphia

MANUFACTURED BY THE HADDON CRAFTSMEN, INC.
CAMDEN, NEW JERSEY, U. S. A.

# CONTENTS

# FOREWORD

This book is both an essay in Comparative Anatomy and also a guide to the dissection of the Albino rat.

During the past 40 years the use of the Albino or of the Pied rat for laboratory purposes has steadily increased and experiments involving operations have become more and more numerous.

The need for an anatomy of this animal is thus indicated. The present study was begun more than a decade ago. Despite handicaps it has been persistently pursued and so brought to its present form. It represents the most complete anatomical presentation of any of our laboratory animals and further is unique in that both the figures and the text are the work of the author, an unusual combination.

It is hoped that it will prove of value not only as an introduction to the Comparative Anatomy of the mammal but also in those operative studies in which the rat is used as the test animal.

HENRY H. DONALDSON

# PREFACE

At the present time there is no general treatise on the normal anatomy of the rat with the exception of Martin and Moale's "Vertebrate Dissection, Part III, How to Dissect a Rodent" (1884), and Hunt's "Laboratory Manual of the Anatomy of the Rat" (1924). The former is, of course, too elementary to be considered adequate, and the latter is designed solely for student use. A third book, Howell's "Anatomy of the Wood Rat" (1926), while serving as a valuable reference is obviously not directly applicable to the albino rat, and furthermore, treats the subject mainly from the point of view of muscle function, omitting entirely the circulatory, nervous, and endocrine systems. All three books differ radically in aim and scope from the present work, which is designed in no sense as a text book, but as a reference book or atlas for research workers, instructors and advanced students in Comparative Anatomy. It is hoped that it may prove of value either for the dissection of a mammal, for operative work, or for pathological studies. For ordinary laboratory use the superficial dissections will be found sufficient.

With these ends in view, and presupposing a general knowledge of mammalian anatomy on the part of the reader, the author has attempted to show the most important gross anatomical structures by numerous drawings and to avoid lengthy descriptions wherever possible. Structures not visible to the naked eye have been omitted as not coming within the scope of the present work. It will be readily understood, however, that for purposes of careful dissection, especially of the nervous and circulatory systems in so small an animal, some slight magnification was highly desirable. The lowest power obtainable with the binocular dissecting microscope was therefore freely used in making the preparations.

The author has been impressed throughout the work by the close structural similarity of the rat to man and by the fact that only the texts on human anatomy were useful when it came to the identification of details of structure. In many cases relationships which occur as variations in man are found to be the usual condition in the rat. This is undoubtedly due in large measure to the difference in posture. In a few of the cases where such relationships occur, they have been described in more detail. In fact man has been made the basis of comparison throughout. The author is however fully cognizant of the fact that the standard works on human anatomy are based upon large numbers of individuals and that many more rats must be examined before it can be stated finally what is the usual condition and what are the variations. For the present purpose it was deemed sufficient to make some preliminary dissections, then to select not less than ten animals, usually more males than females, and make a thorough dissection of each. This was done for each system and the drawings show the conditions found in the majority of these specimens. Care must be taken, therefore, not to jump to the conclusion that variations from the condition described, or shown, are necessarily anomalies. In general the variations are few and, when they occur, slight.

Treatment of the subject from the standpoint of regional anatomy was considered, but this method of approach is only satisfactory when an anatomical text book is available for reference. It has therefore seemed more appropriate to follow a systematic arrangement, and thus to make the information more easily accessible for other investigators. In view of this the arrangement of chapters follows the order commonly found in the standard text books of human anatomy. Within the chapters the figures follow the pages of description and are arranged, as far as possible, in the order of head, trunk, limbs, and from superficial to deep.

In the index, where reference to several figures is given, the more important ones are shown in heavy type.

The bibliography includes not only anatomical references dealing directly with the rat, but also those dealing in a more general way with comparative anatomy, together with the familiar texts of human anatomy which have served as a constant source of reference in the present study.

The B. N. A. has been used as far as possible. Following Emmel's 1927 edition of this nomenclature, some terms have been anglicized in accordance with the "majority of the standard English and American anatomical texts." In many instances terms taken from comparative anatomy or from the older nomenclature are given in parentheses following the B. N. A. terms.

For a few drawings where size seemed important, for example in the skeletal system and for some of the endocrines, the camera lucida was used. The majority of drawings are freehand and may be considered semi-diagrammatic inasmuch as there is doubtless some slight distortion where parts are drawn aside to show relationships more completely. For permission to use the diagram of the lymph channels, I wish to thank Dr. Thesle T. Job of Loyola University.

As far as practicable, labels have been placed on each figure but where these are too numerous and were in danger of obscuring the details they have been removed to one side or to adjacent key figures.

The specimens used were all albino norway rats (Mus norvegicus albinus) from the rat colony of the Wistar Institute. Some of the specimens used for the study of the circulatory system were injected by Dr. F. D. Lambert, of the Botany Department of Tufts College. The rest of the material used for this purpose was injected by Dr. I. E. Gray at the Supply Department of the Marine Biological Laboratory, Woods Hole, Massachusetts.

The majority of the material was preserved in formalin. It was found desirable to tie the animal in an extended position on coarse wire netting before immersing in formalin to prevent the contraction of the ventral and limb muscles during the hardening process, thus greatly facilitating the dissection.

For the studies of the vascular system either a starch or gelatin injection mass such as is commonly used in most laboratories, was employed. The animals were injected through the heart or the femoral vessels.

In a smaller group of animals a different method was used. The animals were injected through the femoral artery with enbalming fluid followed by a non-bleeding arterial injection mass, Murphy's artery-red lacquer. For such a small animal this was diluted slightly with the usual lacquer thinner. The specimens were then stored in a tank and covered with cloths soaked in 2% carbolic. This gives a flexible specimen which makes for greater ease in dissection but is, in the author's opinion, not so satisfactory for visceral studies as the formalin method which leaves the viscera in a firmer condition.

This work was begun at the suggestion of Dr. Harris H. Wilder of Smith College, by whose death the author lost a most valued friend and teacher. The chapters on muscles and skeleton were done under his direction. These chapters were then submitted to Dr. M. J. Greenman, Director of the Wistar Institute, and at his suggestion it was decided to complete this study with the idea of publishing an anatomical atlas of the albino rat. The remainder of the work was carried on under the guidance of Dr. Henry H. Donaldson to whom more than to anyone else I am deeply indebted for his continued support of the work and for his constant encouragement and advice. To Dr. Greenman and Dr. Donaldson I wish to express my appreciation of their generosity in supplying material from the colony of the Institute, and in extending laboratory privileges at the Institute and at Woods Hole.

I am also under obligation to Dr. Harold D. Senior for invaluable suggestions and criticisms with regard to the circulatory system.

Further acknowledgment is also due Dr. Wayne J. Atwell of the School of Medicine of Buffalo University, and Dr. Henry W. Stiles of the College of Medicine, Syracuse University, for the use of laboratory facilities.

To my husband, Walter F. Greene, without whose generous and unfailing interest this work would never have reached completion, I am especially grateful.

Finally I wish to acknowledge with gratitude the publication of the present work by the American Philosophical Society.

These conditions have made it possible to produce an anatomy of the rat which is more complete than that available for any other of the lower mammals, and the results should have a value not only for the anatomist, but also for an interpretation of the evolutionary development of the mammalian series.

Syracuse                                                        EUNICE CHACE GREENE
    1935

# CHAPTER I

## GENERAL APPEARANCE

### (Figs. 1–5)

Mus norvegicus albinus (Frontispiece) is an albino mutant from the wild gray form (Mus norvegicus) which is represented in the laboratory of the Wistar Institute by several color varieties of which the albino is most common. The albino race differs from the gray race in the relative development of some of the viscera (Donaldson '24). The albino has a span of life of about three years, and is nearly completely grown at six to eight months of age, when the male may weigh 250 grams and the female somewhat less. When fully mature the albino rat weighs from 200–400 grams.

The description which follows applies to the more conspicuous features of the external appearance, as they appear in the mature animal.

The albino is characterized by small and thick ears and the length of the tail is about 85% of the length of the body, being relatively a little longer in the female than in the male. The fact that the tail is somewhat shorter than the body serves to distinguish this species from the house rat (Mus rattus) in which the tail is distinctly longer than the body.

Fraser ('31) who has made a careful histological study of the hair pattern states, "On each surface there are two conspicuous lengths of hair, the longer being the more sparse. The hairs are arranged in two classes of groupings, that with and that without a long hair, which is usually quite centrally located. According to their location within a group, follicles may be divided into three kinds. They may be central; adjacent, grouped with a central; or associated, grouped without a central follicle. Follicles other than the central are mostly of composite nature, that is, there is more than one hair to a follicle. The central follicle is only occasionally composite."

Butcher ('34) who has worked out the hair cycles in the rat, confirms Fraser's account and goes on to state that, "hair growth in the young albino rat is cyclic. These cycles occur approximately every 35 days, the resting period and the growing stage each being about 17 days in length. The elongated follicles of the growth period in the first cycle in the dorsum last from birth until the sixteenth or seventeenth day of life. Quiescence during which the follicles are inactive and short, then continues until the thirty-second to thirty-fourth day when active growth and lengthening begin again. It is remarkable that practically all the follicles are in the same stage at the same time and that the resting condition is established within a very short interval."

"The hair cycles in the prepubertal rat parallel the growth activity in the ovary. The gonad, however, is not responsible for this hair growth, since the hair cycles occur in ovariectomized animals as in normal rats."

"In the rat new hairs in their growth do not normally push out the old hairs.

1

Of course, this results in an additional hair in the follicle with each new growth. Finally, a follicle with several hairs, a composite follicle is established. Composite follicles are not present in the first coat."

It is further shown that "the growth and quiescent stages in the venter are quite similar to those in the dorsum, with the exception that the comparable condition in the venter is 4 or 5 days in advance of the similar stage in the dorsum. In other words growth begins first in the venter and spreads dorsally. Likewise, quiescence in the venter would be observed 4 or 5 days before there was any evidence of it in the dorsum."

In certain regions the bristles are developed as long vibrissae (Fig. 1). In the rodents these are arranged in four groups, and in accordance with Pocock's terminology (1914), may be described as follows. The *buccal* vibrissae are those around the muzzle. These are divided into two groups, the *mystacial* and *submental*. The mystacial in the rat grow in five or six parallel rows, running from the nose backward along the upper lip. The number in a row varies from five to ten. Some of this group are longer than the whole head of the animal. The skin in this region is very thick to accommodate the exceptionally large hair follicles, and forms what is known as the vibrissae or mystacial pad. Inasmuch as the vibrissae serve as tactile organs they are richly supplied with nerves, branches of the maxillary division of the trigeminal. The *submental* group of vibrissae, on the chin, is composed of two rather definite rows and a few smaller more scattered vibrissae. In addition to these there are usually three just above the eye, the *superciliary*, while the *genal* group is doubled and represented by a single bristle a little below the outer corner of the eye, and two more near the corner of the mouth. The *interramal* group consists of an unpaired tuft of three or four vibrissae in the midline between the rami of the mandibles.

The eyelids are well developed, and except for the nictitating membrane (supported by a semilunar cartilage) in the medial corner of the eye, only the cornea is visible. The eyelashes are very fine and short, while the tarsal, or Meibomian glands, are large and readily seen. By rather frequent winking, the eye is kept moist with secretions from the lacrimal and Harderian glands.

The nares, which are shaped like a reversed comma, open laterally and may be closed under water. Unlike the rat, most Murine and Sciurine rodents have the nostrils separated by a vertical groove. In this animal, however, the vertical groove begins just below the nares and forms a cleft in the upper lip, exposing the superior incisors, even when the mouth is closed. Back of the incisors the hairy lip extends inward toward the midline, and the opposite halves practically meet, thus excluding the upper incisors from the mouth cavity proper.

Both fore and hind feet are pentadactyle (Figs. 2, 3, 4, 5). The "thumb" or pollex (Fig. 4) is much reduced in the forefoot. Nevertheless it must be considered a distinct digit since it has a fully developed nail which is peculiarly flattened unlike the nails of the other digits. Furthermore it has two perfectly distinct phalanges as shown in serial sections through the paw. Typical walking pads are evident (Fig. 4), five apical pads on the tips of the digits, three interdigital pads on the palm (the first interdigital seems to have disappeared with the reduction of the pollex), and two at the base of the palm, the hypothenar pad on the ulnar side and the thenar pad on the radial side. The hind foot (Figs. 3, 5) has its full quota, five apical, and four interdigital pads, with two more, one medial and one lateral, in the region of the metatarsals.

The palms and soles are devoid of hair but the backs of the paws are sparsely covered with short thin hair.

The tail, which is actively used as a balancing organ, relative to the body length is shorter than in Mus rattus. Scales overlapping like shingles cover the tail, while the hairs are reduced to three rather short bristles emerging from under the edge of each scale.

In the female the teats of Mus norvegicus are usually twelve in number, three pairs in the pectoral and three pairs in the abdominal region (Fig. 109). Just anterior to the anus is the vagina which is closed by a plug until the time of puberty, which occurs on the average between 70 and 90 days. Anterior to the vagina is the orifice of the urethra at the base of the clitoris which is enclosed in a small prepuce (Fig. 109).

In the male the testes descend between the 30th and 40th day into the thin-walled scrotal sacs. Externally the sacs are not distinctly separated, but beneath the skin the division is complete. On the tip of the scrotal sacs the hair is sparse and fine. The penis is enclosed in a loose sheath, the prepuce, and upon being extruded is found to have a single cartilaginous or bony process, the *os penis* (Figs. 92, 280), on its ventral wall.

## LIST OF FIGURES

Frontispiece—Mus Norvegicus Albinus

### I. GENERAL APPEARANCE

Fig.

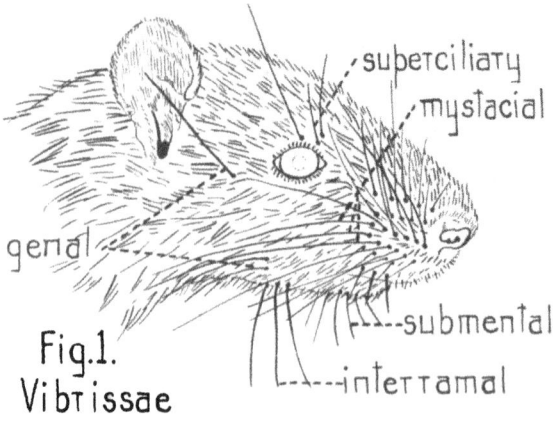

Fig.1.
Vibrissae

superciliary
mystacial
genal
submental
interramal

flat
nail
plx

Fig.4.
Drawing of
Right forepaw

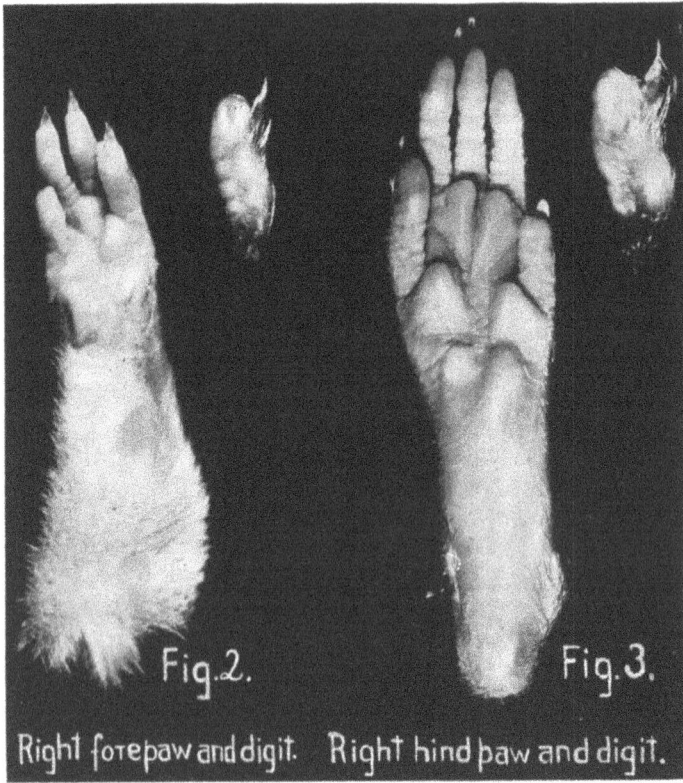

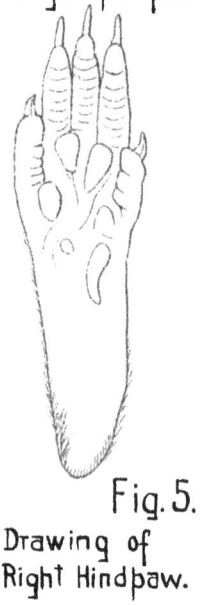

Fig.5.
Drawing of
Right Hindpaw.

Fig.2.
Fig.3.

Right forepaw and digit.    Right hind paw and digit.

# CHAPTER II

## SKELETON
### (Figs. 6–50)

The following order has been followed:

Axial skeleton
Skull
Vertebral column
Thorax

Appendicular skeleton
Anterior appendage
Posterior appendage

Descriptions and figures are based on adult but not senile specimens. In senile material some fusion occurs as in man, although in the long bones there is scarcely a time when the line of diaphysis and epiphysis is not visible.

An attempt has been made to show as many of the essential features as possible by drawings without resorting to lengthy descriptions. A few points will, however, need further elucidation, especially where the structure or arrangement differs markedly from human anatomy.

In all figures, names of bones are capitalized to distinguish them from names of features.

## AXIAL SKELETON
### SKULL
### (Figs. 7–20)
#### TABLE OF COMPARISON OF THE BONES OF THE RAT AND HUMAN SKULL

| *Rat* | *Human* |
|---|---|
| 2 Nasals (Figs. 7–10) | 2 Nasals |
| 2 Premaxillaries (Figs. 9–12) | |
| 2 Maxillaries (Figs. 7–12) | 2 Maxillaries |
| 2 Zygomatics (Figs. 7–12) | |
| 2 Palatines (Figs. 11–12) | 2 Palatines |
| 1 Vomer (Fig. 14) | 1 Vomer |
| 2 Lacrimals (Figs. 7–10) | 2 Lacrimals |
| 1 Ethmoid (Figs. 13–18) | 1 Ethmoid |
| 2 Frontals (Figs. 7–10) | 1 Frontal |
| 1 Basisphenoid (Figs. 9–12) | |

5

| *Rat* | *Human* |
|---|---|
| 1 Presphenoid (Figs. 9–12) | 1 Sphenoid |
| 2 Parietals (Figs. 7–10) | 2 Parietals |
| 1 Occipital (Figs. 7–12) | 1 Occipital |
| 1 Interparietal (Figs. 7–10) | |
| 2 Squamosals (Figs. 7–12) | |
| 2 Periotic capsules (Figs. 9, 10, 12) | 2 Temporals |
| 2 Tympanic bulla (Figs. 9–12) | |
| 6 Auditory ossicles (Fig. 154) | 6 Auditory ossicles |
|    2 Malleus |    2 Malleus |
|    2 Incus |    2 Incus |
|    2 Stapes |    2 Stapes |
| 4 Turbinates (Fig. 15) | 2 Turbinates |
|    2 Naso-turbinals | |
|    2 Maxillo-turbinals | |
| 1 Hyoid (Fig. 21) | 1 Hyoid |
| 2 Mandibles (Figs. 19, 20) | 1 Mandible |

The *craniopharyngeal canal* (Figs. 11–13) pierces the basisphenoid bone. According to Donaldson, ''the canal is patent, in the albinos, in from 3 to 5 percent of the cases. In addition, the foramina can be located in another 3 to 5 percent, and these might be considered patent if you could get bristles slender enough, and at the same time strong enough, to be pushed through the canal. There are all gradations from this to complete disappearance of foramina, both endo-cranially and on the outside.''

VERTEBRAL COLUMN
(Figs. 6, 22–27)

The vertebral column consists of seven cervical, thirteen thoracic, six lumbar, four sacral, and from twenty-seven to thirty caudal vertebrae. Since a typical vertebra of each region has been shown, a few features only, deserve mention here. The spinous process of the second thoracic vertebra is longer than that of any other vertebra and is further extended by a small triangular piece, which articulates with it.

The *sixth* cervical vertebra displays a marked peculiarity. Extending ventrally and slanting caudally from the transverse process is a thin plate of bone, quite thin medio-laterally but in its antero-posterior dimension almost as broad as the centrum and with its ventral edge somewhat thickened. It lies directly ventral to the vertebrarterial foramen. This *carotid* or *Chassaignac's tubercle* is apparently an enlarged costal element and would seem to correspond to the anterior tubercle of the transverse process in man. No hint of such a process appears in either of the adjacent vertebrae. Howell (1926), calls this an inferior lamella and states that, ''the longus colli muscle extends both ways from this process and may be chiefly responsible for the variation which it exhibits, but the latter is believed to be phylogenetic as well.''

In the *seventh cervical* vertebra a vertebrarterial foramen may be small or lacking.

The *sacro-iliac joint* is made by the first two sacral vertebrae with the auricular surface of the ilium.

*Chevron bones* are present as far caudad as the fifth or sixth vertebra from the tip of the tail (Fig. 6).

## THORAX

Numerous bony elements make up the thoracic basket (Figs. 31, 32). It is conical in shape, smaller at the anterior end, and larger posteriorly. In cross section it is practically circular. Dorsally it is composed of the thirteen thoracic vertebrae and the dorsal extremities of the ribs, while the dorsal segments of the ribs form the lateral walls, and the ventral segments, with the sternum, constitute the ventral elements.

Beginning with the sixth thoracic vertebra there is apparently a slight differentiation of the transverse process into (1) tubercular articular process for the rib articulation, and (2) a more dorsally located process, a metapophysis, which in the vertebrae of the posterior region of the thorax becomes more and more closely associated with the anterior surface, until, in the last three thoracic vertebrae, this process and the anterior articular process are indistinguishable.

The dorsal segments of the ribs are, in the case of the rat, completely ossified, while the ventral elements are calcified, thus the animal has no true costal cartilages.

## APPENDICULAR SKELETON

### ANTERIOR APPENDICULAR SKELETON

#### (Figs. 28–30, 33–40)

*The clavicle and associated elements* (Figs. 29, 30). Attaching the lateral end of the clavicle to the acromion process of the scapula is a thin disc of cartilage, usually considered the distal portion of the primitive procoracoid, and called by Parker the mesoscapular segment. At the medial end is a disc of cartilage which is also a vestige of the procoracoid cartilage. There thus extends between the sternum and the scapula a chain of skeletal elements: omosternum, proximal procoracoid piece, clavicle, distal procoracoid piece (mesoscapular segment of Parker). Of these the clavicle and omosternum are ossified in the rat, the others remaining cartilaginous. At this point I wish to acknowledge information received from Dr. Alden B. Dawson based on his histological studies, verifying the above statement. Dawson finds that, "the omosternum begins to ossify at three months and at the end of the fourth month, ossification has proceeded to considerable extent."

The *humerus, ulna,* and *radius* (Figs. 33–38), present no marked peculiarities or variations from the general mammalian type.

The *carpus* (Figs. 39, 40). A series of nine rather nodular bones, with more or less irregular surfaces and flattened articular areas, forms the carpus. Some fusion has taken place and certain modifications from the more generalized primitive condition have occurred. The following table will serve as a comparison of the carpal bones of the rat with the more generalized form of primitive vertebrates and with that of man.

| | Primitive | Rat | Human | B. N. A. |
|---|---|---|---|---|
| Proximal row | Radiale | Navicular ⎫ Fused (Scapho- | Navicular | Cs naviculare manus |
| | Intermedium | Lunate ⎬ lunar) | Lunate | Cs lunatum |
| | Ulnare | Triangular | Triangular | Os triguetrum |
| | Carpale I | Greater multangular | Greater multangular | Os multangulum majus |
| Distal row | Carpale II | Lesser multangular | Lesser multangular | Os multangulum minus |
| | Carpale III | Capitate | Capitate | Os capitatum |
| | Carpale IV | Hamate | Hamate ⎫ Fused | Os hamatum |
| | Carpale V | Centrale | Centrale ⎭ | Os centrale |
| | Centrale | | | |
| | | Pisiform ⎫ Sesamoids | Pisiform Sesamoid | Os pisiforme |
| | | Falciform ⎬ | | Os falciformis |
| | | Ulnar sesamoid | | |

If we except the sesamoids, the proximal row is reduced to two bones by the fusion of the navicular and the lunate, which may best be designated as scapholunar. The centrale is present in the rat, interposed between the proximal and distal rows. It does not extend through to the flexor side of the carpus. The pisiform has a shallow depression which rests on a rounded articular surface of the triangular. On the ventral surface of the wrist it has a prominent knob-like projection which helps to form the almost complete circle of bone which encloses the tendons of the wrist. The falciform is a narrow slightly curved bone, extending obliquely across the wrist, thus forming a bar across the flexor surface, under which the flexor tendons of the digits pass to their insertion. There is a second ulnar sesamoid in the rat, imbedded in the thick ulnar pad of the palmar surface. Ordinarily this is lost in dissection, but by rendering the specimen transparent with the use of caustic potash and glycerine, and staining the bones in situ with alizarine crimson (Schultze's method), this sesamoid remains in place and is shown in the accompanying camera lucida drawing (Fig. 40).

On the flexor surface of each metacarpophalangeal joint are two sesamoids. Each digit with the exception of the first, is composed of three segments, the phalanges. On the flexor surface, at the distal end of the middle phalanx, is a single sesamoid. The first digit is made up of only two phalanges, as is the case in typical mammals, and the sesamoid lies at the base of the terminal phalanx (Fig. 40).

### POSTERIOR APPENDICULAR SKELETON
(Figs. 41–50)

The *femur* (Figs. 45, 46). On the flexor surface of the distal end of the femur, just proximal to the condyles, are two small but well-defined articular surfaces for the sesamoid bones, the lateral and medial fabellae, which are developed in connection with the tendons of origin of the heads of the triceps surae. The semilunar cartilages of the knee joint are also ossified in the rat.

The *tibia and fibula* (Figs. 47, 48). It is necessary to consider the bones of the lower leg, the tibia and fibula, together, as they are partially fused, even in very young specimens. The epiphysis of the proximal end of the tibia, even in specimens where it is completely fused with the diaphysis, is easily observed as the line of demarcation remains always distinct. The fibula articulates proximally with the tibia. Distally it fuses with the latter for a short distance, but beyond the line of complete fusion the distal extremity of the fibula regains complete freedom and its distal end is quite distinct, projecting slightly beyond the tibia. Distally where the bones separate, a slight ridge appears on the lateral surface of the fibula and ends in a sharp projection, the lateral malleolus, under which the tendons of the peroneal muscles pass to the foot. The distal end of the tibia forms the medial malleolus.

The *tarsus* (Figs. 49, 50). Eight irregular bones, varying greatly in size, and arranged in two parallel rows, comprise the tarsus. Very little fusion has taken place, and the principal modification is the result of the enlargement of the two bones of the proximal row, the talus and the calcaneus.

The following table will serve as a means of comparison with the more generalized condition of the tarsus of the primitive vertebrates and with that of man:

| | *Primitive* | *Rat* | *Human* | *B. N. A.* |
|---|---|---|---|---|
| Proximal row | Tibiale | Tibiale | | |
| | Intermedium | Talus | Talus | Talus |
| | Fibulare | Calcaneus | Calcaneus | Calcaneus |
| Interposed | Centrale | Navicular | Navicular | Os naviculare pedis |
| Distal row | Tarsale I | 1st Cuneiform | 1st Cuneiform | Os cuneiforme primum |
| | Tarsale II | 2nd Cuneiform | 2nd Cuneiform | Os cuneiforme secundum |
| | Tarsale III | 3rd Cuneiform | 3rd Cuneiform | Os cuneiforme tertium |
| | Tarsale IV | Cuboid | Cuboid | Os cuboideum |
| | Tarsale V | | | |

On the flexor surface of each metatarsophalangeal joint are two sesamoids, and one sesamoid lies on the flexor surface at the distal end of each middle phalanx, except in the case of the first digit. Here, where one phalanx is lacking, the sesamoid is found at the base of the terminal phalanx. Thus correspondence is very nearly complete between the hand and the foot, even to full correspondence of the sesamoids of the digits. The most important difference lies in the fact that the navicular and lunate in the hand, corresponding to the tibiale and talus in the foot, are fused in the former but not in the latter.

There are in the rat four additional sesamoids developed in connection with the tendons of the foot, one on the extensor surface of the calcaneus, one on the flexor surface of the tibiale, one just medial to the first cuneiform, and the fourth on the flexor surface at the base of the fourth and fifth metatarsals. Schultze's method was used here as well as for the fore foot, thus making it possible to show these sesamoids in situ (Fig. 50).

## LIST OF FIGURES

### II. SKELETON

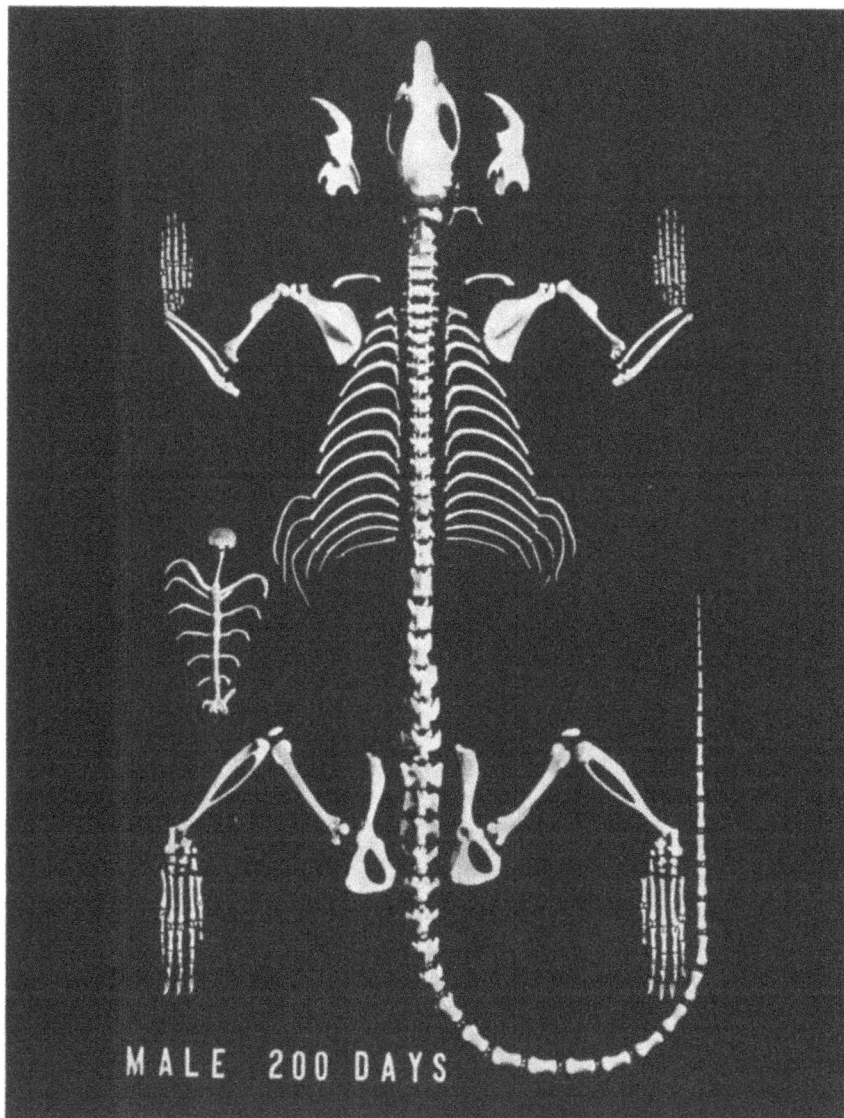

Fig.6.    Skeleton of Albino Rat.
Based on a preparation in the Museum of the Wistar Institute.

Fig. 7.   Dorsal Aspect of the Skull

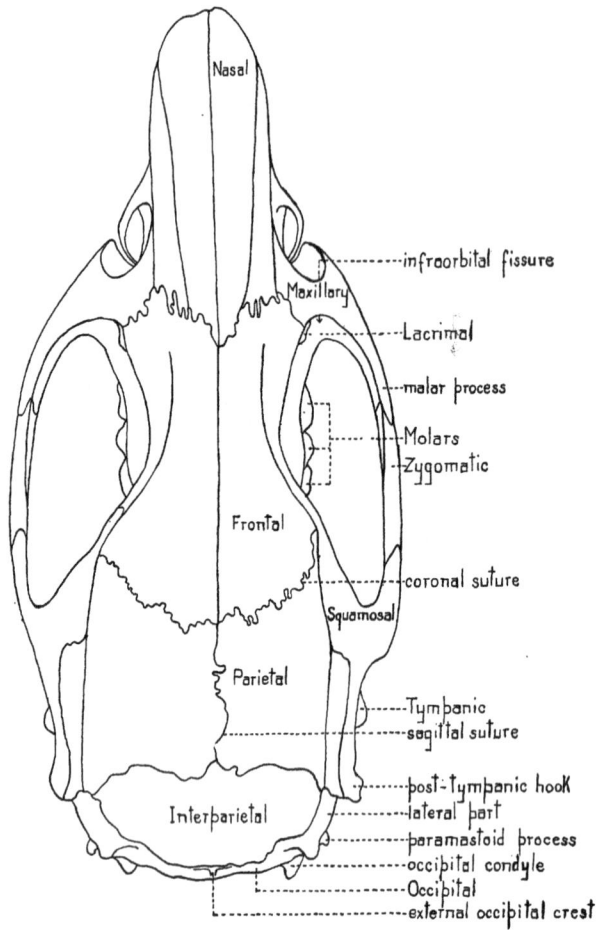

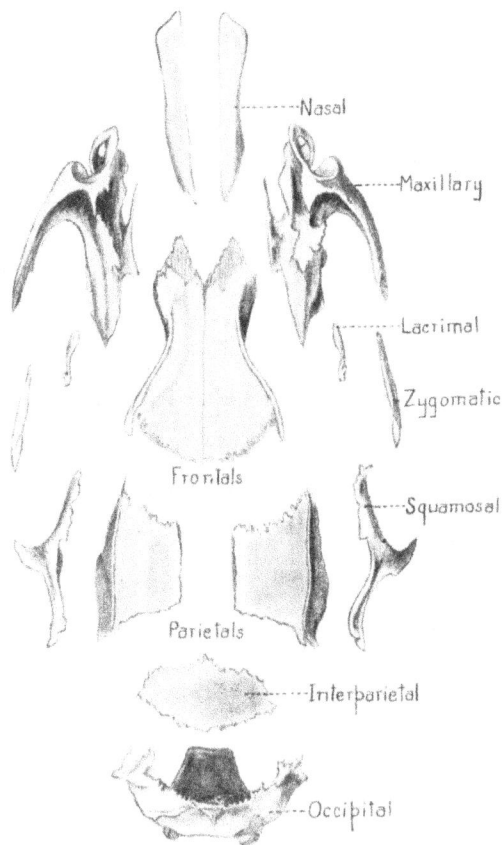

Fig. 8. Dorsal Aspect of Disarticulated Skull.

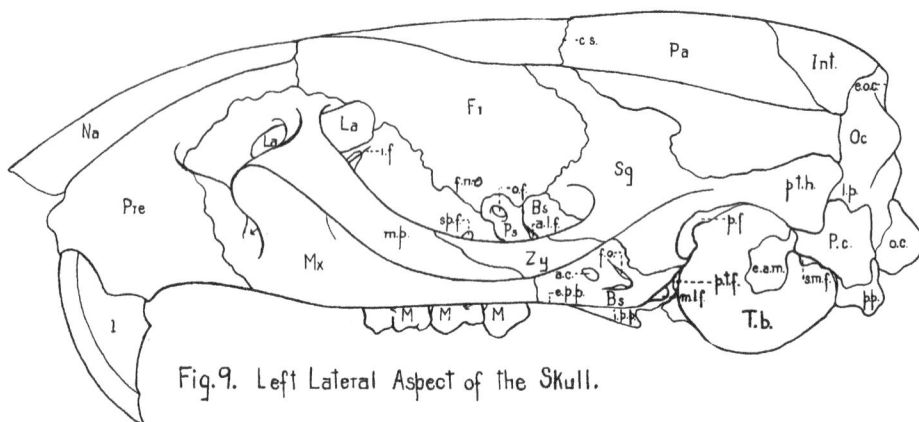

Fig. 9. Left Lateral Aspect of the Skull.

| BONES | | |
|---|---|---|
| Bs. | basisphenoid | |
| Fr. | frontal | |
| I. | incisor | |
| Int. | interparietal | |
| La. | lacrimal | |
| M. | molar | |
| Mx. | maxillary | |
| Na. | nasal | |
| Oc. | occipital | |
| Pa. | parietal | |
| P.c. | periotic capsule | |
| Ps. | presphenoid | |
| Pre. | premaxillary | |
| Sq. | squamosal | |
| T.b. | tympanic bulla | |
| Zy. | zygomatic | |

FEATURES

| | |
|---|---|
| a.c. | alisphenoid canal for internal carotid artery |
| a.l.f. | anterior lacerated foramen—foramen rotundum for III, IV, VI, and ophthalmic and maxillary divisions of V |
| c.s. | coronal suture |
| e.a.m. | external acoustic meatus |
| e.o.c. | external occipital crest |
| e.p.p. | external pterygoid process |
| f.n. | anterior ethmoidal foramen for nasociliary branch of V |
| f.o. | foramen ovale for mandibular division of V |
| i.f. | infra-orbital fissure for infra-orbital branch of V |
| i.p.p. | internal pterygoid process |
| l.p. | lateral part of occipital |
| m.l.f. | middle lacerated foramen for internal maxillary artery |
| m.p. | malar process |
| o.c. | occipital condyle |
| o.f. | optic foramen for II |
| p.f. | post-glenoid foramen for vein from the transverse sinus |
| p.p. | paramastoid process |
| p.t.f. | petrotympanic fissure for exit of pterygopalatine branch of internal carotid artery, entrance of chorda tympani, and vein to pterygoid plexus |
| p.t.h. | post-tympanic hook |
| sm.f. | stylo-mastoid foramen for exit of VII |
| sp.f. | spheno-palatine foramen for sphenopalatine blood vessels and nerves to the nasal cavity |

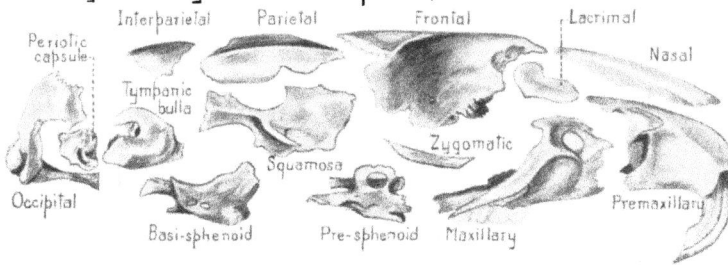

Fig. 10. Right Lateral Aspect of Disarticulated Skull.

Fig.11. Basal Aspect of the Skull.

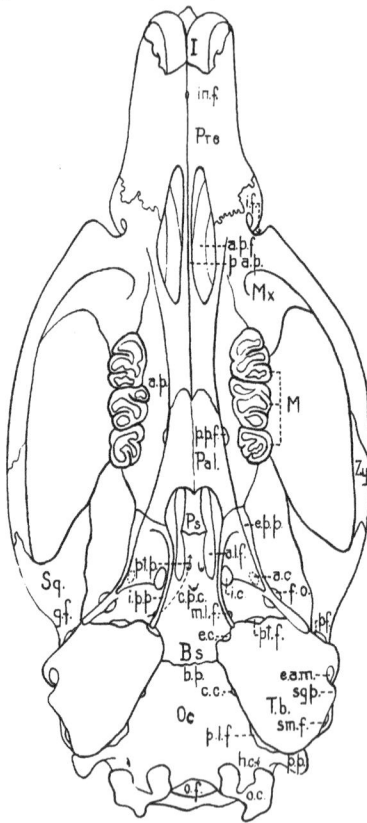

VENTRAL ASPECT

**BONES**

| | | | |
|---|---|---|---|
| Bs. | basisphenoid | Pre. | premaxillary |
| I. | incisor | Ps. | presphenoid |
| M. | molar | Sq. | squamosal |
| Mx. | maxillary | T.B. | tympanic bulla |
| Oc. | occipital | Zy. | zygomatic |
| Pal. | palatine | | |

**FEATURES**

a.c. alisphenoid canal for internal carotid artery

a.l.f. anterior lacerated foramen—foramen rotundum for III, IV, VI, ophthalmic and maxillary divisions of V, and for ophthalmic vessels

a.p. alveolar process

a.p.f. anterior palatine foramen for nasopalatine branch of V to soft palate

b.p. basilar part of occipital

c.c. carotid canal

c.p.c. cranio-pharyngeal canal

e.a.m. external acoustic meatus

e.c. Eustachian canal

e.p.p. external pterygoid process

f.o. foramen ovale for mandibular division of V

g.f. glenoid fossa for articulation with mandible

h.c. hypoglossal canal for XII

i.c. interpterygoid foramen

i.f. infra-orbital fissure for zygomatic branch of V

in.f. incisive foramen for palato-labial artery

i.p.p. internal pterygoid process

m.l.f. middle lacerated foramen for pterygopalatine branch of internal carotid artery

o.c. occipital condyle

oc.f. occipital foramen

pa.p. palatine process

p.f. post-glenoid foramen for vein from the transverse sinus to internal maxillary

p.l.f. posterior lacerated foramen for nerves IX, X, and XI, entrance of pterygopalatine branch of internal carotid artery and int. jug. vein

p.p. paramastoid process

p.p.f. posterior palatine foramen for palatine art. and palatine branch of V

pt.f. petro-tympanic fissure for exit of pterygopalatine branch of internal carotid artery, entrance of chorda tympani nerve, and vein to the pterygoid plexus

sm.f. stylo-mastoid foramen for exit of VII

sg.p. stylo-glossal process

Fig. 12.
Basal Aspect of
Disarticulated Skull

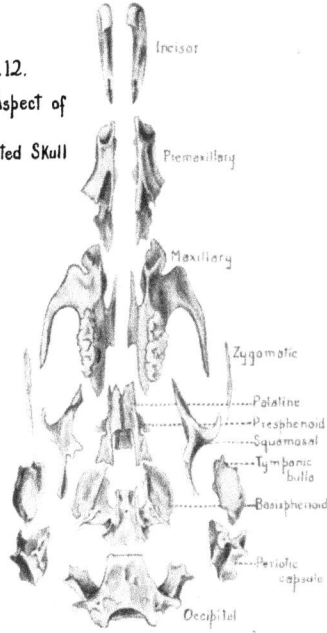

Incisor

Premaxillary

Maxillary

Zygomatic

Palatine
Presphenoid
Squamosal
Tympanic bulla

Basisphenoid

Periotic capsule

Occipital

Fig. 13.
Floor of Cranial Cavity

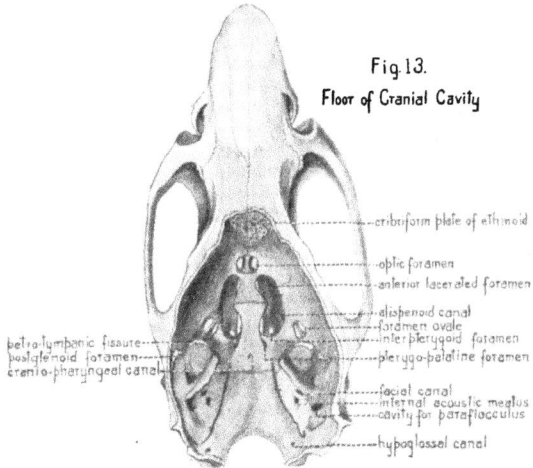

cribriform plate of ethmoid

optic foramen
anterior lacerated foramen

alisphenoid canal
foramen ovale
inter-pterygoid foramen
pterygo-palatine foramen

petro-tympanic fissure
postglenoid foramen
cranio-pharyngeal canal

facial canal
internal acoustic meatus
cavity for paraflocculus

hypoglossal canal

Fig.14. Median Sagittal Section of Skull.

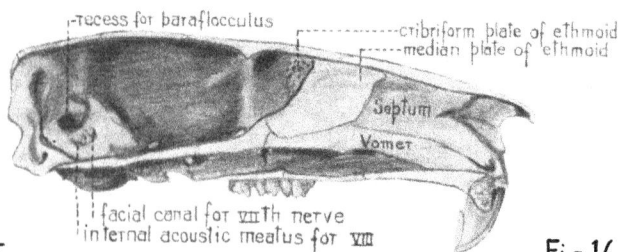

recess for baraflocculus — cribriform plate of ethmoid — median plate of ethmoid

Septum

Vomer

facial canal for VIIth nerve
internal acoustic meatus for VIII

Fig.15.
Lateral Wall of Nasal Cavity, septum removed

Fig.16.
Left Lateral Aspect of Ethmoid

Fig.17. Anterior Aspect of Ethmoid

Fig.18.
Posterior Aspect of Ethmoid

Fig.19.
Lateral Aspect of Mandible

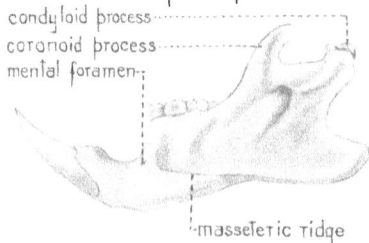

condyloid process
coronoid process
mental foramen

masseteric ridge

Fig.20.
Medial Aspect of Mandible

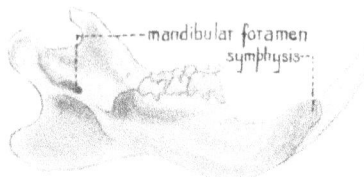

mandibular foramen
symphysis

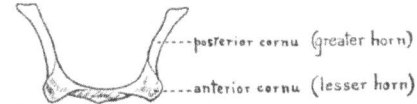

Fig.21.
**Hyoid bone**

- - - posterior cornu (greater horn)

- - - anterior cornu (lesser horn)

Fig. 22.  Left Lateral Aspect of Atlas and Epistropheus

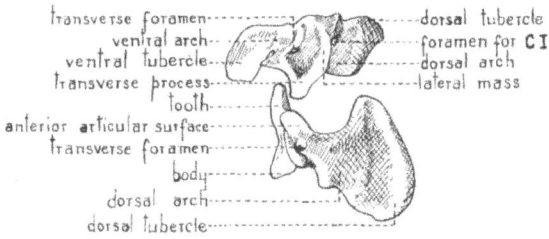

transverse foramen
ventral arch
ventral tubercle
transverse process
tooth
anterior articular surface
transverse foramen
body
dorsal arch
dorsal tubercle

dorsal tubercle
foramen for C I
dorsal arch
lateral mass

Fig.23.  Anterior Aspect of Fifth Cervical Vertebra

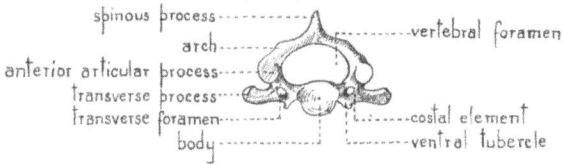

spinous process
arch
anterior articular process
transverse process
transverse foramen
body

vertebral foramen

costal element
ventral tubercle

Fig.24.  Anterior Aspect of Seventh Thoracic Vertebra

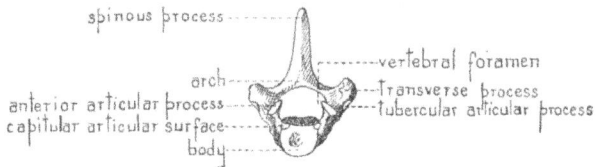

spinous process

arch
anterior articular process
capitular articular surface
body

vertebral foramen
Transverse process
tubercular articular process

Fig. 25. Left Lateral Aspect of Fourth Lumbar Vertebra

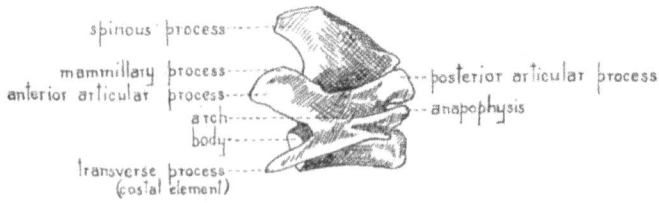

spinous process
mammillary process
anterior articular process
arch
body
transverse process
(costal element)
posterior articular process
anapophysis

Fig. 26. Dorsal Aspect of Sacrum

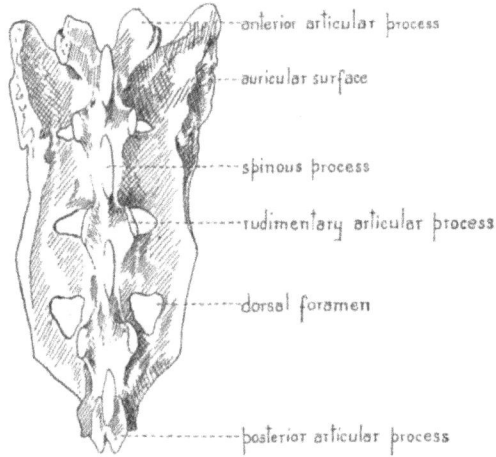

anterior articular process
auricular surface
spinous process
rudimentary articular process
dorsal foramen
posterior articular process

Fig. 27. Dorsal Aspect of Fourteenth Caudal Vertebra

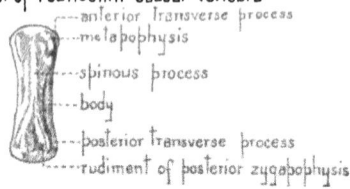

anterior transverse process
metapophysis
spinous process
body
posterior transverse process
rudiment of posterior zygapophysis

Fig.28. Lateral aspect of right scapula

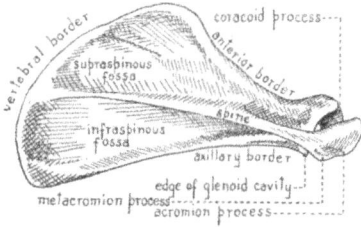

Fig.30.
The Clavicle and Associated Elements

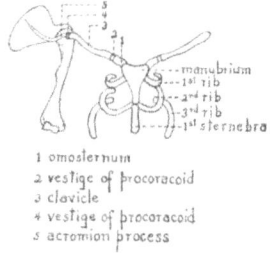

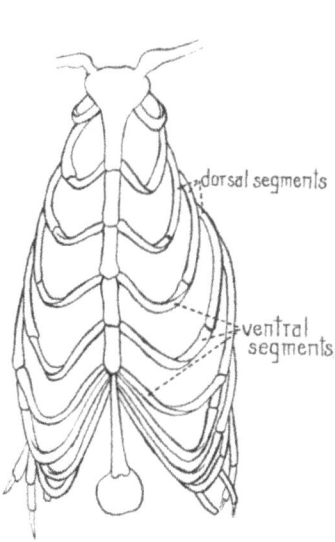

vertebral border
coracoid process
anterior border
subraspinous fossa
spine
infraspinous fossa
axillary border
edge of glenoid cavity
metacromion process
acromion process

Fig.29.
Ventral aspect of right clavicle

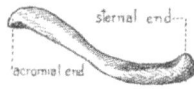

sternal end
acromial end

manubrium
1st rib
2nd rib
3rd rib
1st sternebra

1 omosternum
2 vestige of procoracoid
3 clavicle
4 vestige of procoracoid
5 acromion process

clavicle
manubrium
dorsal segments
sternebra
vertebral column
ventral segments
xiphoid process

Fig. 31
Ventral Aspect of Thoracic Basket
Camera Lucida Drawing of Dried Specimen

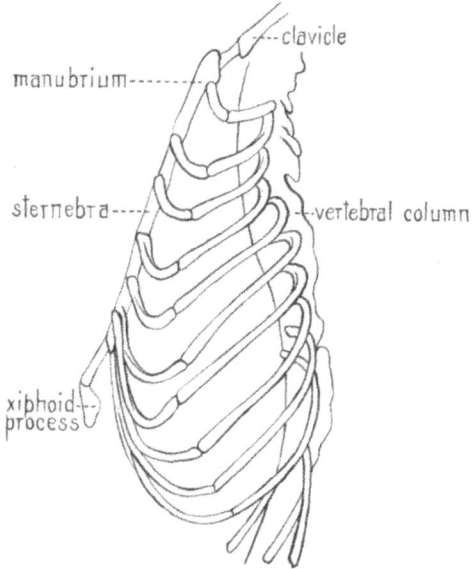

Fig.32. Lateral Aspect of Thoracic Basket
Camera Lucida Drawing of Dried Specimen

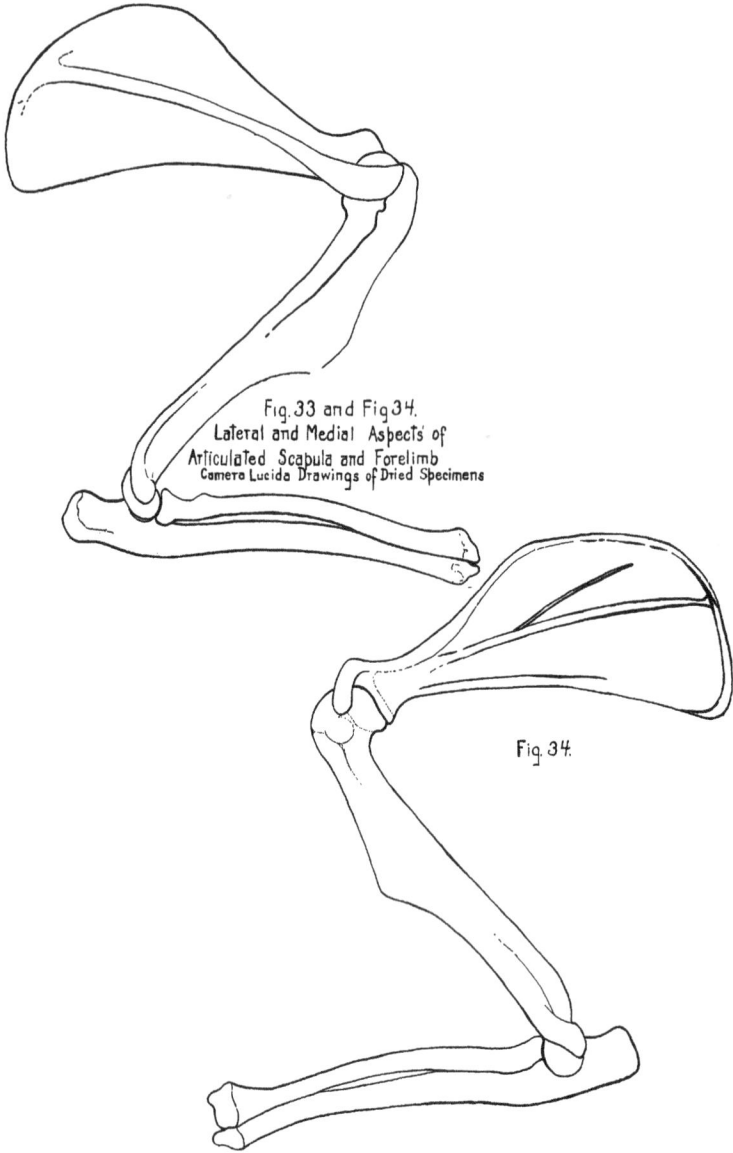

Fig. 33 and Fig 34.
Lateral and Medial Aspects of
Articulated Scapula and Forelimb
Camera Lucida Drawings of Dried Specimens

Fig. 34.

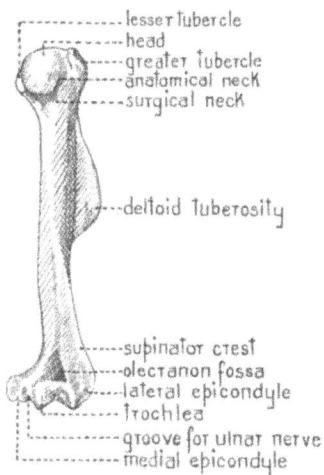

lesser tubercle
head
greater tubercle
anatomical neck
surgical neck

deltoid tuberosity

supinator crest
olecranon fossa
lateral epicondyle
trochlea
groove for ulnar nerve
medial epicondyle

Fig. 35.
Extensor surface of right humerus.

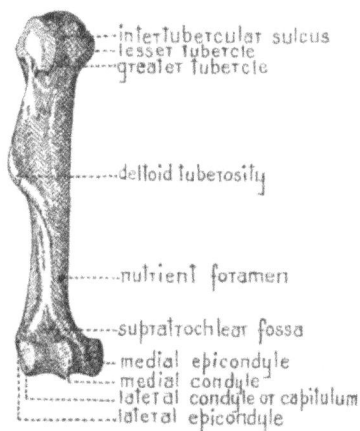

intertubercular sulcus
lesser tubercle
greater tubercle

deltoid tuberosity

nutrient foramen

supratrochlear fossa
medial epicondyle
medial condyle
lateral condyle or capitulum
lateral epicondyle

Fig. 36.
Flexor surface of right humerus.

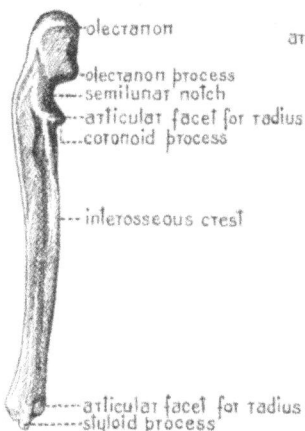

olecranon
olecranon process
semilunar notch
articular facet for radius
coronoid process

interosseous crest

articular facet for radius
styloid process

Fig. 37. Lateral aspect of right ulna.

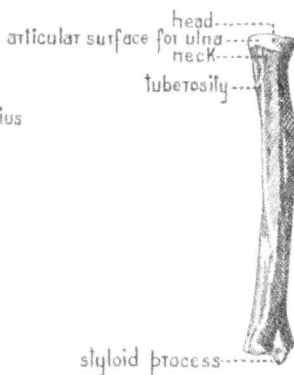

head
articular surface for ulna
neck
tuberosity

styloid process

Fig. 38.
Extensor surface of right radius.

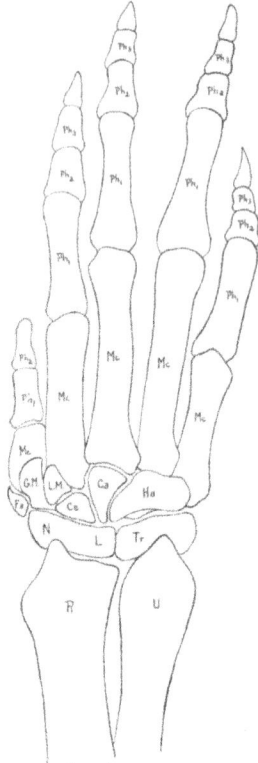
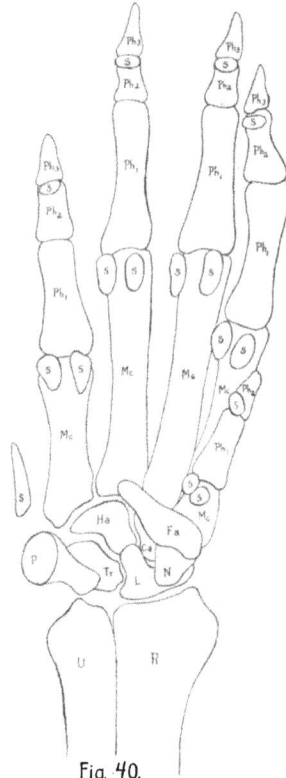

Fig. 39.                                    Fig. 40.

Extensor and Flexor Surfaces of Right Manus.

R = radius                    Ce = centrale
U = ulna                      Ca = capitate
N = navicular                 Ha = hamate
L = lunate                    Mc = metacarpal
Tr = triangular               Ph = phalanx
G.M. = greater multangular    Fa = falciformis
L.M. = lesser multangular     P = pisiform
            S = sesamoid

Dorsal

Ventral

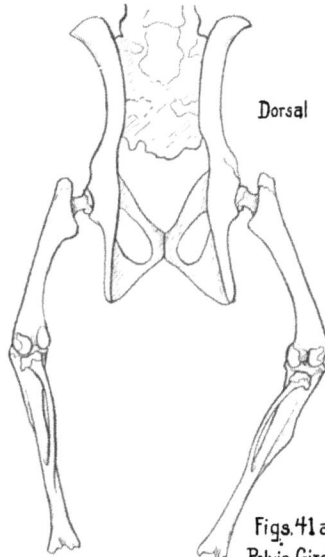

Figs. 41 and 42.
Pelvic Girdle and
Posterior Appendage
Camera Lucida Drawings of
Dried Specimens.

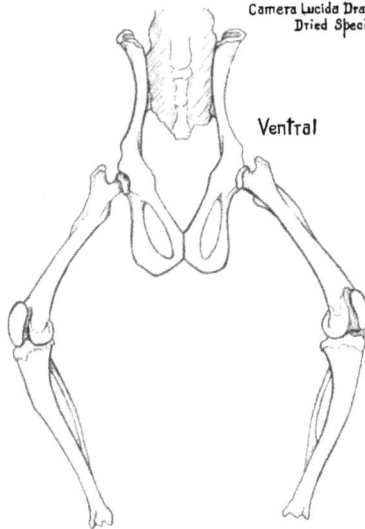

Fig. 43. Ventral Aspect of Ossa Coxae

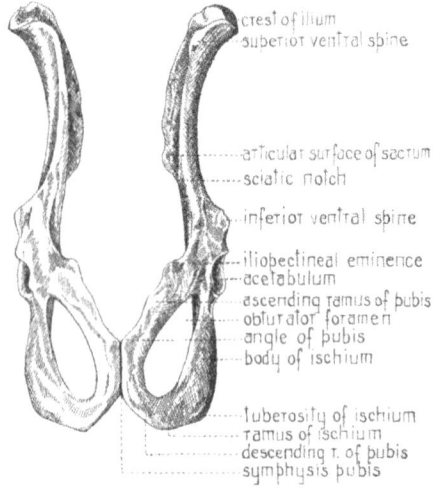

crest of ilium
superior ventral spine

articular surface of sacrum
sciatic notch

inferior ventral spine

iliobectineal eminence
acetabulum
ascending ramus of pubis
obturator foramen
angle of pubis
body of ischium

tuberosity of ischium
ramus of ischium
descending r. of pubis
symphysis pubis

Fig. 44
Lateral Aspect of Right Os Coxae

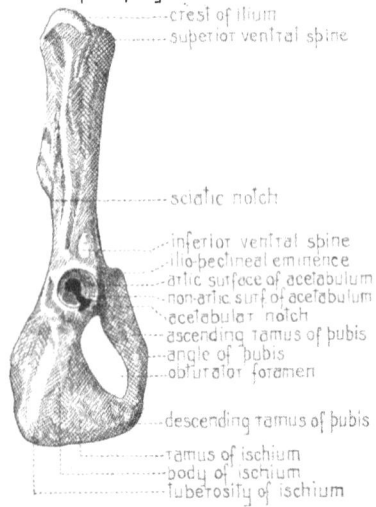

crest of ilium
superior ventral spine

sciatic notch

inferior ventral spine
ilio-bectineal eminence
artic. surface of acetabulum
non-artic. surf of acetabulum
acetabular notch
ascending ramus of pubis
angle of pubis
obturator foramen

descending ramus of pubis

ramus of ischium
body of ischium
tuberosity of ischium

Fig. 45. Extensor Surface of Right Femur

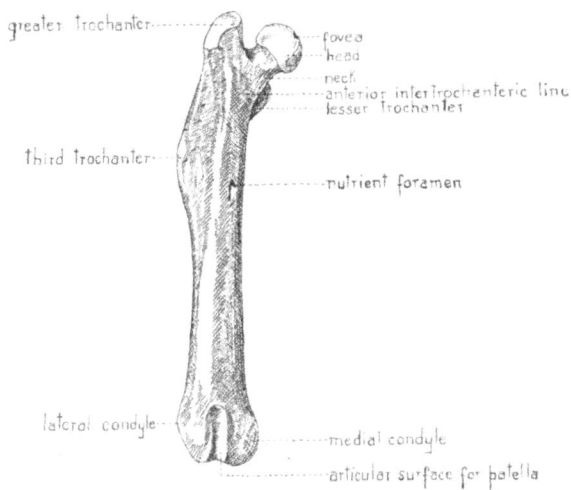

greater trochanter

fovea
head
neck
anterior intertrochanteric line
lesser trochanter

third trochanter

nutrient foramen

lateral condyle

medial condyle

articular surface for patella

Fig. 46. Flexor Surface of Right Femur

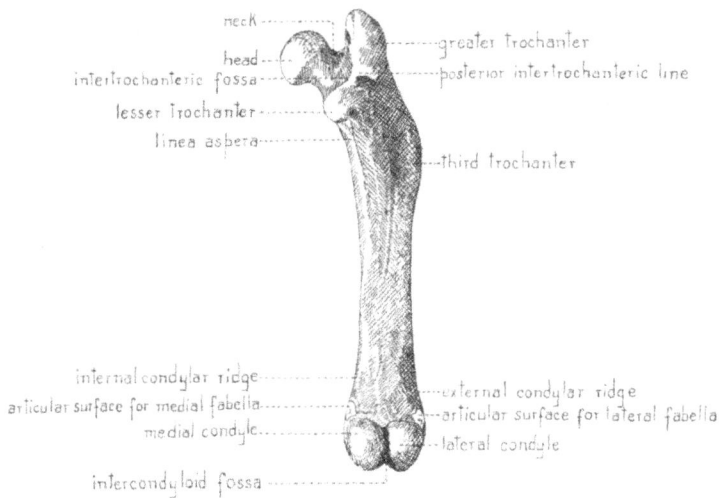

neck

head

intertrochanteric fossa

lesser trochanter

linea aspera

greater trochanter

posterior intertrochanteric line

third trochanter

internal condylar ridge

articular surface for medial fabella

medial condyle

external condylar ridge

articular surface for lateral fabella

lateral condyle

intercondyloid fossa

Fig. 47.
Extensor Surface of Tibia and Fibula

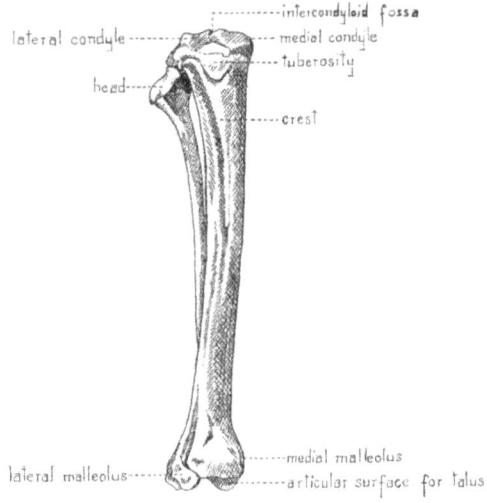

Fig. 48. Flexor Surface of Tibia and Fibula

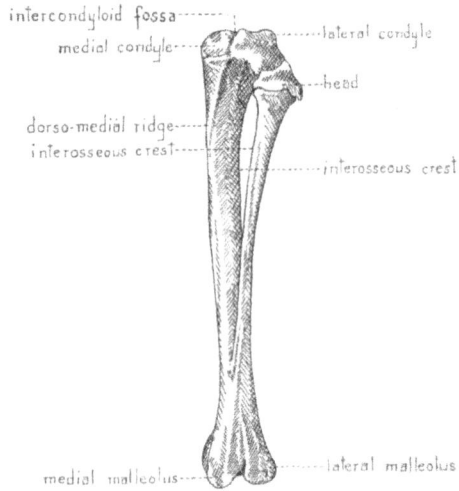

Bones of the Right Pes

Ta = talus
C = calcaneus
Ti = tibiale
N = navicular
$C_1$ = 1st cuneiform
$C_2$ = 2nd cuneiform
$C_3$ = 3rd cuneiform
Cd = cuboid
Mt = metatarsal
Ph = phalanx
S = sesamoid

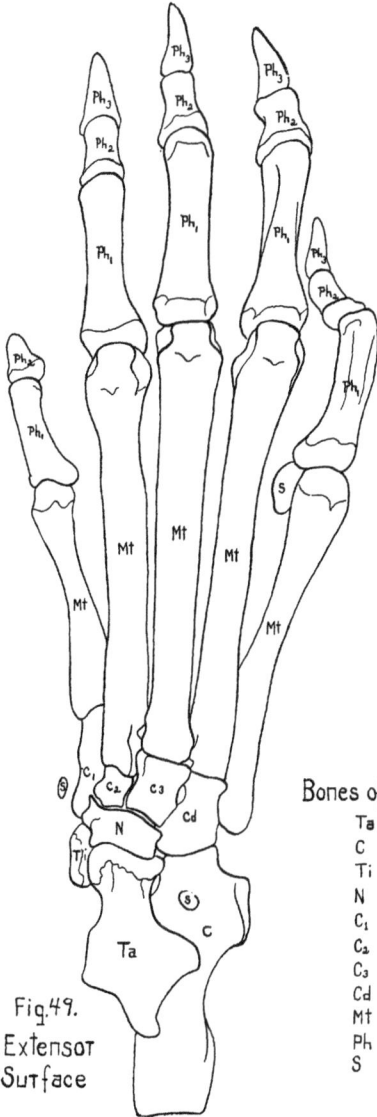

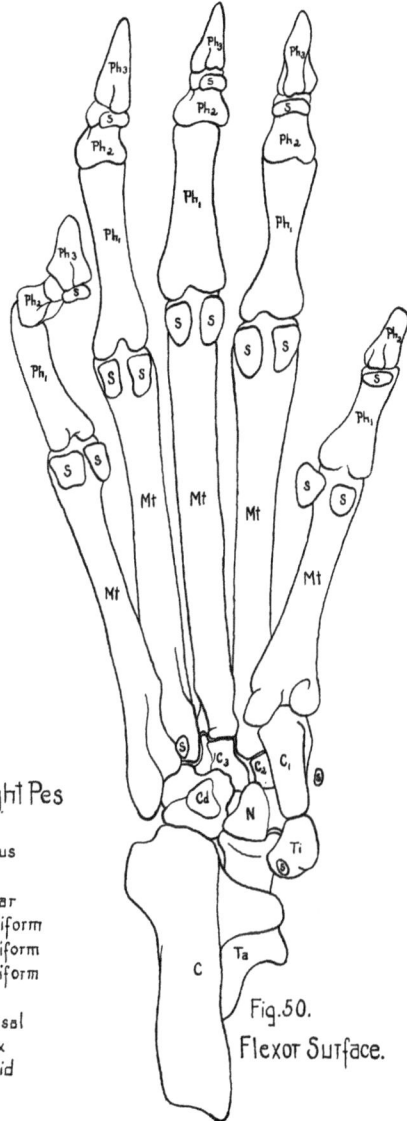

Fig.49.
Extensor
Surface

Fig.50.
Flexor Surface.

# CHAPTER III

## MUSCLES

### (Figs. 51–102)

Inasmuch as human anatomy has been used as the basis of comparison, and since the textbooks on this subject present the only standardized arrangement of muscles, this same order has been adopted, with few exceptions, for the present work. The principal exception is in the head where the rat does not, of course, show the extensive development of facial muscles which constitute such a large part of the head musculature in man. In view of this, only the more important and readily distinguished head muscles are described.

At the risk of incompleteness an exact description of some muscles has been omitted altogether as not being essential for the present purpose, for example, the intrinsic muscles of the vertebral column. The nerve and blood supply of the muscles has been given so far as determined.

The muscles have been described in the following order:

Muscles of the head
  Superficial muscles of the head
  Muscles of mastication
Muscles of the antero-lateral region of the neck
  Superficial cervical
  Lateral cervical
  Suprahyoid and infrahyoid
  Anterior vertebral
  Lateral vertebral
Muscles of the trunk
  Cutaneous muscle of the trunk
  Muscles of the thorax
  Muscles of the abdomen
  Caudal muscles
Muscles of the anterior appendage
  Muscles connecting the anterior appendage to the vertebral column; superficial muscles of the back
  Muscles connecting the anterior appendage to the thoracic walls; muscles of the pectoral region
Muscles of the shoulder
Muscles of the brachium
Muscles of the antibrachium
Muscles of the manus
Muscles of the posterior appendage
  Muscles of the thigh
    Anterior femoral muscles
    Medial femoral muscles
    Muscles of the gluteal region
    Posterior femoral muscles (hamstring muscles)
  Muscles of the leg
    Anterior crural muscles
    Posterior crural muscles
    Lateral crural muscles
  Muscles of the pes
    Dorsal muscle of the pes
    Plantar muscles of the pes

## MUSCLES OF THE HEAD

### SUPERFICIAL MUSCLES OF THE HEAD

Inasmuch as the platysma is so extensive in this form and covers a large part of the lateral surface of the head as well as the neck, it will be discussed with the head muscles.

The superficial muscles of the head are:

| | |
|---|---|
| Levator auris longus | Orbicularis oculi |
| Interscutularis | Orbicularis oris |
| Frontalis | Buccinator |
| Levator labii superioris | Platysma |
| Zygomaticus | Auriculolabialis |
| Dilator naris | Sternofacialis |

**M. Levator auris longus** (Figs. 51, 53) consists of two portions, cranial and caudal. The *cranial* part arises from the spines of the first four cervical vertebrae and runs forward to its insertion on the anterior part of the base of the auricle. The *caudal* part, from the spines of the fourth and fifth cervical vertebrae, runs nearly parallel to the cranial portion and is inserted into the posterior part of the base of the auricle.

*Nerve-supply.*—Posterior auricular branch of facial nerve.

*Blood-supply.*—Posterior auricular artery.

**M. Interscutularis** (Fig. 53) crosses the parietal region of the skull connecting the base of the ears.

**M. Frontalis** (Fig. 51), a portion of epicranius or occipito-frontalis of human anatomy, arises from the skin over the upper border of orbicularis oculi, which is only slightly developed in the rat, and inserts on the Galea aponeurotica.

*Nerve-supply.*—Temporal branches of facial nerve.

*Blood-supply.*—Anterior auricular artery.

**M. Levator labii superioris** (Figs. 51, 59), corresponding to the infraorbital head of quadratus labii superioris in man, arises as a thin slip from the premaxilla just above the root of the zygoma and from the anterior margin of orbicularis oculi and extending obliquely downward attaches in the mystacial pad of the upper lip.

*Nerve-supply.*—Facial nerve.

*Blood-supply.*—Angular artery.

**M. Zygomaticus** (Fig. 59) arises from the masseteric sheath upon the surface of the zygomatic arch and runs as an exceedingly delicate muscle to the upper corner of the mouth where it blends with the orbicularis oris.

*Nerve-supply.*—Facial nerve.

*Blood-supply.*—Angular artery.

**M. Dilator naris** (Fig. 58) is a delicate muscle slip running, under cover of levator labii superioris, to the wing of the nose.

*Nerve-supply.*—Facial nerve.

*Blood-supply.*—Angular artery.

**M. Orbicularis oculi and orbicularis oris** are both poorly defined, present no peculiarities, and need no description.

*Nerve-supply.*—Facial nerve.

*Blood-supply.*—External maxillary artery.

**M. Buccinator** (Fig. 55), arising from the maxilla and premaxilla along the ridge formed by the upper incisor, and just anterior to the anterior deep masseter, runs downward and blends with orbicularis oris.

*Nerve-supply.*—Lower zygomatic branches of the facial nerve.

*Blood-supply.*—Superficial temporal; internal and external maxillary arteries.

**M. Panniculus carnosus** is divisible into two main portions, namely that of the head and neck, the *platysma* and *sternofacialis*, and that of the trunk, the *cutaneous maximus*. The latter will be considered under trunk musculature.

**M. Platysma** (Figs. 51–53, 58) is an extremely thin sheet of muscle of the head and neck region, closely associated with the subcutaneous fascia, in fact, in young animals it is with difficulty separated from this. It is divisible into *cranial*, and *superficial* and *deep cervical* portions. The cranial portion has its origin from the mid-dorsal line of the neck, where its attachment is partially covered by levator auris longus. It arises from the base of the ear and from the parietal region under cover of frontalis. It runs forward and downward past the base of the ear, covering the angle of the jaw before splitting just posterior to the corner of the mouth into two slips which attach to the orbicularis oris, one in the region of the lower and one in the region of the upper lip. Howell in his work on the wood rat (1926) finds the fibers which arise from the ear and from the parietal region quite distinct from the platysma and calls this muscle *auriculolabialis*. In the albino rat this is indistinguishable from platysma.

The superficial portion of the cervical panniculus arises by decussating fibers from the ventral region of the neck from the pectoral region to the symphysis mandibuli. From its origin it runs antero-laterally beneath the cranial portion of the platysma to attachment on the fascia of the masseter. The deep portion, sometimes called *sternofacialis*, arising from the manubrium of the sternum runs upward over sternomastoideus and under cover of the superficial platysma to spread over the masseter where it divides into three slips ending on the ear cartilage and adjacent fascia.

*Nerve-supply.*—Cranial portion by inframandibular branch of facial nerve. "Cervical portion by the superficial cervical and facial nerves." (Parsons 1894–96.)

*Blood-supply.*—By muscular branches of the external carotid, and by cutaneous, submental, inferior labial, and lateral nasal branches of the external maxillary artery. Also by branches from the posterior auricular, and by sternomastoid and superficial cervical branches of the cervical trunk.

### MUSCLES OF MASTICATION

The muscles of this group are four:

| | |
|---|---|
| Masseter | Pterygoideus externus |
| Temporalis | Pterygoideus internus |

**M. Masseter** (Figs. 54, 55) is more or less distinctly separable into four portions, anterior and posterior superficial, and anterior and posterior deep. This is in accord with Parsons' findings (1894-96) in his study of sciuromorphine, hystricomorphine, and myomorphine rodents.

*The anterior superficial part* (Fig. 54) arises by a strong slender tendon from the lateral surface of the maxilla just posterior to the suture of the maxilla and premaxilla and inserts along the lower margin and internal surface of the angular process of the mandible, adjacent to the insertion of internal pterygoid. *The posterior superficial part* (Fig. 54) arising from the whole length of the zygomatic arch inserts on the lower part of the lateral surface of the mandible and into the masseteric ridge which extends along the lateral surface and lower margin of the bone. This large portion of the masseter shows in the direction of its fibers some indication of a possible division into a more superficial posterior portion and a deeper anterior. Parsons (1894) found this separation quite distinct in Sciuromorphs and speaks of the anterior deep masseter and posterior superficial masseter. In Hystricomorphs, he found what he called the anterior deep masseter passing through the infraorbital foramen, but in myomorphine rodents he found both these so-called anterior deep divisions were present simultaneously. He therefore concludes that the anterior deep muscle of the Sciuromorphs is a forward extension of the posterior superficial while the infraorbital muscle of the Hystricomorphs is the true anterior deep muscle. *The anterior deep part* (Fig. 55) arises from the fossa just anterior to the infraorbital fissure and converges to a slender muscle which passes through the infraorbital fissure and inserts upon the lateral surface of the mandible at the anterior end of the masseteric ridge, nearly covered by the posterior deep masseter. *The posterior deep part* (Fig. 55) takes its origin from the lower border and inner surface of the zygoma and inserts upon the lateral surface of the ramus of the mandible; it is with difficulty separated from the posterior superficial muscle.

*Nerve-supply.*—Masseteric branch of the mandibular division of the trigeminal nerve.

*Blood-supply.*—Masseteric and muscular branches of the internal and external maxillary; and by superficial temporal and transverse facial arteries.

**M. Temporalis** (Fig. 57). Its fibers arise from the post-tympanic hook of the temporal bone, from the temporal ridge of the parietal, and from the inner surface of the zygoma. As the fibers from the temporal and parietal reach the root of the zygoma, which they use as a pulley, they change their course, and with the remaining fibers of the muscle insert into both surfaces of the mandible, as well as into the anterior border of the coronoid process and the groove between it and the molar teeth.

*Nerve-supply.*—Deep temporal branch of the mandibular division of the trigeminal nerve.

*Blood-supply.*—Middle temporal branch of the superficial temporal, and anterior deep temporal branch of the internal maxillary artery.

**M. Pterygoideus externus** (Fig. 56) arises from the external pterygoid ridge and inserts into the medial surface of the condylar process of the mandible.

*Nerve-supply.*—Pterygoid branch of the mandibular division of the trigeminal nerve.

*Blood-supply.*—Pterygoid branch of the internal maxillary, and ascending palatine branch of the external carotid artery.

**M. Pterygoideus internus** (Fig. 56) arises from the lateral surface of the feebly marked internal pterygoid ridge and inserts on the medial surface of the mandible into the ridge which runs posteriorly from the molar teeth below the mandibular foramen.

*Nerve-supply.*—Pterygoid branch of the mandibular division of the trigeminal nerve.

*Blood-supply.*—"Tonsillar" branch of the external maxillary, ascending palatine branch of external carotid, and pterygoid branch of internal maxillary artery.

## MUSCLES OF THE ANTERO-LATERAL REGION OF THE NECK

### SUPERFICIAL CERVICAL MUSCLES

The *Platysma* has been described above under the head musculature.

### LATERAL CERVICAL MUSCLES

Clavotrapezius
Sternomastoideus
Cleidomastoideus

The *trapezius* muscle in the rat is made up of three portions, *clavotrapezius*, *acromiotrapezius*, and *spinotrapezius*. Each of these is such a distinct muscle that the clavotrapezius may well be treated with the lateral cervical group, while acromiotrapezius and spinotrapezius are included with the superficial muscles of the back.

**M. Clavotrapezius** (cleido-occipitalis cervicalis) (Figs. 60–62, 69) is a thin band which arises from the superior nuchal line and the external occipital protuberance, passes laterally and posteriorly and is inserted into the outer third of the posterior border of the clavicle.

*Nerve-supply.*—Spinal accessory and third and fourth cervical nerves through the subtrapezial plexus.

*Blood-supply.*—Transverse cervical artery of the cervical trunk.

**M. Sternomastoideus** (Figs. 60–62) arises from the anterior border of the manubrium of the sternum and is inserted into the mastoid and paroccipital processes of the skull.

*Nerve-supply.*—Spinal accessory and a branch from the third and fourth cervical nerves through the subtrapezial plexus.

*Blood-supply.*—Superficial cervical artery of cervical trunk.

**M. Cleidomastoideus** (Figs. 60–62) arises from the middle third of the clavicle partly covered by clavotrapezius, and extends as a narrow ribbon between sternomastoideus and cleido-occipitalis to its insertion into the mastoid process and occipital and temporal ridges.

*Nerve-supply.*—Spinal accessory plus a branch from the third and fourth cervical nerves through the subtrapezial plexus.

*Blood-supply.*—Occipital artery and superficial cervical branch of the cervical trunk.

SUPRAHYOID AND INFRAHYOID MUSCLES

The suprahyoid muscles are:

Digastricus<br>
Stylohyoideus<br>
Mylohyoideus<br>
Geniohyoideus<br>
Transversus mandibularis

**M. Digastricus** (Figs. 60, 62–64) is attached posteriorly to the paroccipital process of the skull and anteriorly into the ventral border of the mandible just posterior to the symphysis. The two bellies of the muscle are separated by a tendon of insertion which attaches them to the hyoid. Between the tendons of opposite sides stretches a fibrous arch and from its convexity which is directed forward, additional fibers arise to contribute to the anterior belly. At an angle formed by the two mandibles the digastric muscles separate a trifle and expose the transverse mandibular muscle.

*Nerve-supply.*—Anterior belly (and mylohyoideus) by mylohyoid branch of the inferior alveolar, posterior belly (and stylohyoideus) by the digastric branch of the facial nerve.

*Blood-supply.*—Anterior belly by a muscular branch of the external carotid and by the submental artery; posterior belly by the ascending pharyngeal and occipital arteries.

**M. Stylohyoideus** (Figs. 63, 65) arises from the base of the skull just medial to the paroccipital process and is inserted into the epihyal cartilage close to the hyoid bone. It passes under digastricus.

*Nerve-supply.*—Stylohyoid branch of the facial nerve.

*Blood-supply.*—Occipital artery and submental branch of the external maxillary artery.

**M. Mylohyoideus** (Figs. 63, 65) arises from the whole extent of the mylohyoid line of the mandible, and by means of the tendinous arch of the digastricus is inserted into the hyoid bone.

*Nerve-supply.*—Mylohyoid branch of the inferior alveolar branch of the trigeminal nerve.

*Blood-supply.*—Sublingual branch of the lingual artery from the external carotid and by the submental branch of the external maxillary.

**M. Geniohyoideus** (Fig. 65) arises from the median surface of the mandible close to the symphysis. From opposite sides the two muscles extend somewhat posteriorly converging toward the midline where they insert by means of a median fibrous raphe which extends from the symphysis to the hyoid bone. Posteriorly the two muscles coalesce and attach to the body and posterior cornu of the hyoid.

*Nerve-supply.*—Hypoglossal nerve.

*Blood-supply.*—Sublingual branch of the lingual artery from the external carotid.

**M. Transversus mandibularis** (Fig. 63) is a short band of muscle fibers between the two mandibles slightly posterior to the symphysis. This muscle is present in all rodents in which the two mandibles form a movable arthrosis.

*Nerve-supply.*—Mylohyoid branch of inferior alveolar nerve.

*Blood-supply.*—Submental branch of the external maxillary artery.

The infrahyoid muscles are:

> Sternohyoideus
> Sternothyreoideus
> Thyreohyoideus
> Omohyoideus

**M. Sternohyoideus** (Figs. 60, 62) arises from the manubrium of the sternum and is inserted into the lower border of the body of the hyoid.

*Nerve-supply.*—From the first two cervical nerves via ansa hypoglossi.

*Blood-supply.*—Ascending pharyngeal artery.

**M. Sternothyreoideus** (Fig. 64) takes its origin from the posterior surface of the manubrium and from the edge of the first and second costal cartilages, and is inserted into the thyroid cartilage.

*Nerve-supply.*—From the first, second, and third cervical nerves through the ansa hypoglossi.

*Blood-supply.*—Ascending pharyngeal branch of the external carotid artery.

**M. Thyreohyoideus** (Fig. 64) is a small muscle extending from the thyroid cartilage to the angle between the body and the posterior cornu of the hyoid.

*Nerve-supply.*—Hypoglossal nerve.

*Blood-supply.*—Ascending pharyngeal branch of the external carotid artery.

**M. Omohyoideus** (Figs. 60, 62) extends from the anterior border of the scapula to the hyoid bone.

*Nerve-supply.*—From the first, second, and third cervical nerves via ansa hypoglossi.

*Blood-supply.*—Ascending pharyngeal branch of the external carotid artery.

### ANTERIOR VERTEBRAL MUSCLES

> Longus colli (Fig. 66)
> Longus capitis (Fig. 66)
> Rectus capitis (Fig. 67)

The muscles of this group are not described.

*Nerve-supply.*—Anterior rami of the cervical nerves.

*Blood-supply.*—Descending branch of the occipital, ascending pharyngeal, cervicalis profunda, and the ascending cervical arteries.

> Scalenus anterior
> Scalenus medius
> Scalenus posterior

**M. Scalenus anterior.** In his "Account of the Myology of the Myomorpha," (1896), Parsons gives the name of scalenus anticus only to a muscle inserted into the first rib "in front of," i.e., ventral to, the subclavian artery and brachial plexus. This muscle when present arises by a tendon from the basioccipital in front of and internal to the levator claviculae. Such a muscle is not, as a rule, present in the rat.

**M. Scalenus medius** (Figs. 67, 74, 78) is the most extensive of the scalenes, arising from the transverse processes of the second to the seventh cervical vertebrae and inserting by means of digitations on the anterior surface of the first to sixth ribs.

*Nerve-supply.*—Branches from the cervical nerves.

*Blood-supply.*—Ascending cervical branch of the cervical trunk.

**M. Scalenus posterior** (Figs. 71, 74) arises from the transverse processes of the second to seventh cervical vertebrae and is then identified as two slips which interdigitate with serratus anterior and are inserted into the fourth and fifth ribs.

*Nerve-supply.*—Second to seventh cervical nerves.

*Blood-supply.*—Ascending cervical branch of the cervical trunk.

## MUSCLES OF THE TRUNK

The trunk musculature will be considered under the following heads:

> Cutaneous muscle of the trunk
> Muscles of the thorax
> Muscles of the abdomen
> Caudal muscles

### CUTANEOUS MUSCLE OF THE TRUNK

The portion of panniculus carnosus which covers the trunk is called **M. Cutaneous maximus** (Figs. 51, 53). Part of its fibers take origin from the lesser tubercle of the humerus and unite with fibers arising from the surface of the pectoralis. The muscle emerges from beneath the arm as a rather stout muscle which then radiates posteriorly over the dorsal, lateral, and ventral surfaces of the trunk. The dorsal fibers cover the shoulder and extend to the mid-dorsal line; the lateral mass becomes much thinner before inserting into the skin of the gluteal region, beyond which point the fibers again converge to an attachment at the base of the tail. Finally, of those fibers which spread out ventrally, some end in the fascia of the extensor and medial surfaces of the thigh, while some extend to the mid-ventral line.

*Nerve-supply.*—Lateral and ventral parts by the medial anterior thoracic nerve from C VIII and Th I through the brachial plexus. Cervical part "by superficial cervical and facial nerves." (Parsons, 1894.)

*Blood-supply.*—Mainly by branches of the lateral thoracic artery, circumflexo-subscapular trunk, and thoracodorsal artery.

MUSCLES OF THE THORAX

Of these only three muscles seem sufficiently important to be described, namely:

Serratus posterior superior
Serratus posterior inferior
Diaphragm

**M. Serratus posterior superior** (Fig. 71) arises from the ligamentum nuchae, the spines of the first two thoracic vertebrae, and the dorsolumbar fascia and is inserted into the ribs from the fourth to the ninth.

**M. Serratus posterior inferior** (Fig. 71) arises from the spines of the posterior thoracic and lumbar vertebrae by means of the lumbar fascia and is inserted into the ribs from the ninth to the thirteenth.

*Nerve-supply.*—Posterior divisions of the spinal nerves.

*Blood-supply.*—From subscapular artery by the dorsal thoracic branch.

**The Diaphragm,** according to Morrell, (1872), takes its origin from "the cartilages of all the false ribs and the last true rib, and from the superior (internal) surface of the ensiform cartilage, and is inserted between the ribs and the spine. It presents a foramen dextrum for the vena cava inferior; foramen sinistrum for the œsophagus and pneumogastric nerve; close to the vertebral column two large muscle processes, of which the right is longest and largest, termed crura, form an inverted arch beneath the aorta (hiatus aorticus) through which the thoracic duct and aorta pass."

*Nerve-supply.*—Phrenic nerve from C IV and V.

*Blood-supply.*—Superior and inferior phrenic arteries.

MUSCLES OF THE ABDOMEN

The abdominal muscles are divided into two groups; (1) the ventro-lateral muscles: (2) the posterior muscles.

*Ventro-lateral Muscles of the Abdomen*

Obliquus externus abdominis
Obliquus internus abdominis
Cremaster
Transversus abdominis
Rectus abdominis
Pyramidalis

The extent of the three thin muscles which form the lateral walls of the abdomen, namely, obliquus externus and internus abdominis, and transversus abdominis, have not been shown in sufficient entirety to warrant any reference to figures.

**M. Obliquus externus abdominis** arises by digitations, alternating with slips of serratus magnus, from the fourth to the twelfth ribs and from the lumbar fascia.

Spreading out as a thin sheet over the lateral and ventral surface of the body its insertion is rather extensive. Passing ventrally and posteriorly, its fibers are inserted on the crest of the ilium, from which they bridge the inguinal region as Poupart's ligament attaching to the anterior part of the body of the pubis. Another portion of the muscle is also inserted into the anterior part of the body of the pubis, but it is separated from the portion mentioned above, by a large triangular region, the external abdominal ring. Anterior to this latter portion the fibers pass ventral to the rectus to the linea alba, while still more anteriorly the fibers blend with the rectus.

*Nerve-supply.*—Lower intercostal nerves.

*Blood-supply.*—Superior and inferior epigastric, musculophrenic, deep circumflex iliac, and ascending branch of lateral femoral circumflex arteries.

**M. Obliquus internus abdominis** is so closely blended with transversus abdominis, which lies beneath it, that it is extremely difficult to distinguish their attachments.

The muscle arises from the dorso-lumbar fascia, the crest of the ilium, and from Poupart's ligament, and extending ventro-anteriorly, is inserted into the cartilages of the false ribs, and into the linea alba as far as the symphysis pubis. In the inguinal region, fibers of the internal oblique form the *cremaster muscle* (Fig. 90) which encloses the testis.

*Nerve-supply.*—Lower intercostals, anterior branch of the iliohypogastric, and sometimes from the ilioinguinal nerve.

*Blood-supply.*—Superior and inferior epigastric, musculophrenic, pubic branch of the obturator, and deep circumflex iliac arteries.

**M. Transversus abdominis** lying directly under obliquus internus, arises from the inner surfaces of the posterior ribs, from the dorso-lumbar fascia, from the crest of the ilium and from Poupart's ligament, and is inserted into the linea alba by an aponeurosis which passes beneath rectus abdominis.

*Nerve-supply.*—Lower intercostals, some fibers from anterior branch iliohypogastric, and sometimes from ilioinguinal nerve.

*Blood-supply.*—Superior and inferior epigastric, musculophrenic, and deep circumflex iliac arteries.

**M. Rectus abdominis** (Figs. 74, 266, 267) is a flat thin band extending the whole length of the ventral surface of the trunk from symphysis pubis to the anterior end of the sternum. It is inserted into the first rib, the medial third of the clavicle and into the manubrium (Fig. 74). In the abdominal region it is separated from the corresponding muscle of the opposite side by the linea alba. The muscle of each side arises by two, occasionally three, slips which cross the mid-line alternating with corresponding slips from the rectus abdominis of the opposite side, giving a striking interdigitation. One of the left slips crosses most superficially (Figs. 266, 267 a-d).

*Nerve-supply.*—Anterior divisions of the fifth to thirteenth thoracic nerves.

*Blood-supply.*—Superior, inferior, and deep epigastric arteries.

**M. Pyramidalis** (Fig. 91) takes its origin from the pubic crest partly covered by adductor longus. Its fibers spread fan-wise to a more extensive insertion on the sheath of rectus abdominis from the symphysis to the suspensory ligament of the penis.

*Nerve-supply.*—From the sacral plexus.

*Blood-supply.*—Inferior external pudendal artery.

### *Posterior Muscles of the Abdomen*

Psoas major
Psoas minor
Iliacus
Quadratus lumborum

**M. Psoas major** (Figs. 90, 98) takes its origin from the lateral surface of the centra and the ventral surface of the transverse processes of all the lumbar vertebrae. The muscle is quite distinctly separated into an inner and outer portion by some of the branches of the lumbar plexus. Insertion is on the lesser trochanter of the femur.

*Nerve-supply.*—Branches of the second and third lumbar via the femoral nerve.

*Blood-supply.*—Lumbar; iliolumbar; superficial circumflex iliac, and muscular branch of the deep circumflex iliac artery.

**M. Psoas minor** (Figs. 90, 98) arises from the lateral surface of the centra of the second to sixth lumbar vertebrae and is inserted into the ilio-pectineal eminence.

*Nerve-supply.*—Branch of the first lumbar nerve.

*Blood-supply.*—Lumbar, iliolumbar, superficial circumflex iliac, and muscular branch of the deep circumflex iliac artery.

**M. Iliacus** (Figs. 90, 98) arises from the fifth and sixth lumbar vertebrae and is inserted with the psoas major into the lesser trochanter of the femur.

*Nerve-supply.*—By branches of the second and third lumbar nerves.

*Blood-supply.*—Superficial circumflex iliac, muscular branch of the deep circumflex iliac, and by the ascending and descending branches of the lateral femoral circumflex artery.

**M. Quadratus lumborum** (Figs. 90, 99) is made up of a medial and a lateral portion, separated by the transverse processes of the lumbar vertebrae.

The medial portion is composed of five bundles taking their origin as follows:

From Vertebra Th10 to L2⎫
           Th11 to L3⎪ and inserting into the up-
           Th12 to L4⎬ per half of the margin
           L1   to L5⎪ of the acetabulum.
           L2   to L6⎭

The lateral portion arises from the two lowest ribs and is inserted into the border of the ilium.

*Nerve-supply.*—Mainly by the first lumbar nerve.

*Blood-supply.*—Iliolumbar and lumbar arteries.

### CAUDAL MUSCLES

**M. Flexor caudae brevis** (externus) (Figs. 90, 256) arises "from the ventral surface of all the sacral and the first four or five caudal vertebrae. The several tendons developing extend caudad with those of flexor longus,'' (Howell '26) and insert upon the flexor surface of the caudal vertebrae.

**M. Flexor caudae longus** (internus) (Figs. 90, 256) takes its origin ''from the ventral surfaces of the centra and diapophyses of the vertebrae caudad from and including the fifth lumbar. Numerous tendons are developed which are inserted onto the caudal vertebrae.'' (Howell '26).

*Nerve-supply.*—From the sacral plexus.

*Blood-supply.*—Mainly from the superior gluteal and internal pudendal arteries, and by small branches from the obturator artery.

**M. Abductor caudae internus** (Fig. 256) takes its origin from the lower half of the medial surface and ventral border of the ilium and inserts by tendon along the ventral surface of the tail with the flexor caudae.

**M. Abductor caudae externus** (Fig. 256) arising from the medial surface of the ascending ramus of the pubis and pubic symphysis, is inserted along the lateral surface of the base of the tail.

*Nerve-supply.*—By branches from the sacral plexus.

*Blood-supply.*—From the superior gluteal and the internal pudendal arteries, and by small branches from the obturator.

**M. Extensor caudae externus** (lateralis). According to Howell '26, ''its fibers have deep origin from the diapophyses of the sacral and caudal vertebrae. The long slender tendons which are developed from them pass to the dorsal part of the caudal vertebrae farther caudad.''

**M. Extensor caudae internus** (medialis) takes its origin ''from the spinous processes of the sacral and a few of the caudal vertebrae. Tendons developing are inserted into the dorsa of the vertebrae further caudad.'' (Howell '26). Neither of these extensor muscles are shown in drawings.

*Nerve-supply.*—By branches from the sacral plexus.

*Blood-supply.*—By branches of the caudal arteries.

## MUSCLES OF THE ANTERIOR APPENDAGE

Muscles connecting the anterior appendage to the vertebral column; superficial muscles of the back.

Muscles connecting the anterior appendage to the thoracic walls; muscles of the pectoral region.

Muscles of the shoulder.

Muscles of the brachium.

Muscles of the antibrachium.

Muscles of the manus.

| | | |
|---|---|---|
| Acromiotrapezius | Rhomboideus major | Levator scapulae |
| Spinotrapezius | Rhomboideus minor | Levator claviculae |
| Latissimus dorsi | Occipitoscapularis | |

**M. Acromiotrapezius** (dorso-scapularis superior, anterior trapezius, or trapezius superior) (Figs. 69, 72) is a thin triangular muscle arising from the spines of all the cervical and the first four thoracic vertebrae. The fibers converge laterally and are inserted into the acromion process and the spine of the scapula.

*Nerve-supply.*—From the spinal accessory and second and third cervical nerves through the subtrapezial plexus.

*Blood-supply.*—Superficial cervical, and descending branch of the occipital artery.

**M. Spinotrapezius** (dorso-scapularis, inferior posterior trapezius, or trapezius inferior) (Figs. 69, 72) arises from the spines of the vertebrae from the fourth thoracic to the third lumbar. Its fibers converge laterally and insert into the proximal third of the spine of the scapula.

*Nerve-supply.*—From the spinal accessory and from the second and third cervical nerves through the subtrapezial plexus.

*Blood-supply.*—By subscapular branch from axillary, and by muscular branches of the superficial cervical artery.

**M. Latissimus dorsi** (Figs. 69, 83) arises from the spines of the eighth to twelfth thoracic vertebrae and from the dorso-lumbar fascia as far as the level of the third lumbar vertebra. From this extensive origin its fibers converge and enter the axilla between the extensor and flexor muscles, giving a superficial contour known as the dorsal tendon of the axilla. The muscle is then continued as a thin narrow tendon which passes medial to the shaft of the humerus and is inserted into the upper third of the medial surface of the deltoid ridge under cover of the biceps brachii.

*Nerve-supply.*—By the thoracodorsal (long subscapular) nerve from the sixth and seventh cervical.

*Blood-supply.*—Subscapular branch of the axillary artery.

**M. Rhomboideus major** (Fig. 70) is a flat quadrangular muscle which arises from the spines of the fourth to the seventh cervical vertebrae and is inserted into the vertebral border of the scapula.

*Nerve-supply.*—Dorsal (posterior) scapular nerve from the fifth cervical.

*Blood-supply.*—Ascending cervical branch of the cervical trunk, and by a branch of the superficial cervical artery.

**M. Rhomboideus minor** (Fig. 70) is a thin band arising from the spines of the first three cervical vertebrae and is inserted into the vertebral border of the scapula at the terminus of the spine, overlapped by rhomboideus major.

*Nerve-supply.*—Dorsal (posterior) scapular nerve from the fifth cervical.

*Blood-supply.*—Ascending cervical branch of the cervical trunk, and by a branch of the superficial cervical artery.

**M. Occipitoscapularis** (levator scapulae dorsalis, rhomboideus occipitalis, or rhomboideus capitis) (Fig. 70) is a thin ribbon-like muscle which arises from the lambdoidal ridge and is inserted into the end of the spine of the scapula extending onto the vertebral border.

*Nerve-supply.*—Dorsal (posterior) scapular nerve from the fifth cervical.

*Blood-supply.*—By branches of the ascending and superficial cervical arteries.

**M. Levator scapulae** (levator anguli scapulae (Figs. 71, 74) is a flat quadrilateral muscle which arises from the lateral processes of the last four cervical vertebrae and is inserted into the medial surface of the scapula close to the vertebral border anterior to serratus magnus with which it is practically continuous.

*Nerve-supply.*—Special branch from the third cervical, and branch of the dorsal scapular nerve from the fifth cervical.

*Blood-supply.*—Superficial cervical, ascending and descending branches of the transverse cervical artery and by the ascending cervical branch of the cervical trunk.

**M. Levator claviculae** (levator scapulae ventralis) (Figs. 67, 69) arises from the transverse process of the atlas and from the ventral surface of the basioccipital with longus capitis. It extends as a narrow ribbon to its insertion into the metacromion process.

In man this muscle occasionally occurs as a separated outer part of levator anguli scapulae.

*Nerve-supply.*—By a branch from the third cervical nerve.

*Blood-supply.*—Superficial cervical artery.

MUSCLES CONNECTING THE ANTERIOR APPENDAGE TO THE THORACIC WALLS (MUSCLES OF THE PECTORAL REGION)

> Pectoralis major
> Pectoralis minor
> Subclavius
> Serratus anterior

**M. Pectoralis major** (Fig. 75) is made up of two portions, a superficial and a deeper portion. The *superficial portion* is a flat muscle arising from the anterior half of the manubrium of the sternum and inserting into the deltoid ridge just proximal to the deeper portion of the muscle.

The *deeper portion* arises from the manubrium of the sternum and from the first and second sternebrae. It is a relatively broad and thick muscle until close to its insertion, where it passes beneath the deltoid muscles and is inserted into the distal portion of the deltoid ridge.

*Nerve-supply.*—Superficial portion by the upper branch of the lateral anterior thoracic nerve, deeper portion by the lower branch of the lateral anterior thoracic from the fifth and sixth cervical nerves.

In man the pectoralis major is mainly supplied by the lateral anterior thoracic arising from the fifth, sixth, and seventh cervical nerves, but it receives in addition a filament from the medial anterior thoracic from the eighth cervical and first thoracic by means of a loop between the lateral and medial anterior thoracic nerves. This loop is not present in the rat and pectoralis major therefore receives its innervation entirely from the lateral anterior thoracic.

*Blood-supply.*—Acromiodeltoid artery, thoracoacromial branch of the axillary, and lateral (long) thoracic artery.

**M. Pectoralis minor** (Figs. 75–77) is divisible into three portions. The *first portion*, which is flat and triangular arises from the second to the fifth sternebrae and is inserted by two slips. The anterior slip is tendinous and inserts on the coracoid process of the scapula. The posterior slip is inserted into the lesser tuberosity of the humerus.

The *second portion* arises from the fifth sternebra and is inserted by a short tendon into the proximal end of the deltoid ridge.

The *third portion*, sometimes known as xiphihumeralis, arises from the xiphoid process. It is a thin ribbon-like muscle crossing beneath the first and second portions of pectoralis minor to its insertion into the coracoid process of the scapula just beneath the insertion of the anterior slip of the first portion.

*Nerve-supply.*—Lateral and medial anterior thoracic nerves.

*Blood-supply.*—Thoracoacromial branch of the axillary, and by the lateral (long) thoracic artery.

**M. Subclavius** (Fig. 78) is a small cylindrical muscle arising from the first rib and inserted into the under surface of the middle of the clavicle.

*Nerve-supply.*—By a special individual nerve from the sixth cervical.

*Blood-supply.*—Acromiodeltoid artery.

**M. Serratus anterior** (serratus magnus) (Figs. 71, 74) is a very irregular flat muscle arising by digitations from the first to the seventh ribs and inserting into the medial surface of the scapula close to its vertebral border just posterior to the levator anguli scapulae with the border of which it is nearly continuous.

*Nerve-supply.*—Long (posterior) thoracic from the sixth, seventh, and eighth cervical nerves.

*Blood-supply.*—Ascending cervical and descending branch of the transverse cervical, and by the thoracodorsal branch of the subscapular artery.

MUSCLES OF THE SHOULDER

Acromiodeltoideus
Spinodeltoideus
Subscapularis
Supraspinatus
Infraspinatus
Teres major
Teres minor

**M. Acromiodeltoideus** (Figs. 69, 72) is a triangular muscle covering the shoulder. It arises from the lateral half of the clavicle and the acromion process and is inserted into the deltoid ridge of the humerus.

*Nerve-supply.*—By a muscular branch of the axillary (circumflex) from the sixth and seventh cervical nerves.

*Blood-supply.*—Acromiodeltoid branch of the cervical trunk, deltoid and acromial branches of the thoracoacromial, the posterior humeral circumflex, and by the deltoid branch of the profunda brachii.

**M. Spinodeltoideus** (Figs. 69, 72, 80, 81) is a flat, rather triangular muscle which arises from the infraspinous fascia and anterior two-thirds of the spine of the scapula and converges toward its insertion into the deltoid ridge of the humerus.

*Nerve-supply.*—By muscular branch of the axillary nerve from the sixth and seventh cervical.

*Blood-supply.*—Subscapular and posterior humeral circumflex arteries.

**M. Subscapularis** (Figs. 68, 79) arises from the whole subscapular fossa and is inserted by a tendon into the dorsal border of the lesser tuberosity of the humerus.

*Nerve-supply.*—Upper (short) and lower (middle) subscapular nerves, the former arising from the fifth and sixth cervical, the latter from the sixth and seventh cervical nerves.

*Blood-supply.*—Subscapular branch of transverse scapular, subscapular artery from the axillary, and by dorsal (circumflex) scapular branch of thoracodorsal.

**M. Supraspinatus** (Figs. 79, 81) is a thick muscle occupying the whole supra-spinous fossa of the scapula. It arises from the anterior margin of the scapula, from its vertebral border as far as the spine, from the superior surface of the spine, and from the whole supraspinous fossa. It is inserted by a tendon into the anterior margin of the head of the humerus.

*Nerve-supply.*—Suprascapular nerve from the fifth and sixth cervical.

*Blood-supply.*—Supraspinous branch of transverse scapular and supraspinous branch of superficial cervical artery.

**M. Infraspinatus** (Figs. 80, 81) occupies the infraspinous fossa of the scapula. It arises from the ventral surface of the spine of the scapula, from the vertebral border below the spine, and from the distal third of the axillary border. It is inserted into the larger tubercle of the humerus.

*Nerve-supply.*—Suprascapular nerve from the fifth and sixth cervical.

*Blood-supply.*—Infraspinous branch of transverse scapular artery and by a branch of the circumflexo-subscapular trunk.

**M. Teres major** (Figs. 79–82) arises from the proximal two-thirds of the axillary border of the scapula and is inserted into the medial surface of the shaft of the humerus.

*Nerve-supply.*—Posterior division of axillary (circumflex) from the sixth and seventh cervical nerves.

*Blood-supply.*—Subscapular artery from axillary.

**M. Teres minor** (Fig. 80) arises from the ventral third of the axillary border of the scapula and is inserted into the larger tubercle of the humerus just distal to infra-spinatus. It is so closely associated with infraspinatus that it would be called part of it if the former were not supplied by the axillary nerve.

*Nerve-supply.*—Posterior division of the axillary nerve from the sixth and seventh cervical.

*Blood-supply.*—Posterior humeral circumflex artery.

MUSCLES OF THE BRACHIUM

Coracobrachialis
Biceps brachii
Brachialis
Triceps brachii
Dorso-epitrochlearis brachii
Anconeus

**M. Biceps brachii** and **coracobrachialis** (Fig. 84). Biceps brachii takes its origin by two heads, the short head and the long head. The *short head* of biceps brachii and *coracobrachialis* arise by a common tendon from the coracoid process of the scapula. In addition, coracobrachialis derives some fibers from the extensor surface of the short head of biceps. The musculocutaneous nerve frequently perforates the muscle between its two parts. Half way down the humerus the two muscles separate. The short head of the biceps unites with the main portion of the long head and is inserted into the tuberosity of the radius, while coracobrachialis is inserted into the distal half of the shaft of the humerus. The *long head* of biceps arises from the anterior edge of the glenoid cavity. Its tendon passes through the intertubercular sulcus and the muscle divides into two parts. The main portion with that of the short head is inserted into the tuberosity of the radius. A secondary portion is inserted by a fibrous band (the lacertus fibrosus, semilunar or bicipital fascia), into the fascia covering pronator teres.

*Nerve-supply.*—Musculocutaneous from the fifth, sixth, and seventh cervical nerves. Coracobrachialis may also receive a branch from the radial nerve.

*Blood-supply.*—Radial artery.

**M. Brachialis** (brachialis anticus) (Fig. 85) arises from the larger tubercle and neck of the humerus, curves about the shaft, and is inserted by a tendon which passes with the tendon of the biceps brachii between extensor carpi radialis and pronator teres to the medial surface of the ulna just distal to the coronoid process.

*Nerve-supply.*—Musculocutaneous nerve from the fifth, sixth, and seventh cervical, and by a small branch of the radial (musculospiral). In man its innervation is mainly from musculocutaneous but the portion arising from the intermuscular septum is innervated by the radial (musculospiral) nerve.

*Blood-supply.*—Muscular branch of a. profunda brachii.

**M. Triceps brachii** (Figs. 83–86) is made up of three very distinct parts, the long head, the lateral head, and the medial head.

The *long head* (Fig. 85) arises from the ventral third of the axillary border of the scapula by a broad tendinous attachment.

The *lateral head* (Fig. 86) arises from the larger tubercle of the humerus.

The *medial head* (Fig. 84) arises from the proximal two-thirds of the shaft of the humerus.

All three parts are inserted into the olecranon of the ulna.

*Nerve-supply.*—Radial (musculospiral) nerve from the sixth, seventh and eighth cervical.

*Blood-supply.*—Scapular circumflex (dorsal scapular) branch of the subscapular, and the posterior humeral circumflex arteries.

**M. Epitrochleoanconeus** (epitrochlearis, extensor antibrachii longus, or extensor parvus antibrachii) (Fig. 83) is a thin muscle arising from the tendon of insertion of latissimus dorsi and inserting into the medial epicondyle of the humerus. This muscle occurs in about eighteen to twenty percent of cases in man, but was found in all the rats examined.

*Nerve-supply.*—Ulnar nerve.

*Blood-supply.*—Muscular branch of a. profunda brachii.

**M. Anconeus** (Fig. 83) is a very short muscle arising from the lateral epicondyle of the humerus and inserting into the olecranon of the ulna.

*Nerve-supply.*—Radial (musculospiral) nerve.

*Blood-supply.*—Dorsal interosseous recurrent and ulnar collateral arteries.

#### MUSCLES OF THE ANTIBRACHIUM

The muscles of the forearm are divisible into a *volar* and a *dorsal* group.

*Volar Muscles of the Antibrachium*

Superficial
      Pronator teres
      Flexor carpi radialis
      Palmaris longus
      Flexor carpi ulnaris
      Flexor digitorum sublimis
Deep
      Flexor digitorum profundus
      Pronator quadratus

**M. Pronator teres** (Figs. 87, 88) takes its origin from the medial epicondyle of the humerus on the radial side of the flexor carpi radialis, separated from extensor carpi radialis by the insertion of brachialis and biceps brachii. It inserts in the middle of the medial surface of the shaft of the radius.

*Nerve-supply.*—Musculocutaneous nerve.

*Blood-supply.*—Ulnar collateral artery.

**M. Flexor carpi radialis** (Figs. 87, 88) arises from the medial epicondyle of the humerus in association with the radial border of the superficial head of flexor digitorum profundus. Its tendon passes under the radial sesamoid (falciformis) to insert on the proximal end of the third metacarpal.

*Nerve-supply.*—Musculocutaneous nerve.

*Blood-supply.*—Ramus anastomoticus radialis.

**M. Palmaris longus** (Fig. 88) arises from the medial epicondyle of the humerus in association with the ulnar border of the superficial head of flexor digitorum profundus and inserts in the palmar fascia covering the ventral surface of the manus.

*Nerve-supply.*—Median nerve.

*Blood-supply.*—Ulnar collateral artery.

**M. Flexor carpi ulnaris** (Figs. 82, 88) arising from the medial surface of the olecranon and medial epicondyle of the humerus inserts on the pisiform bone.

*Nerve-supply.*—Ulnar nerve.

*Blood-supply.*—Ulnar collateral artery.

**M. Flexor digitorum sublimis** (Fig. 88) has its origin on the medial epicondyle of the humerus in association with the ulnar head of flexor digitorum profundus. It divides into four tendons which pass superficial to the tendon of flexor digitorum profundus to reach the manus and the ventral faces of digits two to five, where each tendon divides, in the region of the metacarpals, into two slips which pass, one on either side of the proximal phalanx, around the tendon of flexor digitorum profundus, to be inserted into the proximal end of the middle phalanx.

*Nerve-supply.*—Median nerve.

*Blood-supply.*—Musculo-anastomotic ramus of the ulnar artery.

**M. Flexor digitorum profundus** (Fig. 88) arises by four heads; (a) *superficial head* from the medial epicondyle of the humerus; (b) *ulnar head* from the medial epicondyle of the humerus; (c) *radial head* from the proximal part of the flexor surface of the radius; (d) *middle head* from the flexor surface of the ulna. Its tendon expands into a broad sheath lying immediately dorsal to the tendon of flexor digitorum sublimis; from this sheath four tendons are given off one to each of the digits from two to five, each tendon passing along the ventral face of the digit and inserting into its distal phalanx. Each tendon passes between the two slips into which the corresponding tendon of flexor digitorum sublimis is divided.

*Nerve-supply.*—Ulnar and volar interosseous branch of the median nerve.

*Blood-supply.*—Musculo-anastomotic ramus of the ulnar artery.

**M. Pronator quadratus** under cover of the long flexor muscles, arises from the flexor surface of the distal fourth of the ulna, and inserts on the flexor surface of the distal fourth of the radius. This small deep muscle is not shown in any drawing.

*Nerve-supply.*—Volar interosseous branch of the median nerve.

*Blood-supply.*—Volar interosseous artery.

*Dorsal Muscles of the Antibrachium*

Superficial

| | |
|---|---|
| Brachioradialis (Supinator longus) lacking. | Extensor digiti quarti |
| Extensor carpi radialis longus | Extensor digiti quinti proprius |
| Extensor carpi radialis brevis | Extensor carpi ulnaris |
| Extensor digitorum communis | |

Deep
    Supinator
    Abductor pollicis
    Extensor pollicis brevis
    Extensor pollicis longus
    Extensor indicis proprius

**M. Extensor carpi radialis longus** (Fig. 87) takes its origin from the lateral epicondyle of the humerus and inserts on the radial side of the distal end of the second metacarpal.

**M. Extensor carpi radialis brevis** (Fig. 87) arises from the lateral epicondyle of the humerus closely associated·with longus and inserts on the radial side of the distal end of the third metacarpal. The tendons of longus and brevis pass together under the tendon of extensor pollicis longus and brevis.

*Nerve-supply.*—Radial nerve.

*Blood-supply.*—A. transversa cubiti.

**M. Extensor digitorum communis** (Fig. 87) arises from the lateral epicondyle of the humerus. The muscle divides into four slips which pass over into tendons in the carpal region. These tendons pass under the annular ligament to their insertion at the base of the distal phalanx of digits two to five.

*Nerve-supply.*—Radial nerve.

*Blood-supply.*—Dorsal interosseous artery.

**M. Extensor digiti quarti** (Fig. 87) takes its origin from the lateral epicondyle of the humerus just lateral to extensor communis digitorum. In the region of the carpus its tendon divides into two slips which pass through a perforation in the annular ligament to the base of the distal phalanx of digit four.

*Nerve-supply.*—Radial nerve.

*Blood-supply.*—Interosseous recurrent artery.

**M. Extensor digiti quinti proprius** (Fig. 87) arises from the lateral epicondyle of the humerus just lateral to extensor communis digitorum and runs to the base of the proximal phalanx of the fifth digit.

*Nerve-supply.*—Radial nerve.

*Blood-supply.*—Interosseous recurrent artery.

**M. Extensor carpi ulnaris** (Fig. 87) springs from the lateral epicondyle of the humerus just lateral to extensor quarti and quinti digiti, and passing along the ulnar border of the forearm is inserted into the base of the fifth metacarpal.

*Nerve-supply.*—Radial nerve.

*Blood-supply.*—Interosseous recurrent artery.

**M. Supinator** arises from the dorsal region of the lateral epicondyle and supinator ridge of the humerus, and inserts upon the medial surface of the shaft of the radius slightly distal to the insertion of pronator teres. This muscle is concealed by extensor carpi radialis and hence is not shown with other muscles of the group.

*Nerve-supply.*—Deep ramus of the radial nerve.

*Blood-supply.*—Interosseous recurrent artery.

**M. Abductor pollicis** (Fig. 89) has its origin on the falciform and inserts on the radial side of the base of the first phalanx of the pollex.

*Nerve-supply.*—Muscular branch of the radial nerve.

*Blood-supply.*—Mediano-radial artery.

**M. Extensor pollicis brevis** (Fig. 87) or abductor pollicis longus, takes its origin from the ulnar surface of the radius and from the interosseous membrane just distal to extensor pollicis longus, and inserts on the base of the first phalanx of the pollex.

*Nerve-supply.*—Branch of the radial nerve.

*Blood-supply.*—From the anastomosis between dorsal interosseous and mediano-radial arteries.

**M. Extensor pollicis longus** (Fig. 87) from the extensor surface of the shaft of the ulna and from the interosseous membrane just proximal to the origin of extensor indicis, crosses tendons of extensor carpi radialis to the base of the second or distal phalanx of the pollex.

*Nerve-supply.*—Radial nerve.

*Blood-supply.*—From the anastomosis between dorsal interosseous and mediano-radial arteries.

**M. Extensor indicis proprius** (Fig. 87) arises from the extensor surface of the ulna just distal to the origin of extensor pollicis. Its tendon divides into two slips which cross beneath the tendon of the extensor digitorum communis and pass under the annular ligament to insert at the base of the distal phalanx of the second and third digits.

*Nerve-supply.*—Radial nerve.

*Blood-supply.*—Dorsal interosseous artery.

### MUSCLES OF THE MANUS

Lateral volar muscles (muscles of the pollex)
    Flexor pollicis brevis
    Adductor pollicis
Medial volar muscles (muscles of the fifth digit)
    Abductor digiti quinti
    Flexor digiti quinti brevis
    Opponens digiti quinti
Intermediate muscles (interosseous)
    Lumbricales
    Interossei

*Lateral Volar Muscles of the Pollex*

**M. Flexor pollicis brevis** (Fig. 89) arises from the annular ligament distal to abductor pollicis and inserts on the outer side of the base of the first phalanx of the pollex.

*Nerve-supply.*—Median nerve.

*Blood-supply.*—Mediano-radial artery.

**M. Adductor pollicis** (Figs. 88, 89) arising from the os magnum and from the bases of the second and third metacarpals, inserts upon the inner side of the base of the first phalanx of the pollex.

*Nerve-supply.*—Ulnar nerve.

*Blood-supply.*—Mediano-radial artery.

### Medial Volar Muscles of the Hand

**M. Abductor digiti quinti** (Fig. 88, 89) takes its origin from the pisiform and inserts upon the base of the first phalanx of the fifth digit.

*Nerve-supply.*—Ulnar nerve.

*Blood-supply.*—Ulnar artery.

**M. Flexor digiti quinti brevis** (Fig. 89) arises from the unciform and inserts upon the inner side of the base of the first phalanx of the fifth digit. Parsons ('96) speaks of this as the ulnar slip of the interosseous muscle to the fifth digit.

*Nerve-supply.*—Ulnar nerve.

*Blood-supply.*—Ulnar artery.

**M. Opponens digiti quinti** (Fig. 89) takes its origin from the pisiform and triangular bones and inserts on the ulnar margin of the distal end of the fifth metacarpal.

*Nerve-supply.*—Ulnar nerve.

*Blood-supply.*—Ulnar artery.

### Intermediate Muscles of the Manus or Hand

**Mm. Lumbricales** (Fig. 88) take their origin from the ventral surface of the tendon of flexor digitorum profundus at the place where it divides into slips for digits two to five. They insert on the proximal end of the first phalanx on the preaxial side of digits two, three, four and five.

*Nerve-supply.*—Those portions of the muscles which insert on the second and third digits are innervated by the median, while those which insert on the fourth and fifth digits are supplied by the ulnar nerve.

*Blood-supply.*—Volar metacarpal arteries from the volar arch.

**Mm. Interossei** (Fig. 89) are small stout muscles lying on the palmar surfaces of the metacarpals of digits two, three, and four. They are divided into *interossei dorsales* and *interossei volares*.

The *interossei dorsales* arise from the ventro-lateral surface of the second, third, and fourth metacarpals with an additional slip on the preaxial side of the fifth, taking its origin from the fourth metacarpal. The muscle on the preaxial side of the third arises from two slips from the second and third metacarpals. The muscle on the postaxial side of the second arises from the third metacarpal. The muscle on the preaxial side of the second arises from the first metacarpal. Near the distal end of the metacarpal each muscle divides into two slips which pass to the lateral surface of the metacarpal, leaving the ventral surface of the distal end exposed, and insert on the lateral surface of the base of the phalanx and onto its sesamoids and partly by slender tendons which continue dorsally to join the extensor tendon of the digit.

The *volares* are three in number, extending the entire length of the palmar surface of the metacarpals, ventral to the dorsales. From the tendon of origin the fibers diverge at the distal portion of the muscle and insert onto the sesamoids at the metacarpophalangeal joint with the tendons of the dorsales.

*Nerve-supply.*—Ulnar nerve.

*Blood-supply.*—Metacarpal arteries.

## MUSCLES OF THE POSTERIOR APPENDAGE

### MUSCLES OF THE THIGH

#### *Anterior Femoral Muscles*

**M. Extensor quadriceps femoris** is the name given to a group of four muscles which are quite distinct at their origin but unite in a common tendon (the ligamentum patellae in which the patella is imbedded), and become inserted on the crest of the tibia. The muscles of the group are *rectus femoris, vastus lateralis, vastus medialis,* and *vastus intermedius.*

**M. Rectus femoris** (Figs. 98, 99) has two heads arising by a bifurcated tendon, the *posterior (reflected) head* from the anterior border of the acetabulum, and the *anterior (straight) head* from the inferior ventral spine of the ilium. Its insertion is by the ligamentum patellae into the tuberosity of the tibia.

*Nerve-supply.*—Posterior division of the femoral nerve.

*Blood-supply.*—Superficial circumflex iliac, ascending and descending branches of lateral femoral circumflex, and lateral superior genicular arteries.

**M. Vastus lateralis** (vastus externus) (Figs. 94–97) arises from the greater trochanter and the third trochanter and is inserted by the ligamentum patellae into the tuberosity of the tibia.

*Nerve-supply.*—Posterior division of the femoral nerve.

*Blood-supply.*—Ascending and descending branches of lateral femoral circumflex, superficial circumflex iliac, superior muscular branch of popliteal, and lateral superior genicular arteries.

**M. Vastus medialis** (vastus internus) (Fig. 98) arises from the neck and proximal end of the shaft of the femur and is inserted by the ligamentum patellae into the tuberosity of the tibia.

*Nerve-supply.*—Posterior division of the femoral nerve.

*Blood-supply.*—Superficial circumflex iliac, descending branch of lateral femoral circumflex, and muscular branch of genu suprema.

**M. Vastus intermedius** (crureus) (Figs. 98, 99) arises from the whole length of the extensor surface of the shaft of the femur and is inserted by the ligamentum patellae into the tuberosity of the tibia.

*Nerve-supply.*—Posterior division of the femoral nerve.

*Blood-supply.*—Superficial circumflex iliac, descending branch of lateral femoral circumflex, muscular branch of genu suprema, and by lateral superior genicular arteries.

### *Medial Femoral Muscles*

**M. Gracilis anticus** (Figs. 98, 99) arises from the posterior half of the symphysis pubis and is inserted into the upper part of the crest and medial border of the tibia covering the insertion of gracilis posticus.

*Nerve-supply.*—Anterior division of the obturator nerve.

*Blood-supply.*—Muscular branches of the femoral artery.

**M. Gracilis posticus** (Figs. 98, 99) arises from the ramus of the ischium and is inserted by a tendon into the tuberosity of the tibia beneath the insertion of gracilis anticus, and proximal to the insertion of semitendinosus.

*Nerve-supply.*—Anterior division of the obturator nerve.

*Blood-supply.*—Muscular branch from the femoral and by the inferior external pudendal artery.

**M. Adductor longus** (Figs. 98, 99) arises from the anterior end of the ascending ramus of the pubis and is inserted into the shaft of the femur proximal to adductor brevis.

*Nerve-supply.*—Anterior division of the obturator nerve.

*Blood-supply.*—Muscular branch of the femoral, pubic branch of pudic-epigastric trunk and by the medial femoral circumflex arteries.

**M. Adductor magnus** (Figs. 98, 99) arises from the posterior region of the ascending ramus of the pubis and from the pubic symphysis, overlapping the origin of the adductor brevis, and is inserted into the tuberosity of the tibia proximal to the insertion of gracilis anticus.

*Nerve-supply.*—By the posterior division of the obturator nerve.

*Blood-supply.*—Medial femoral circumflex, lateral superior genicular, and muscular branches of the femoral and genu suprema.

**M. Adductor brevis** (Figs. 98, 99) arises from the ascending ramus of the pubis, from the symphysis pubis, where it is overlapped by adductor magnus, and from the medial half of the ramus of the ischium. Its insertion is into the third trochanter and into the flexor surface of the distal half of the shaft of the femur from lateral to medial ridge.

*Nerve-supply.*—Posterior division of the obturator nerve.

*Blood-supply.*—Medial femoral circumflex, muscular branch of femoral, muscular branch of genu suprema, lateral superior and middle genicular.

**M. Pectineus** (Figs. 98, 99) takes its origin from the pubic arch and the iliopectineal tubercle and is inserted into the medial ridge of the shaft of the femur proximal to the adductor longus.

*Nerve-supply.*—Femoral nerve.

*Blood-supply.*—Pubic branch of pudic-epigastric trunk, superficial circumflex iliac, and medial femoral circumflex arteries.

## Muscles of the Gluteal Region

**M. Sartorius** is absent in the rat as a distinct muscle. "Gluteus maximus, tensor fasciae femoris (latae) and sartorius are so closely united in Rodents that they form practically one sheet." (Parsons.)

**M. Tensor fasciae latae** (tensor vaginae femoris) (Figs. 94–97) is a thin fan-shaped muscle of the extensor and lateral surface of the thigh. It arises from the crest of the ilium and spreads out to be inserted into the fascia lata of the thigh. Its posterior edge is almost inseparably blended with the gluteus maximus.

*Nerve-supply.*—Superior gluteal nerve.

*Blood-supply.*—Iliolumbar, deep branch of the gluteal, and ascending branch of the lateral femoral circumflex arteries.

**M. Gluteus maximus** (Figs. 94–96) is a thin muscle arising by fascia from the dorsal border of the ilium, from the last three sacral and the first caudal vertebrae. Anteriorly it becomes inseparably united with tensor fasciae latae; posteriorly its origin is covered by the anterior head of the biceps femoris. It is inserted by a tendon slightly distal to the greater trochanter of the femur into the third trochanter.

*Nerve-supply.*—Inferior branch of the superior gluteal, and by the inferior gluteal nerve.

*Blood-supply.*—Inferior branch of superior gluteal, inferior gluteal, medial femoral circumflex, and ascending branch of lateral femoral circumflex arteries.

**M. Gluteus medius** (Figs. 96, 97) is covered by gluteus maximus. It arises from the dorsal border, anterior and lateral crest, and ventral border of the ilium, and by fascia from the sacrum, and is inserted into the greater trochanter.

*Nerve-supply.*—Superior gluteal nerve.

*Blood-supply.*—Superior and inferior branches of superior gluteal, inferior gluteal, medial femoral circumflex, ascending and descending branches of lateral femoral circumflex arteries.

**M. Gluteus minimus** (Fig. 97) is covered by gluteus medius. It arises from the dorsal border of the ilium, posterior to the sciatic notch, from the lateral crest, and from the lateral surface ventral to the lateral crest. Its insertion is into the greater trochanter.

*Nerve-supply.*—Superior gluteal nerve.

*Blood-supply.*—Ascending and descending branches of lateral femoral circumflex, and inferior branch of superior gluteal artery.

**M. Piriformis** (Figs. 97, 284) arises from the sacrum and is practically continuous with gluteus medius. In fact the separation is difficult to distinguish except in specimens muscularly well developed, and then only at its origin. Insertion is with gluteus medius into the greater trochanter.

*Nerve-supply.*—By a branch from the lumbo-sacral plexus. This nerve also supplies obturator internus.

*Blood-supply.*—Inferior gluteal, superior and inferior branches of superior gluteal, and anterior branch of obturator artery.

**M. Obturator externus** (Figs. 96, 97) is a short rather thick muscle which arises from the lateral aspect of the obturator membrane and from the bony margin of the obturator foramen. It is inserted into the trochanteric fossa between quadratus femoris and obturator internus.

*Nerve-supply.*—Posterior division of the obturator nerve.

*Blood-supply.*—Anterior and posterior branches of obturator artery, and medial femoral circumflex.

**M. Obturator internus** (Figs. 96, 97, 283) arises from the medial surface of the ischium, passes over into a tendon which crosses the dorsal border of the ischium and is inserted into the trochanteric fossa of the femur. The muscle does not cover the obturator membrane as in man.

*Nerve-supply.*—By a special nerve from the sacral plexus.

*Blood-supply.*—Obturator and inferior gluteal arteries.

**Mm. Gemelli superior and inferior.** (Fig. 284). These are two fan-shaped muscles situated on either side of the tendon of obturator internus. *Gemellus superior* arises from the anterior part of the dorsal border of the ischium, closely associated with gluteus minimus, and joins the tendon of obturator internus to be inserted into the trochanteric fossa. Gemellus superior has sometimes been mistaken for piriformis. *Gemellus inferior* arises from the postero-dorsal margin of the ischium and is inserted into the trochanteric fossa.

*Nerve-supply.*—Nerves from the sacral plexus.

*Blood-supply.*—To gemellus inferior and superior, by obturator and inferior gluteal arteries; to gemellus superior by ascending branch of lateral femoral circumflex, and by superior branch of superior gluteal.

**M. Quadratus femoris** (Figs. 96, 97) arises from the posterior border of the ischium and is inserted into the posterior side of the shaft of the femur, slightly below the level of the greater trochanter, into the lesser trochanter.

*Nerve-supply.*—Posterior division of the obturator nerve.

*Blood-supply.*—Medial femoral circumflex and anterior branch of obturator artery.

### Posterior Femoral Muscles (Hamstring Muscles)

**M. Semitendinosus** (Figs. 94-97) arises by two heads, the principal and the accessory head. The *principal head* arises from the posterior part of the sciatic tuber. It is joined almost immediately by the *accessory head* which arises from the last sacral and first two caudal vertebrae. The muscle is inserted onto the medial side of the leg by the same fascia as the gracilis, into the distal end of the tuberosity of the tibia.

*Nerve-supply.*—*Principal head* by the tibial division of the sciatic nerve. *Accessory head* by a branch of the lumbo-sacral plexus.

*Blood-supply.*—Inferior external pudendal, medial femoral circumflex, posterior scrotal, and superior muscular branch of popliteal artery.

**M. Semimembranosus** (Figs. 96-99) springs from the posterior edge of the ischium and sciatic tuber and is inserted into the ridge and medial surface of the tibia and into the medial fabella, covered by gracilis anticus and pierced by the internal ligament.

*Nerve-supply.*—Tibial division of the sciatic nerve.

*Blood-supply.*—Medial femoral circumflex and muscular branch of femoral artery.

**M. Caudo-femoralis** (Figs. 95, 96) arises from the posterior sacral and first caudal vertebrae, closely associated with the anterior head of biceps femoris, and is inserted into the flexor surface of the femur from the internal condyle and the internal (medial) fabella, to the lateral condyle and the external (lateral) fabella. This muscle is not present in man. In other comparative works it is sometimes called a second part of semimembranosus.

*Nerve-supply.*—Tibial division of the sciatic nerve.

*Blood-supply.*—Medial femoral circumflex, posterior scrotal, inferior gluteal, muscular branch of femoral, and muscular branch of a. genu suprema.

**M. Biceps femoris** (Fig. 94-96) so-called because of the condition in man from which it was originally named, is here made up of three parts, the *anterior head*, corresponding to the short head in man; the *posterior head* and the *accessory head*, which together correspond to the long head in man.

The *anterior head* arises from the last sacral and the first caudal vertebrae with the caudo-femoralis muscle.

The *posterior head* arises from the sciatic tuber anterior to the accessory head.

The *accessory head* arises from the sciatic tuber with the semitendinosus.

All three heads are inserted by a strong tendinous sheath into the distal end of the femur and the proximal two-thirds of the tibia.

*Nerve-supply.*—In man the nerve to the *anterior head* is a branch of the common peroneal, after the division of the lumbo-sacral trunk into the common peroneal and tibial. In the rat the nerve to the anterior head arises before this division in common with a nerve to gluteus maximus. From its origin, namely, via the lumbo-sacral trunk, and from its relationship, this nerve may be identified as the inferior gluteal. The *posterior* and *accessory* heads are supplied by the tibial division of the sciatic. In man the common peroneal division of the sciatic is derived from the fifth lumbar and first and second sacral nerves, and the tibial division from the first three sacral.

*Blood-supply.*—Medial femoral circumflex, posterior scrotal, inferior gluteal, superior muscular branch of popliteal, lateral superior genicular, and lateral inferior genicular arteries.

### MUSCLES OF THE LEG

#### Anterior Crural Muscles

**M. Tibialis anterior** (Figs. 100, 101) has its origin on the margin of the lateral condyle, tuberosity, and ventral crest of the tibia. Its tendon passes under the annular ligament, crosses to the medial surface and inserts on the first cuneiform and on the proximal end of the first metatarsal.

*Nerve-supply.*—Deep peroneal nerve.

*Blood-supply.*—Anterior tibial recurrent, and muscular branch of anterior tibial arteries.

**M. Extensor digitorum longus** (Fig. 100) arises from the lateral epicondyle of the femur. The muscle divides into four parts; the tendons pass under the annular ligament, lateral to the tendon of tibialis anterior, and insert on the base of the third phalanx of digits two to five.

*Nerve-supply.*—Deep peroneal nerve.

*Blood-supply.*—Muscular branch of anterior tibial artery.

**M. Extensor hallucis longus** (extensor hallucis proprius) (Fig. 100) takes its origin from the distal quarter of the fibula and from the interosseous membrane. Its tendon passes under the annular ligament with the tendons of the tibialis anterior and extensor digitorum longus to the base of the terminal phalanx of the hallux.

*Nerve-supply.*—Deep peroneal nerve.

*Blood supply.*—Muscular branch of anterior tibial artery.

### Posterior Crural Muscles: Superficial

**M. Triceps surae** consists of two muscles, gastrocnemius, and soleus, arising independently but inserting together.

**M. Gastrocnemius** (Figs. 100, 101) is made up of two heads, medial and lateral. The *medial head* arises from the medial epicondyle of the femur and from the medial fabella. The *lateral head* takes its origin from the lateral epicondyle and from the lateral fabella. Their tendons are twisted with that of plantaris. (Parsons '94.)

**M. Soleus** (Figs. 100, 101) arises by a slender tendon from the head of the fibula. Both muscles unite in a strong tendon and insert on the tuber calcanei.

*Nerve-supply.*—Tibial nerve.

*Blood-supply.*—Middle and lateral inferior genicular and external sural arteries, to lateral head; external sural and fibular branch of peroneal to soleus; and internal sural to medial head of gastrocnemius.

**M. Plantaris** (Figs. 100, 101) arises from the lateral epicondyle of the femur, enwrapped by gastrocnemius, from the lateral fabella and medial border of the head of the fibula. Its tendon passes over the tuber calcanei superficial to the triceps surae and becomes continuous with the flexor digitorum brevis.

*Nerve-supply.*—Tibial nerve.

*Blood-supply.*—External sural artery.

### Posterior Crural Muscles: Deep

**M. Popliteus** (Fig. 101) springs from the lateral epicondyle of the femur and inserts on the proximal third of the medial surface of the tibia.

*Nerve-supply.*—Tibial nerve.

*Blood-supply.*—Lateral inferior genicular and posterior tibial arteries.

**M. Flexor hallucis longus** (flexor fibularis) (Fig. 101) arises from the head and medial surface of the shaft of the fibula, interosseous membrane, and flexor surface of the tibia along the dorsolateral crest. Its tendon passes under the medial malleolus and beneath the tendon of flexor digitorum brevis to divide into five slips which, with the exception of the first and fifth, pass between the tendons of flexor digitorum brevis to insert on the proximal end of the third or terminal phalanx of the second, third, and fourth digits. Those of the first and fifth digits pass directly to their insertion on the terminal phalanx of their respective digits.

*Nerve-supply.*—Tibial nerve.

*Blood-supply.*—Posterior tibial artery.

**M. Flexor digitorum longus** (flexor tibialis) (Fig. 101) arises from the tibia below popliteus, and from the head of the fibula. Its tendon passes through a groove under the medial malleolus, gives off a tendon which joins a corresponding one from the flexor hallucis longus to form the flexor tendon of the hallux, then becomes blended with flexor hallucis longus, the two forming tendons which pass between slips of flexor digitorum brevis to the terminal phalanx of the second to fifth digits.

*Nerve-supply.*—Tibial nerve.

*Blood-supply.*—Posterior tibial artery.

**M. Tibialis posterior** (Fig. 101) has its origin from the medial surface of the anterior end of the tibia, from the interosseous ligament and the proximal end of the fibula, and is inserted on the navicular and the first cuneiform.

*Nerve-supply.*—Tibial nerve.

*Blood-supply.*—Posterior tibial artery.

### Lateral Crural Muscles

**M. Peroneus longus** (Fig. 100) arising from the head of the fibula and lateral condyle of the tibia, becomes tendinous and passes under the lateral malleolus and across the flexor surface of the foot to the base of the first metatarsal and to the first cuneiform.

*Nerve-supply.*—Superficial peroneal nerve.

*Blood-supply.*—Anterior tibial recurrent and peroneal arteries.

**M. Peroneus brevis** (Fig. 100) arises from the head and shaft of the fibula and from the interosseous membrane. Its tendon passes with that of peroneus longus under the lateral malleolus to the tuberosity of the fifth metatarsal.

*Nerve-supply.*—Superficial peroneal nerve.

*Blood-supply.*—Anterior tibial recurrent and peroneal arteries.

**M. Peroneus digiti quarti** (Fig. 100) takes its origin from the head of the fibula. Its tendon passes with that of peroneus digiti quinti, peroneus longus, and peroneus brevis under the lateral malleolus to the distal end of the fourth metatarsal.

*Nerve-supply.*—Deep peroneal nerve.

*Blood-supply.*—Fibular branch of peroneal artery.

**M. Peroneus digiti quinti** (Fig. 100) arises from the proximal half of the shaft of the fibula and inserts upon the distal end of the fifth metatarsal.

*Nerve-supply.*—Deep peroneal nerve.

*Blood-supply.*—Fibular branch of peroneal artery.

*Dorsal Muscle of the Foot*

**M. Extensor digitorum brevis** (Fig. 100) is represented by two small muscles arising from the lateral process of the calcaneus and inserting on the base of the second
phalanx of digits two and three.

*Nerve-supply.*—Deep peroneal nerve.
*Blood-supply.*—Superficial sural artery.

*Plantar Muscles of the Foot*

**M. Flexor digitorum brevis** (Fig. 101) arises from the tendon of plantaris as three
slender muscles which pass over into long tendons. Each tendon, at the base of the
first phalanx, divides into two which pass around the tendon of flexor hallucis longus
and insert on the proximal end of the second phalanx of the second, third and fourth
digits.

*Nerve-supply.*—Medial plantar nerve.
*Blood-supply.*—Posterior tibial and lateral plantar arteries.

**M. Abductor digiti quinti** (Figs. 101, 102) has its origin on the calcaneus and its
insertion on the tuberosity of the fifth metatarsal.

*Nerve-supply.*—Lateral plantar nerve.
*Blood-supply.*—Lateral plantar artery.

**M. Quadratus plantae** (accessorius) (Fig. 101) takes its origin from the calcaneus
and inserts on the tendons of flexor digitorum longus and flexor hallucis longus.

*Nerve-supply.*—Lateral plantar nerve.
*Blood-supply.*—Lateral plantar artery.

**Mm. Lumbricales** (Fig. 101). Four muscles arise from the ventral surface of the
tendon of flexor hallucis longus at the place where it divides into slips for digits two
to five. Two additional slips arise from the ventral surface of the tendon of flexor
hallucis longus before it divides. The former insert on the proximal end of the first
phalanx of digits two, three, four, and five. The latter two slips insert on the preaxial
side of flexor digitorum brevis tendons to digits three and five.

*Nerve-supply.*—Those muscles which insert on the proximal end of the first phalanx
of digits three, four and five, are supplied by the lateral plantar nerve. The slip to the
first phalanx of the fifth digit is supplied by the medial plantar, as are also the two slips
inserting on the tendons.

*Blood-supply.*—Dorsal metatarsal artery.

**M. Flexor hallucis brevis** (abductor hallucis, Parsons) (Figs. 101, 102) is composed
of two parts arising from the navicular but inserting separately. One inserts on the
lateral, the other on the medial side of the base of the first phalanx of the hallux. The
latter blends with the interosseous muscle to digit I.

*Nerve-supply.*—Medial plantar nerve.
*Blood-supply.*—Superficial branch of medial plantar artery.

**M. Flexor digiti quinti brevis** (Fig. 102) arises from the flexor surface of the cuboid. At the point where it passes over the base of the fourth and fifth metatarsals the tendon contains a sesamoid bone. It inserts on the lateral side of the base of the first phalanx of the fifth digit. This muscle might readily be considered in the lumbricales series.

*Nerve-supply.*—Lateral plantar nerve.

*Blood-supply.*—Deep branch lateral plantar artery.

**M. Adductor indicis** (Fig. 102) arises from the tendon of the interossei of the third digit and inserts on the sesamoid of the second digit.

*Nerve-supply.*—Medial plantar nerve.

*Blood-supply.*—Second plantar metatarsal artery.

**Mm. Interossei plantares** (Fig. 102) are small slender muscles on the volar surface of the metatarsals. Parsons '96 says, "In no Rodent were any distinct dorsal interosseous muscles found." However, there are two interossei for digits II, III and IV, with one each for digits I and V. The tendons of muscles going to digits I and II arise from the navicular; those for III, IV and V from the cuboid. A small sesamoid is imbedded in the tendons from the cuboid. There is some blending of fibers of one interosseous muscle with another, in such a way as to defy description. An attempt has been made to show this in the figure.

*Nerve-supply.*—Lateral plantar nerve.

*Blood-supply.*—Dorsal metatarsal arteries.

## LIST OF FIGURES

### III. MUSCLES

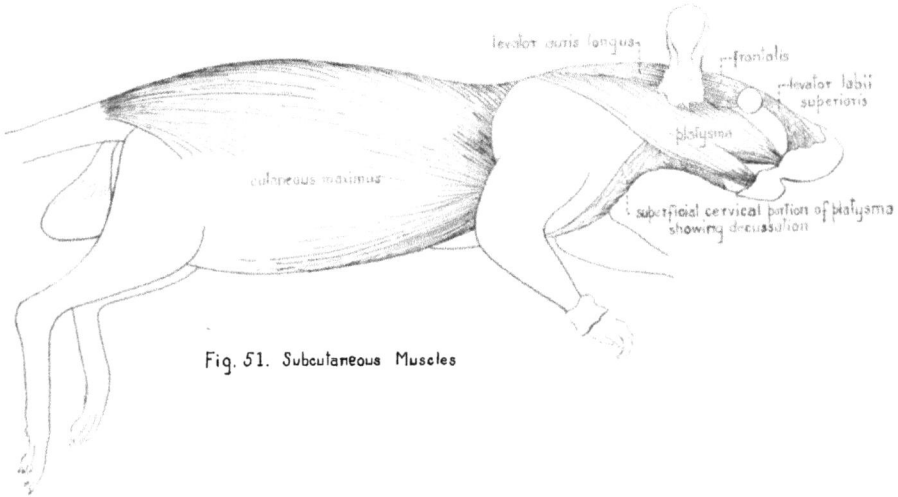

Fig. 51. Subcutaneous Muscles

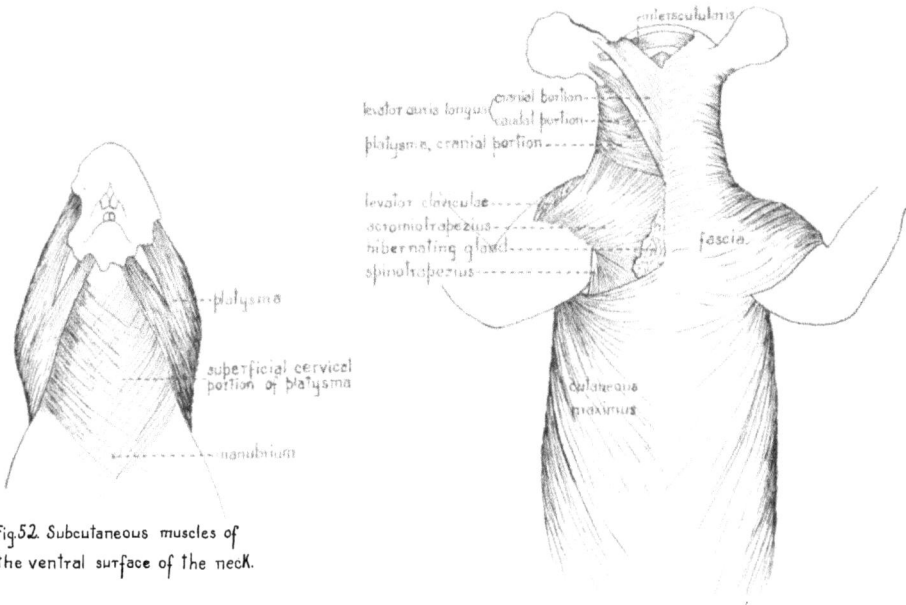

Fig. 52. Subcutaneous muscles of
the ventral surface of the neck.

Fig. 53. Dorsal aspect of subcutaneous muscles
with fascia removed from left side.

Fig. 54.

anterior superficial
posterior superficial } masseter

Fig. 55.

anterior deep masseter

buccinator

posterior deep
anterior deep

Fig. 56.

internal pterygoid
external pterygoid

Fig. 57.

temporalis

Fig. 58.

dilator naris

platysma
levator labii superioris

Fig. 59.

zygomaticus
levator labii superioris

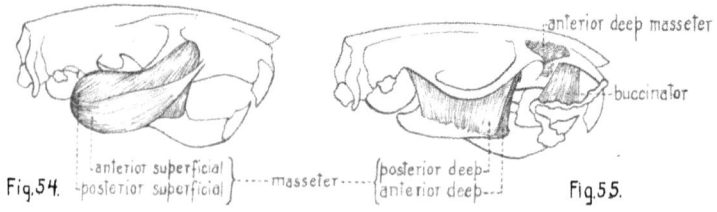

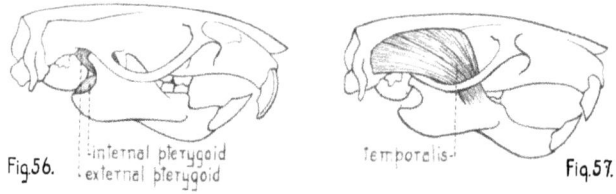

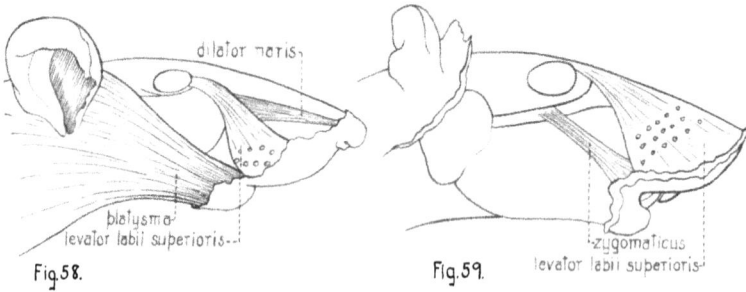

Muscles of Mastication

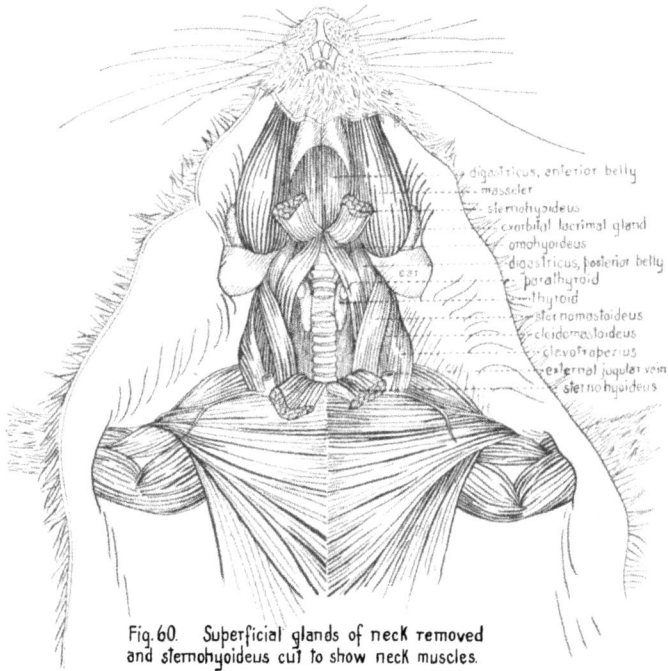

Fig. 60. Superficial glands of neck removed
and sternohyoideus cut to show neck muscles.

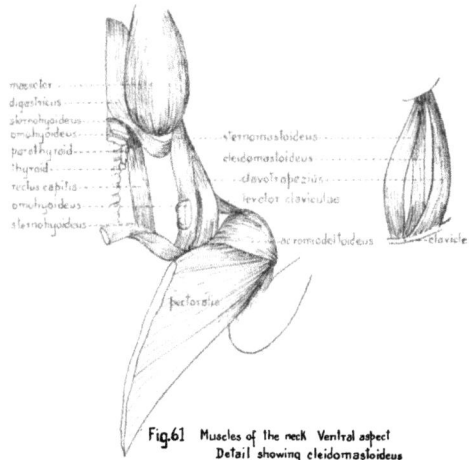

Fig.61 Muscles of the neck Ventral aspect
Detail showing cleidomastoideus

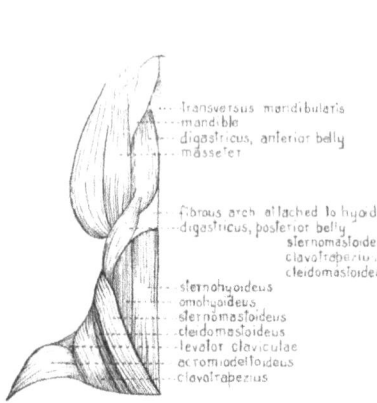

Fig 62    Lateral Cervical Muscles

transversus mandibularis
mandible
digastricus, anterior belly
masseter

fibrous arch attached to hyoid
digastricus, posterior belly
sternomastoideus
clavotrapezius
cleidomastoideus

sternohyoideus
omohyoideus
sternomastoideus
cleidomastoideus
levator claviculae
acromiodeltoideus
clavotrapezius

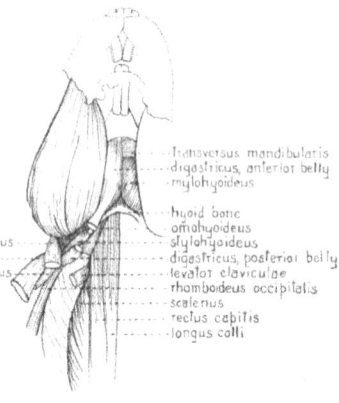

Fig 63.

transversus mandibularis
digastricus, anterior belly
mylohyoideus

hyoid bone
omohyoideus
stylohyoideus
digastricus, posterior belly
levator claviculae
rhomboideus occipitalis
scalenus
rectus capitis
longus colli

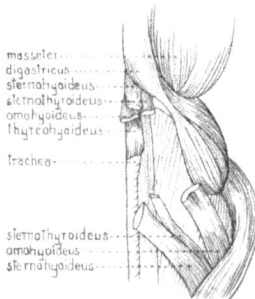

Fig 64.

masseter
digastricus
sternohyoideus
sternothyoideus
omohyoideus
thyreohyoideus

trachea

sternothyroideus
omohyoideus
sternohyoideus

Muscles of the Hyoid

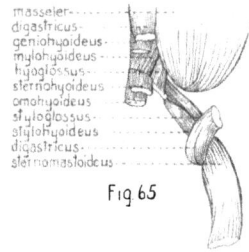

masseter
digastricus
geniohyoideus
mylohyoideus
hyoglossus
sternohyoideus
omohyoideus
styloglossus
stylohyoideus
digastricus
sternomastoideus

Fig 65

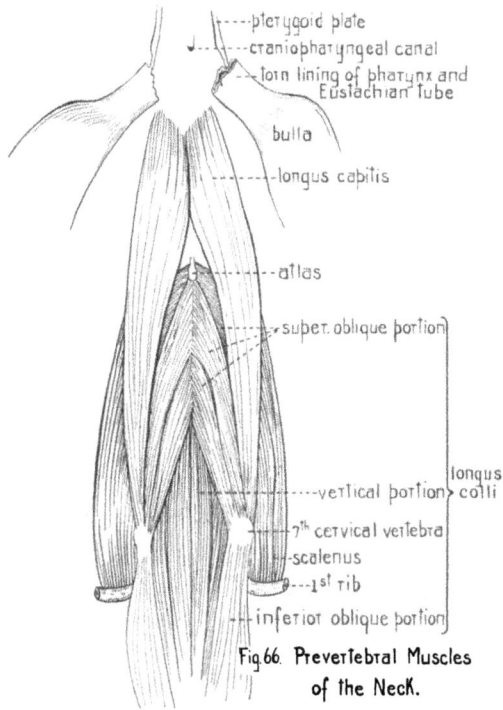

pterygoid plate
craniopharyngeal canal
torn lining of pharynx and
Eustachian tube

bulla

longus capitis

atlas

super. oblique portion

vertical portion ⎱ longus
⎰ colli
7th cervical vertebra
scalenus
1st rib
inferior oblique portion

Fig. 66. Prevertebral Muscles
of the Neck.

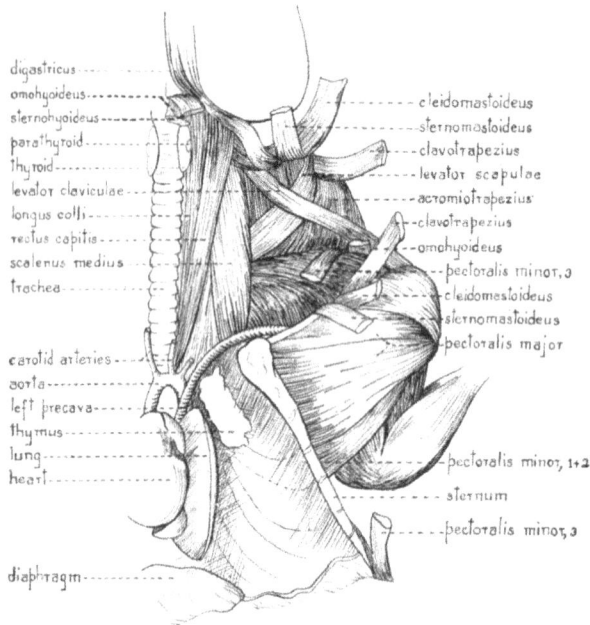

digastricus
omohyoideus
sternohyoideus
parathyroid
thyroid
levator claviculae
longus colli
rectus capitis
scalenus medius
trachea

carotid arteries
aorta
left precava
thymus
lung
heart

diaphragm

cleidomastoideus
sternomastoideus
clavotrapezius
levator scapulae
acromiotrapezius
clavotrapezius
omohyoideus
pectoralis minor, 3
cleidomastoideus
sternomastoideus
pectoralis major

pectoralis minor, 1+2
sternum
pectoralis minor, 3

Fig.67. Deep muscles of the neck   Ventral aspect.

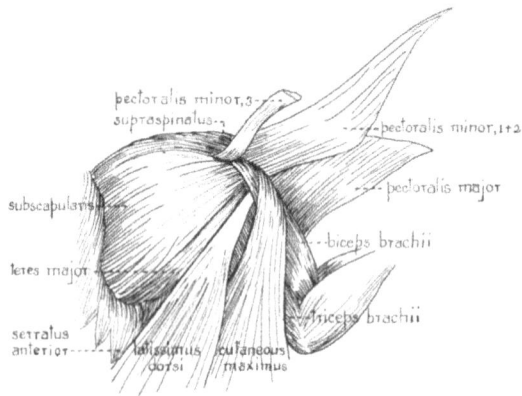

pectoralis minor, 3
supraspinatus
subscapularis
teres major
serratus anterior
latissimus dorsi
cutaneous maximus

pectoralis minor, 1+2
pectoralis major
biceps brachii
triceps brachii

Fig.68. Pectoralis cut and left arm reflected,
showing muscles of the axillary region

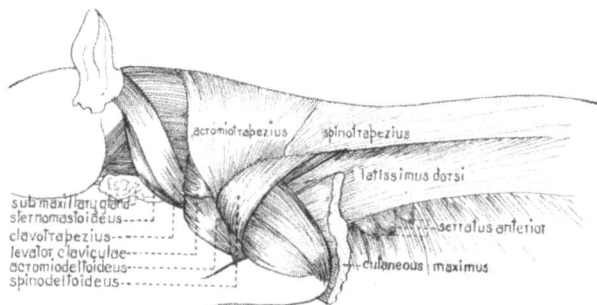

Fig. 69. Superficial muscles of the shoulder and neck. Lateral aspect.

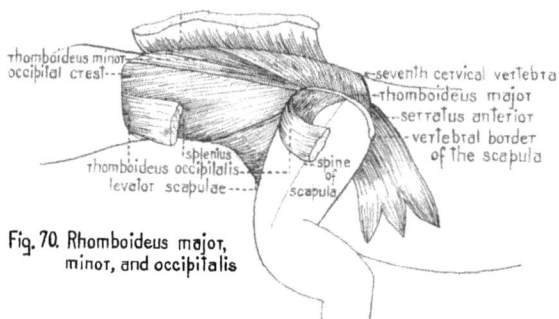

Fig. 70. Rhomboideus major, minor, and occipitalis

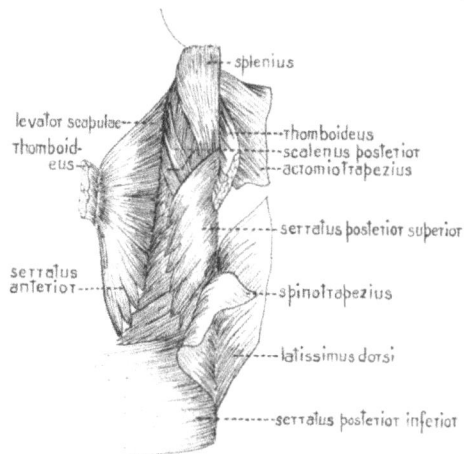

Fig. 71. Deeper muscles of the back. Dorsal aspect. Rhomboideus cut and scapula drawn away from thorax.

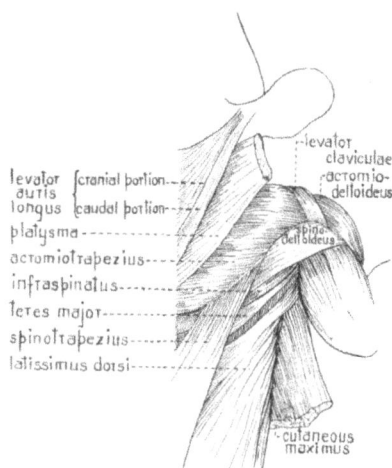

Fig. 72. Superficial muscles of the shoulder and neck, Dorsal Aspect.

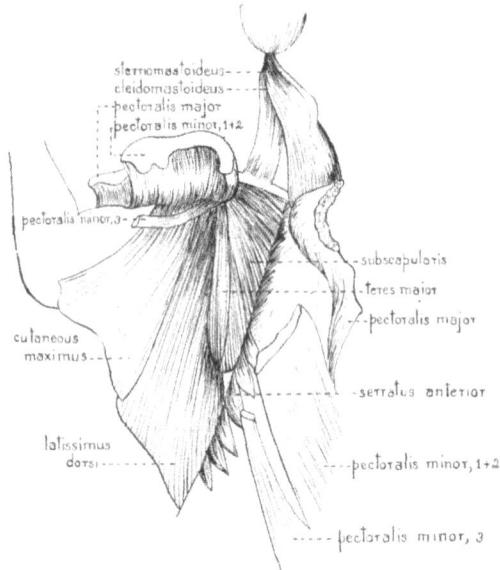

sternomastoideus
cleidomastoideus
pectoralis major
pectoralis minor, 1+2

pectoralis minor, 3

subscapularis

teres major

pectoralis major

cutaneous maximus

serratus anterior

latissimus dorsi

pectoralis minor, 1+2

pectoralis minor, 3

Fig. 73 Muscles of the right axilla

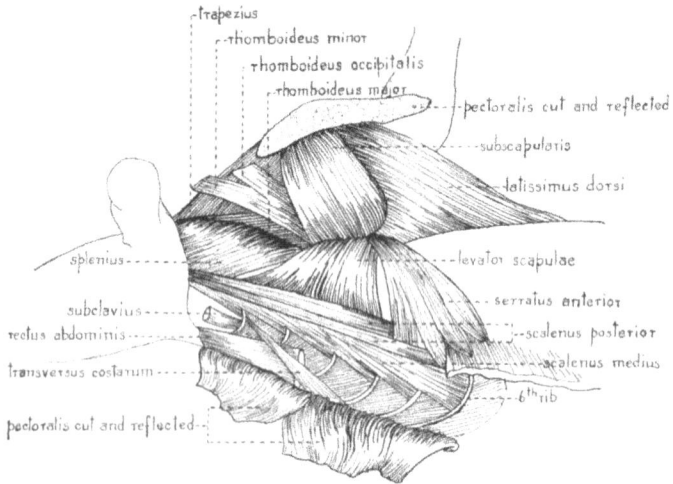

trapezius
rhomboideus minor
rhomboideus occipitalis
rhomboideus major

pectoralis cut and reflected

subscapularis

latissimus dorsi

splenius

levator scapulae

subclavius
rectus abdominis

serratus anterior

scalenus posterior

transversus costarum

scalenus medius

6th rib

pectoralis cut and reflected

Fig. 74. Deep muscles of the thoracic wall   Lateral aspect.
Pectoralis cut, and left arm reflected.

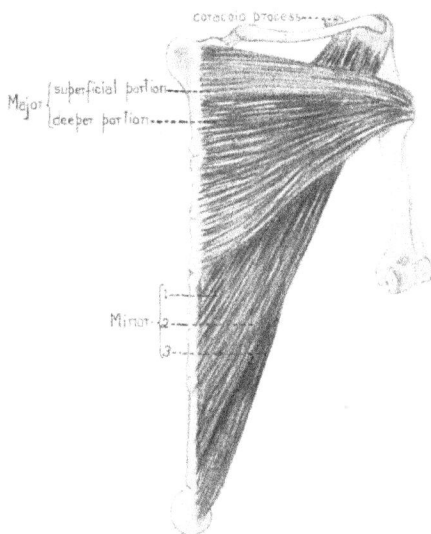

Fig. 75. Pectoralis major and minor

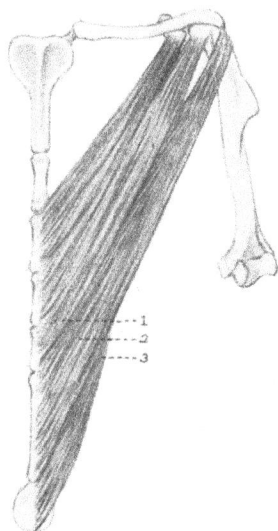

Fig. 76. Pectoralis minor, 3 portions

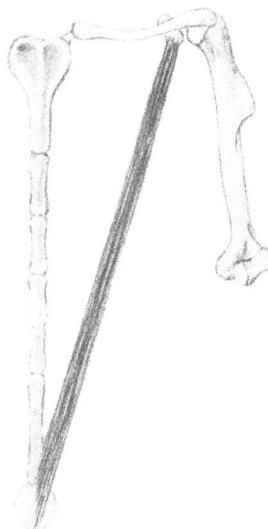

Fig. 77. Pectoralis minor, 3rd portion.

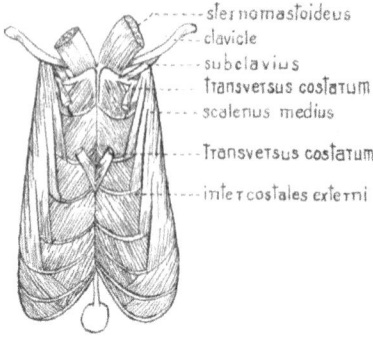

Fig. 78. Muscles of the thorax

- sternomastoideus
- clavicle
- subclavius
- transversus costarum
- scalenus medius
- transversus costarum
- intercostales externi

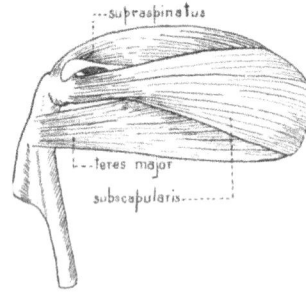

Fig. 79. Muscles of the scapula. Medial aspect.

- supraspinatus
- teres major
- subscapularis

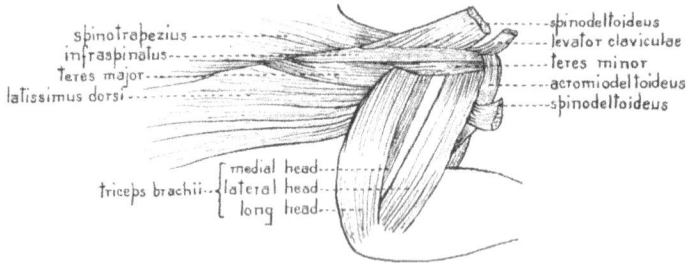

Fig. 80.　Muscles of the shoulder　Lateral aspect.

- spinotrapezius
- infraspinatus
- teres major
- latissimus dorsi
- spinodeltoideus
- levator claviculae
- teres minor
- acromiodeltoideus
- spinodeltoideus
- triceps brachii { medial head, lateral head, long head }

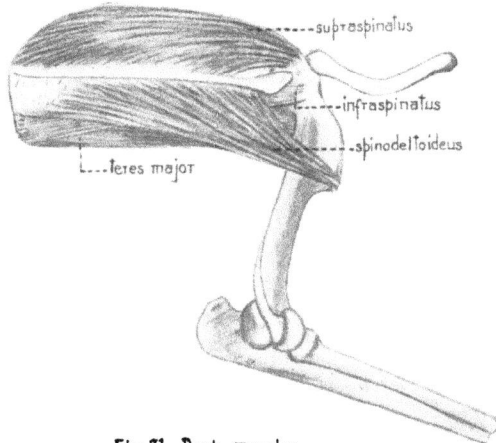

Fig. 81. Deep muscles
of the scapula. Lateral aspect.

- supraspinatus
- infraspinatus
- spinodeltoideus
- teres major

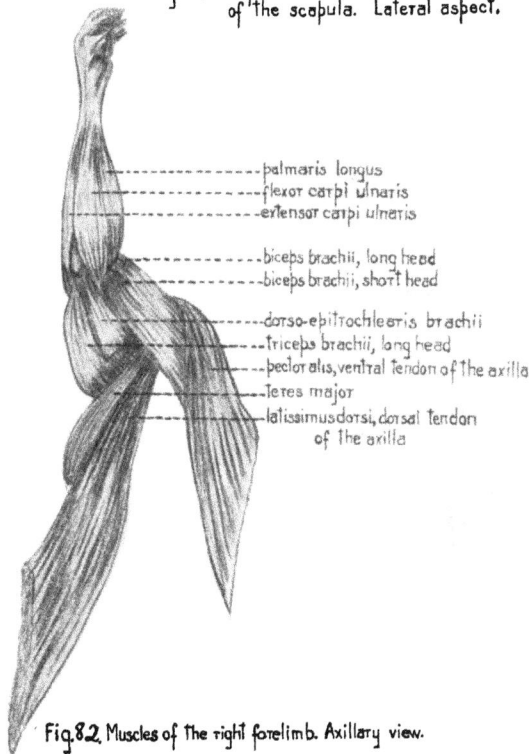

- palmaris longus
- flexor carpi ulnaris
- extensor carpi ulnaris
- biceps brachii, long head
- biceps brachii, short head
- dorso-epitrochlearis brachii
- triceps brachii, long head
- pectoralis, ventral tendon of the axilla
- teres major
- latissimus dorsi, dorsal tendon
  of the axilla

Fig. 82. Muscles of the right forelimb. Axillary view.

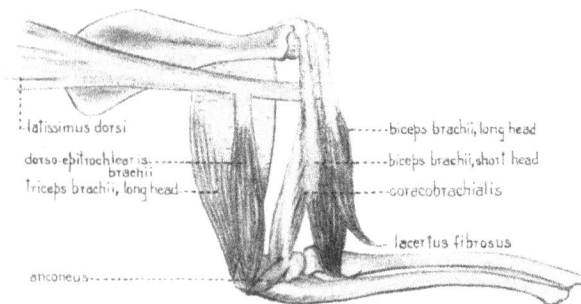

latissimus dorsi

dorso-epitrochlearis brachii

Triceps brachii, long head

anconeus

biceps brachii, long head

biceps brachii, short head

coracobrachialis

lacertus fibrosus

Fig.83. Muscles of the upper arm, biceps brachii, coracobrachialis, dorso-epitrochlearis. and anconeus.

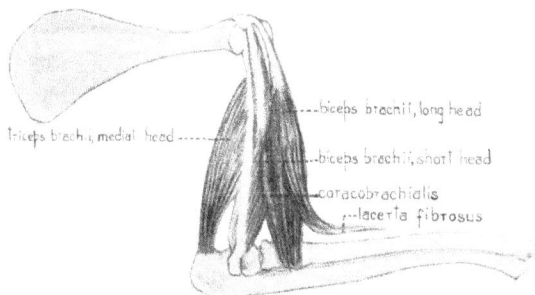

Triceps brachii, medial head

biceps brachii, long head

biceps brachii, short head

coracobrachialis

lacertus fibrosus

Fig.84. Muscles of the upper arm, triceps brachii, medial head

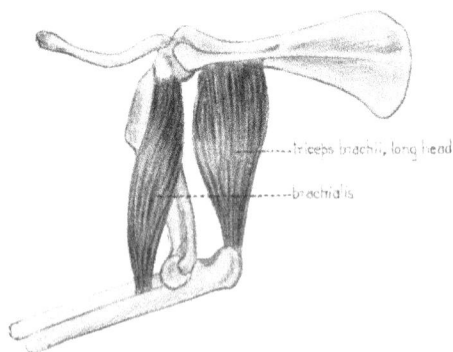

triceps brachii, long head

brachialis

Fig.85. Muscles of the upper arm, brachialis, triceps brachii long head.

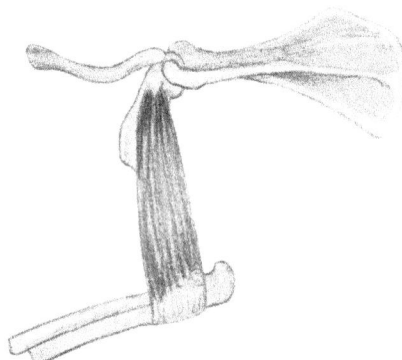

Fig.86. Triceps brachii, lateral head.

biceps brachii
flexor carpi radialis
pronator teres
extensor carpi radialis longus
extensor carpi radialis brevis
extensor digitorum communis
extensor digiti quinti
extensor digiti quarti
extensor carpi ulnaris
extensor indicis proprius
palmaris longus
extensor pollicis longus
extensor pollicis brevis

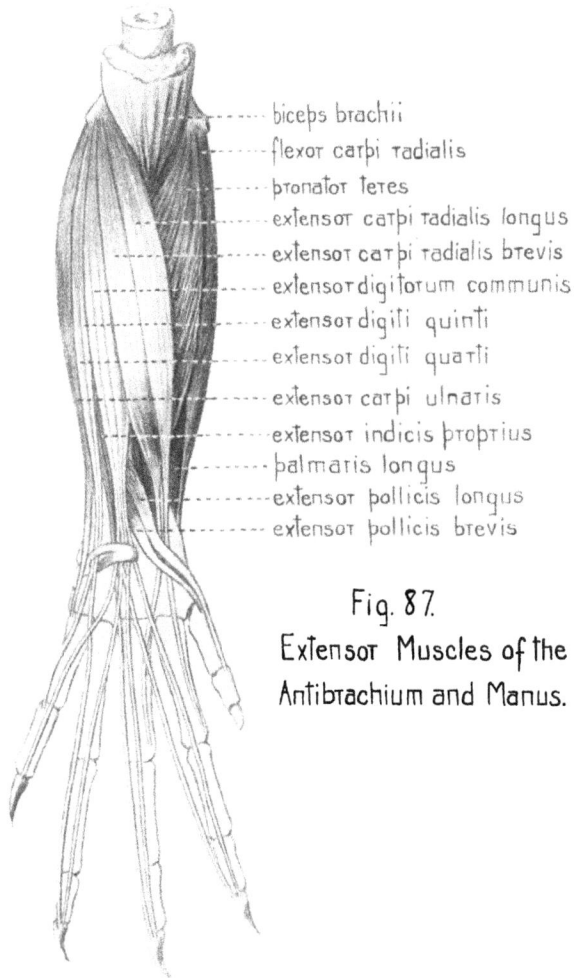

Fig. 87.
Extensor Muscles of the
Antibrachium and Manus.

flexor carpi ulnaris
palmaris longus
flexor digitorum profundus, superficial head
flexor digitorum profundus, radial head
flexor carpi radialis
pronator teres
extensor carpi radialis longus
extensor carpi radialis brevis
flexor digitorum profundus, ulnar head

flexor digitorum sublimis

extensor pollicis brevis
extensor pollicis longus

abductor digiti quinti
adductor pollicis

lumbricales

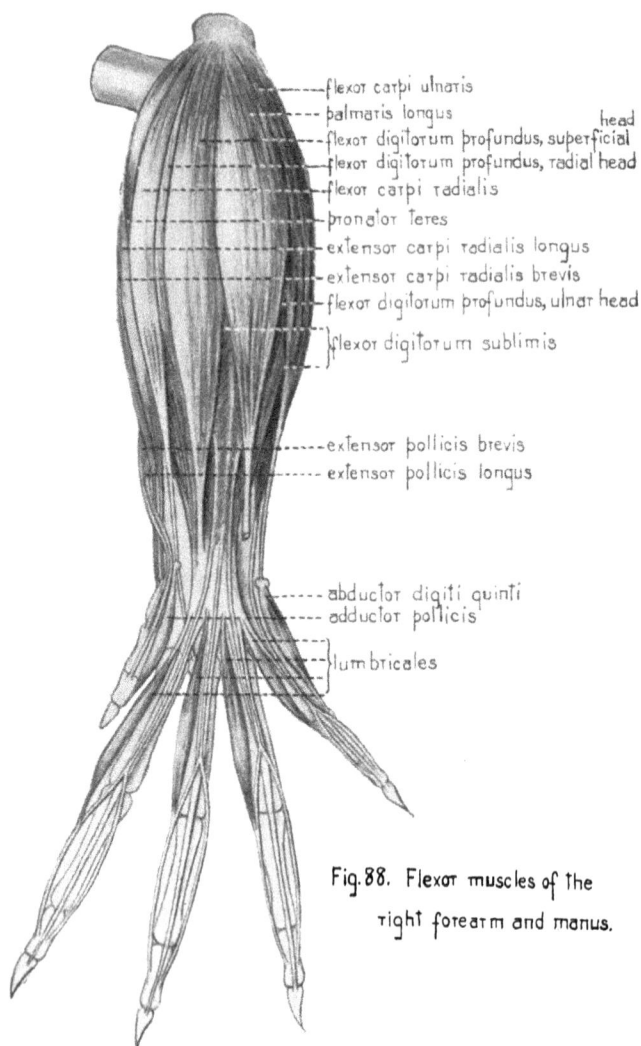

Fig. 88. Flexor muscles of the
right forearm and manus.

Fig.89. Interossei of Manus, and Flexor Muscles of Digits I and V.

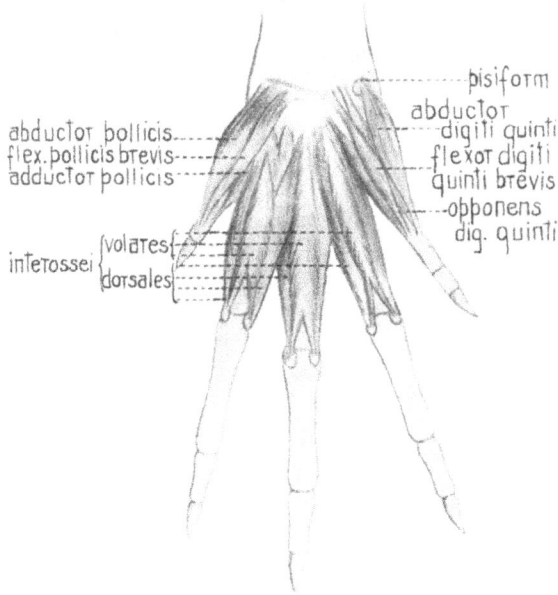

abductor pollicis
flex. pollicis brevis
adductor pollicis

pisiform
abductor digiti quinti
flexor digiti quinti brevis
opponens dig. quinti

interossei { volares
{ dorsales

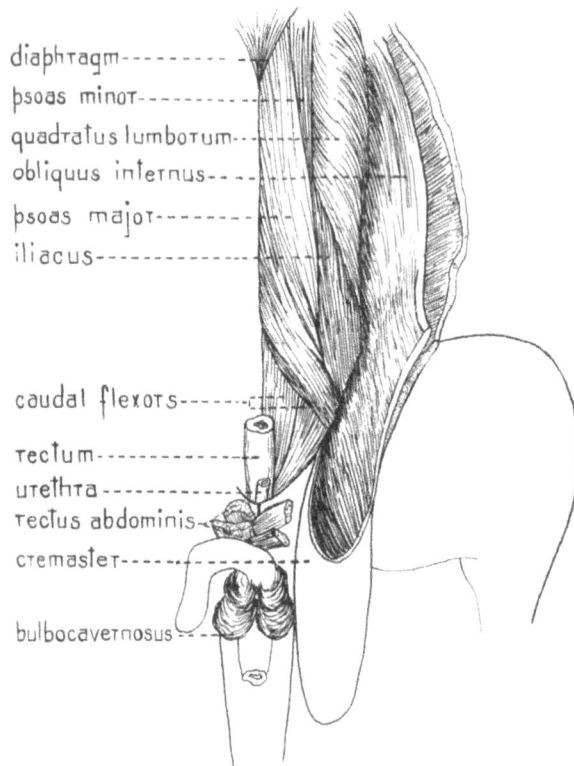

Fig. 90.    Muscles of the Lumbar Region.

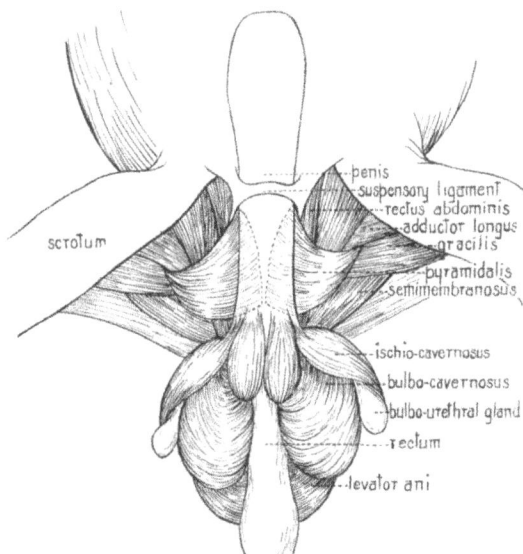

Fig. 91 Pyramidalis, and Muscles of the Penis

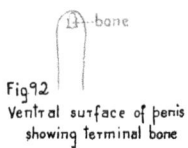

Fig92
Ventral surface of penis
showing terminal bone

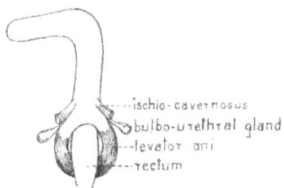

Fig93. Levator ani

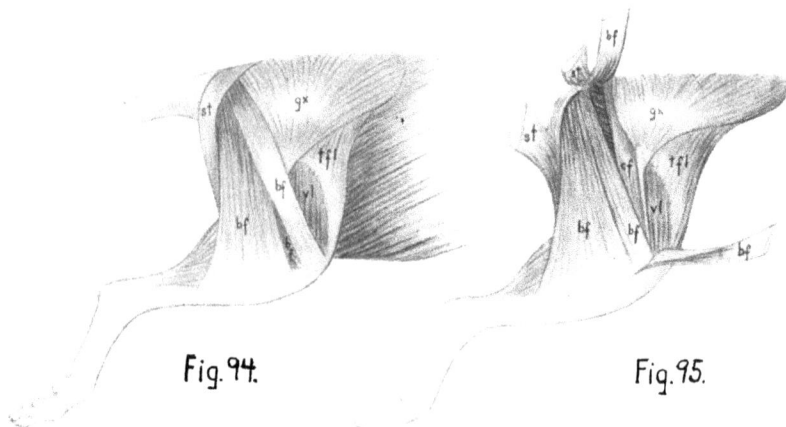

Fig. 94.

Fig. 95.

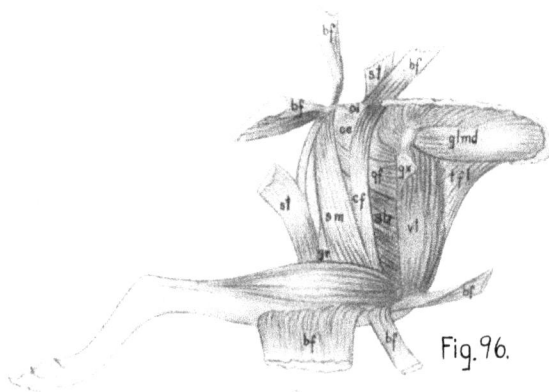

a.b.    adductor brevis
b.f.    biceps femoris
c.f.    caudofemoralis
gl.m.   gluteus medius
gl.mi.   gluteus minimus
gr.    gracilis
g.x.    gluteus maximus
o.e.    obturator externus
o.i.    obturator internus
q.f.    quadratus femoris
sm.    semimembranosus
st.    semitendinosus
t.f.l.   tensor fasciae latae
v.l.    vastus lateralis
p.y.    piriformis

Fig. 96.

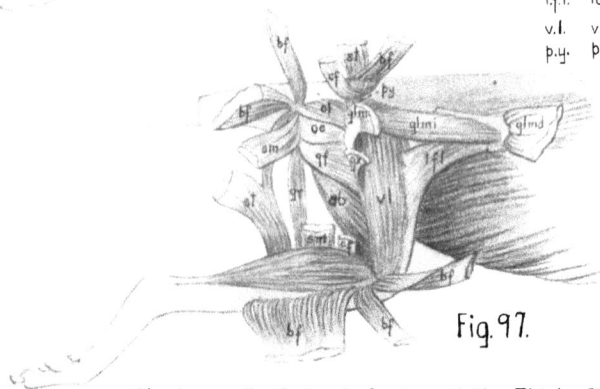

Fig. 97.

Muscles of the Lateral Surface of the Thigh. [Right].

Fig. 98.

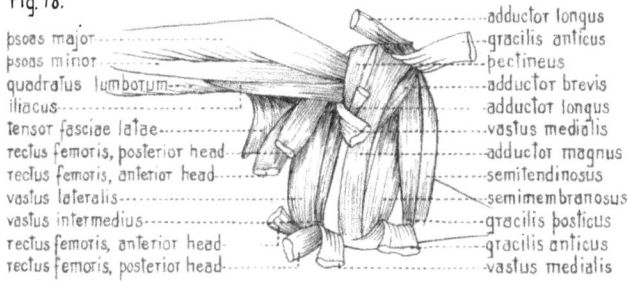

psoas major
psoas minor
quadratus lumborum
iliacus
tensor fasciae latae
rectus femoris, posterior head
rectus femoris, anterior head
vastus lateralis
vastus intermedius
rectus femoris, anterior head
rectus femoris, posterior head

adductor longus
gracilis anticus
pectineus
adductor brevis
adductor longus
vastus medialis
adductor magnus
semitendinosus
semimembranosus
gracilis posticus
gracilis anticus
vastus medialis

Fig. 99.

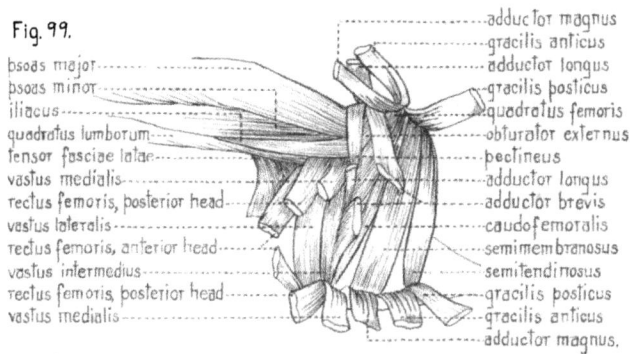

psoas major
psoas minor
iliacus
quadratus lumborum
tensor fasciae latae
vastus medialis
rectus femoris, posterior head
vastus lateralis
rectus femoris, anterior head
vastus intermedius
rectus femoris, posterior head
vastus medialis

adductor magnus
gracilis anticus
adductor longus
gracilis posticus
quadratus femoris
obturator externus
pectineus
adductor longus
adductor brevis
caudofemoralis
semimembranosus
semitendinosus
gracilis posticus
gracilis anticus
adductor magnus.

Muscles of the Medial Surface of the Thigh. [Right]

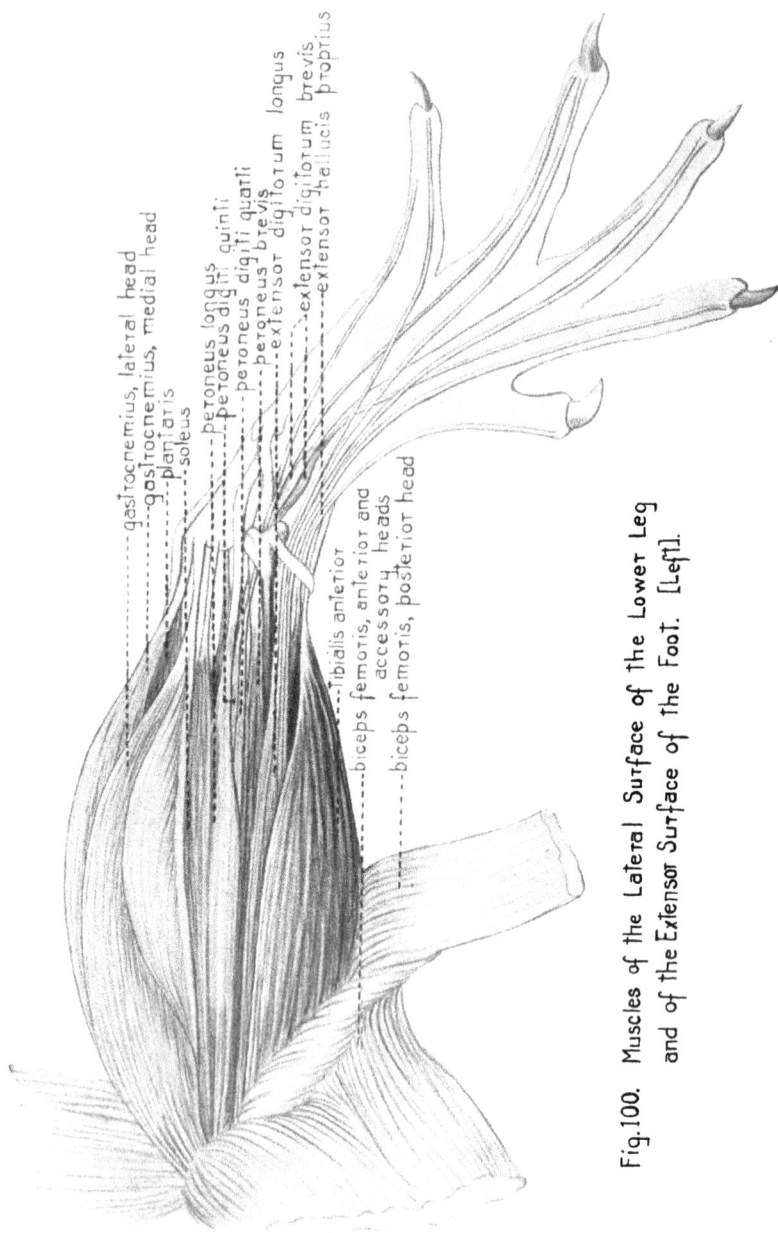

gastrocnemius, lateral head
gastrocnemius, medial head
plantaris
soleus
peroneus longus
peroneus digiti quinti
peroneus digiti quarti
peroneus brevis
extensor digitorum longus
extensor digitorum brevis
extensor hallucis proprius

tibialis anterior
biceps femoris, anterior and accessory heads
biceps femoris, posterior head

Fig. 100. Muscles of the Lateral Surface of the Lower Leg and of the Extensor Surface of the Foot. [Left].

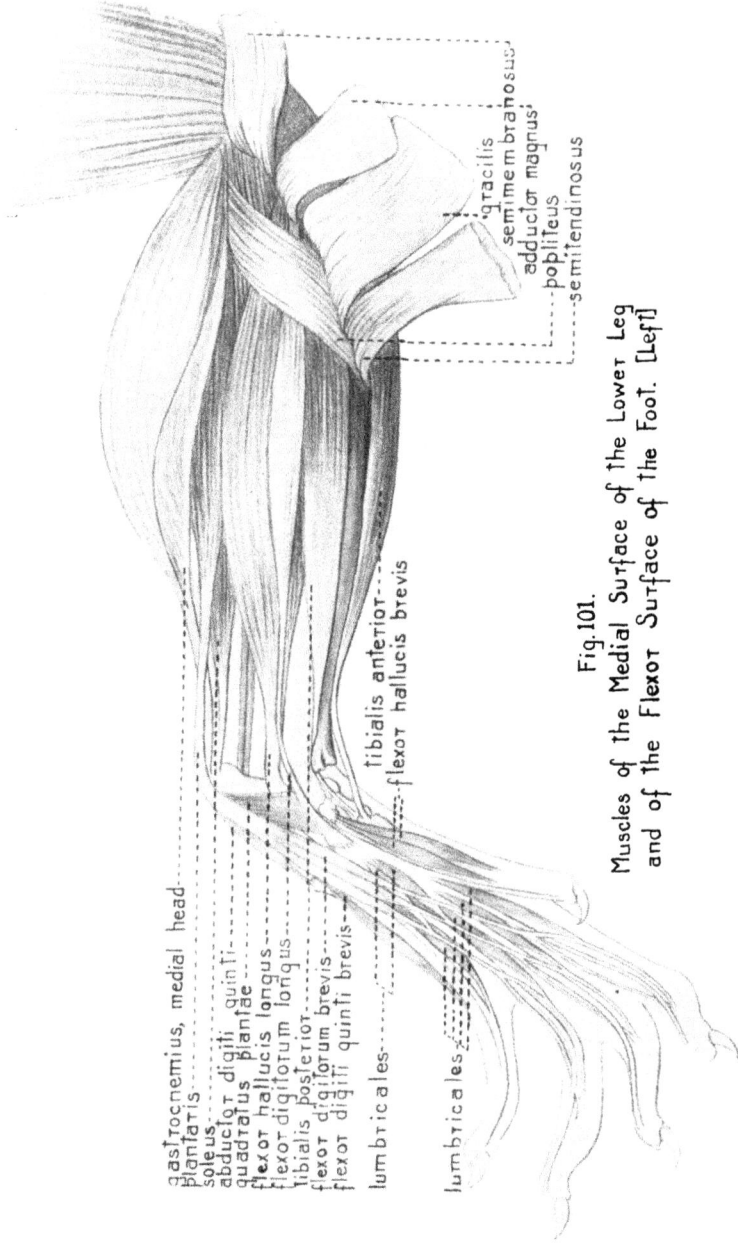

gastrocnemius, medial head
plantaris
soleus
abductor digiti quinti
quadratus plantae
flexor hallucis longus
flexor digitorum longus
tibialis posterior
flexor digitorum brevis
flexor digiti quinti brevis

lumbricales

lumbricales

tibialis anterior
flexor hallucis brevis

gracilis
semimembranosus
adductor magnus
popliteus
semitendinosus

Fig. 101.

Muscles of the Medial Surface of the Lower Leg
and of the Flexor Surface of the Foot. [Left]

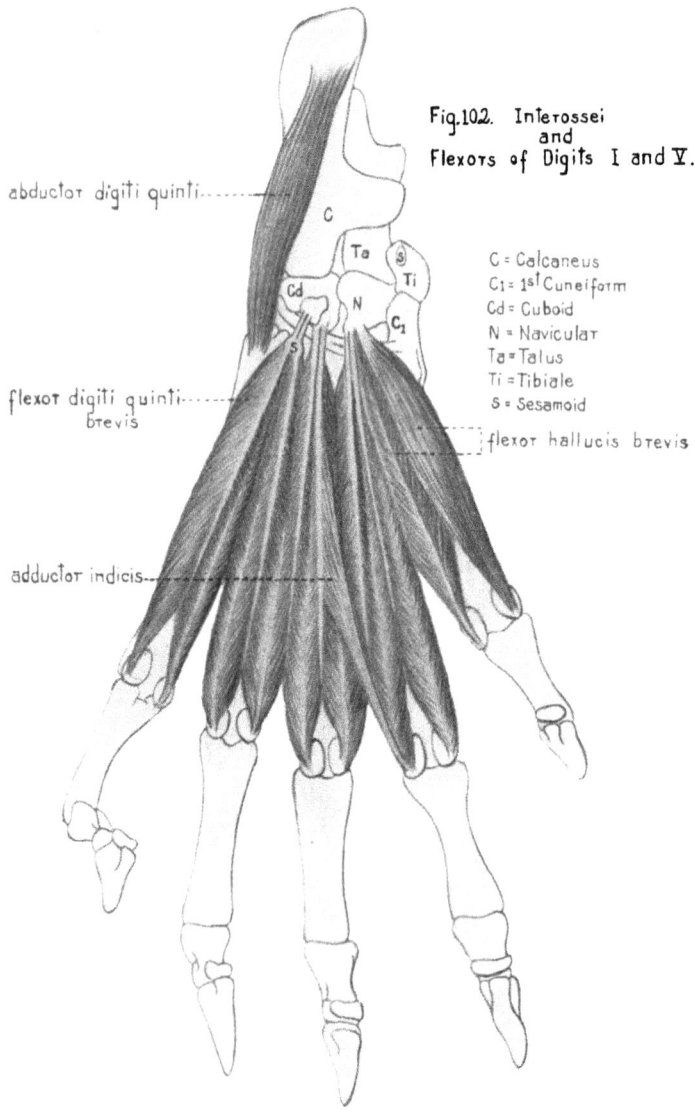

Fig.102. Interossei
and
Flexors of Digits I and V.

C = Calcaneus
C₁= 1ˢᵗ Cuneiform
Cd = Cuboid
N = Navicular
Ta = Talus
Ti = Tibiale
S = Sesamoid

abductor digiti quinti

flexor digiti quinti
brevis

flexor hallucis brevis

adductor indicis

# CHAPTER IV

## VISCERA

(Figs. 103–123)

Here again no attempt at an exhaustive study of the viscera has been made, but such points as seemed important for investigators, or points wherein the rat differs from the familiar laboratory animals, have been emphasized. Among these may be mentioned the lack of tonsils, and of gall bladder, the extremely diffuse pancreas, and the presence of a number of glands not recognized in man.

The viscera will be treated under the following headings:

The mouth cavity
Superficial glands of the head and neck
    Glands of the orbit
    Glands of the neck
Carotid body
Mammary gland
Thoracic viscera
The digestive tract and associated glands
The urogenital system
    Male
    Female

### THE MOUTH CAVITY

Flower 1872, describing the mouth cavity, states that the tonsils are scattered follicular openings, and that the tongue has a median groove on the dorsal surface, marking off the so-called "intermolar eminence." There is no median fraenum, but two lateral ones from near the tip to the band which ties the lower lip to the gum. There are no sublingua but a small pair of salivary papillae close to the median line behind the incisors. A single circumvallate papilla occurs close to the base of the tongue, while the dorsum in front of this is evenly covered with small pointed conical papillae directed backwards, showing more and larger vertical filaments on the intermolar tract. Scattered sparingly over the anterior part of the surface are conspicuous fungiform papillae.

The dental formula for the rat is — I 1/1, C 0/0, P 0/0, M 3/3, with no change of dentition throughout the life of the animal (Fig. 20).

### SUPERFICIAL GLANDS OF THE HEAD AND NECK
### (Figs. 103-105)

This group comprises the *orbital glands*, the *salivary* and *lymphatic glands* of the neck, and the *multilocular adipose tissue* or "hibernating gland."

#### GLANDS OF THE ORBIT
#### (Fig. 105)

Slightly below and just in front of the ear is the *exorbital lacrimal gland*. From its antero-dorsal border the duct leads forward to the dorso-lateral region of the eye where it unites with one from the intraorbital lacrimal gland before opening on the conjunctiva. The temporal vessels cross the duct midway in its course. The *intraorbital lacrimal gland* (infraorbital gland, Loewenthal) is roughly triangular in outline and occupies the posterior angle of the orbit covered by the connective tissue sheath and not extending very deep within the orbit. After removal of the sheath a small fragment of the *Harderian gland* is visible between the intraorbital lacrimal and the eyeball. Upon dissection this is found to be very extensive, occupying a large part of the orbit and opening on the nictitating membrane. As it is traced further it proves to be horseshoe shaped, extending medially and very deep until it encircles the optic nerve and finally ends in the dorso-lateral region of the orbit close to the bony margin, where it just fails to complete the circle. According to Rolleston 1870, this gland is found in most mammals except Chiroptera and Simiadae. Leydig says, "The secretory cells of the Harderian gland have in the centre a small fat globule." Thus they produce a slightly oily secretion. This undoubtedly accounts for the fact that this gland is particularly well developed in swimming mammals.

#### GLANDS OF THE NECK

**Salivary Glands.** The *parotid gland* (Figs. 103-105) is conspicuous in both lateral and ventral views. It is not a compact structure and runs well up behind the ear and down over the ventro-lateral surface of the neck where it lies along the posterior facial vein and is in contact with the lymph glands of the neck. Its posterior extremity reaches the shoulder and covers the outer half of the clavicle. The exorbital lacrimal gland is closely associated with its anterior border and covers the beginning of the *parotid duct* (Stenson's duct). The duct, formed by the union of three principal branches, crosses the masseter muscle parallel with the buccal and mandibular branches of the facial nerve, and opens opposite the molar teeth.

The two large *submaxillary glands* (Figs. 103-106) are the most prominent structures of the ventral cervical region. They are in contact along the midventral line from the level of the hyoid almost to the manubrium.

Closely applied to the latero-anterior surface of the submaxillary is a gland which appears at first to be a lobe of the submaxillary but which was found by Huntington and Schulte 1912, to be the *major sublingual* (Fig. 104). Loewenthal 1912, speaks of this as the Gl. retrolingualis and calls it a second or accessory submaxillary. Huntington and Schulte state that, "In rodents submaxillary and major sublingual ducts running beneath the mylohyoid open by separate parafrenular openings on the plica sublingualis. Minor sublingual glands open by several ducts on the lesser sublingual (Ravinian) ridge." Their description of the *minor sublingual gland* in the muskrat, as a "compact mass in the alveolingual district covering the lateral aspect of the submaxillary and greater sublingual ducts," is also true for the rat (Fig. 106). This gland lies in contact with the mucous membrane of the floor of the mouth just in front of the lingual nerve.

**Lymph Glands** (Figs. 103, 104). Practically imbedded in the parotid, and covering the junction of the anterior facial and external jugular veins there is a lymph node which Job, 1915, calls the *submaxillary node.* Just anterior to the submaxillary gland are two other nodes, sometimes double, which Job apparently does not describe but which appear in all material seen by the author. In addition there are sometimes small accessory ones along the blood vessels of this region.

The *anterior* and *posterior cervical lymph nodes* (Fig. 108) are seen at a deeper level, lateral to the carotid artery and vagus nerve. The anterior cervical node is covered by omohyoideus and lies upon levator claviculae. The posterior cervical node lies deep to sternomastoid and omohyoid, and may be seen by separating these muscles.

*Multilocular adipose tissue,* or *hibernating gland* so-called (Figs. 103, 107), is located chiefly in the region between the scapulae, but sends a small portion into the neck, where it may be seen in the posterior cervical region between the parotid and submaxillary glands, emerging between the sternomastoid and clavotrapezius as a very pink structure somewhat fatty in appearance.

It is with some hesitancy that this structure is listed as a gland, since, as Rasmussen (1922) points out, there is no justification for regarding this as glandular tissue. But inasmuch as it is most commonly known by the name hibernating gland, and secondly because it is encountered among the neck viscera in dissection, the author has ventured to include it at this point for description, listing it also under its correct name of *multilocular adipose tissue* in the index.

As Rasmussen states (1916, 1922) literature on this gland began with C. Gessner in 1551 and since then it has been known under various names. Barchow, in 1846, first gave it the name of hibernating gland, while others have referred to it as adipose gland, oil gland, brown fat, lipoid or cholesterin gland, organ of hibernation, and hibernating mass. A circumscribed fat mass in human foetuses in the dorso-cervical and interscapular regions, which some consider an homologous structure, is called the interscapular gland by Hatai (1902).

The literature reveals that it is found in about 47 species (5 insectivora, 9 chiroptera, 33 rodentia). A large number of species (rats, mice, rabbits, etc.) possess it but do not hibernate, and it is not found in some animals that do hibernate.

Discussing theories of hibernation, Rasmussen (1916) says, "removal of the hibernating gland in white rats is almost always fatal. It modifies the action of toxins, adrenalin, chloroform, tetanus toxin and cobra venom, retarding action of some, accelerating others. It possibly serves as an economizer of proteins by insuring utilization of reserve carbohydrates and fat during hibernation."

This structure may be followed best by removing the skin of the back and shoulders. The multilocular adipose tissue then appears as a small pink fatty structure in the deep depression between the scapulae, and between acromio- and spinotrapezius. With these two muscles cut the gland may be traced through its relatively large extent. It sends one flattened short lobe posteriorly beneath latissimus dorsi, but the main portion extends anteriorly beneath rhomboideus and levator scapulae, curving forward over the shoulder to the neck where its tip emerges between clavotrapezius and sternomastoideus in the ventral region.

The medial portion of the gland is in contact with that of the opposite side so that the spreading wing-like extensions just described give the name "butterfly gland" which is sometimes applied to it. Long (1922) referring to it as the fat gland, says, "it is closely associated with the blood vessels which go to the body muscles. It is brown to salmon in color; the cells are gorged with droplets of lipoid among which the nucleus is imbedded. It appears first in the 17-day embryo."

Quoting rather extensively from Rasmussen's conclusions (1922), "It is clearly evident that no one has gone extensively enough into the physiological significance of this brown multilocular adipose tissue to justify its classification as a gland. The granular character of the cytoplasm, the presence of numerous discrete fatty globules and other structural features of this tissue are not sufficient criteria by which to judge its glandular properties. The prevailing opinion during the past 50-60 years has been that it is merely one type of adipose tissue. The best designation for this tissue at the present time would seem to be multilocular adipose tissue."

"Extirpation experiments are of little value on account of the extensive distribution of the tissue. The effects of its extracts have been investigated practically by only a single worker (Vignes '13), who believes he has evidence from enzyme action that it functions as a protein sparer during winter sleep. The latest theory, that it may be a special source of vitamines, is based on such meager data that it need not be taken seriously."

### THE CAROTID BODY (CAROTID GLAND)

### (Fig. 331)

Located in the bifurcation of the common carotid into external and internal carotid arteries is the *carotid body* or "*carotid gland*" as it is frequently called. This structure has been the object of much research, and some controversy has arisen as to whether it should be classed as an endocrine organ. Smith '24, working on the origin and development of the carotid body, states that "the understanding of this organ largely depends on the analysis of the growth transformations of the region in which it is found. The vicinity of bifurcation of the common carotid artery is very vascular and abundantly supplied with nerves. These latter arise as branches of the glossopharyngeal and vagus trunks and the superior cervical ganglion and are accompanied by sympathetic ganglion cells, and often, by chromaffin cells, though the latter is not true of the rat. This vascular, nervous parenchymatous structure indicates that we are dealing not with a simple organ, but a complex, the components of which have been brought together by processes dependent on the developmental history of the region. It is indeed this very complexity which has caused the present-day confusion as to the status of the carotid body. As one or the other of its elements has been emphasized by different investigators, so have been formulated the various theories concerning its nature and origin, as (1) epithelial, (2) vascular, (3) nervous."

She further points out that "the inclusion of the carotid body in the endocrine system is dependent on its chromaffin nature, as many believe that cells of this type are internally secreting. Some carotid bodies possess no chromaffin cells as in the rat, and some but few. In summary she concludes that, "there is no evidence to warrant the inclusion of the carotid body in the endocrine system."

### Mammary Glands

#### (Figs. 109, 110)

Myers 1916, making a study of the mammary glands of young animals, about 10–14 days of life, describes six separate pairs of mammary glands, three thoracic, one abdominal, and two inguinal. The greatest interval occurs between the third thoracic and the abdominal pair. In observing 100 rats he finds the number of glands varying from 10 to 13 with six pairs as the normal. Only one specimen showed a supernumerary gland. The second thoracic are those most often absent.

After two weeks the nipples are obscured by hair. At the age corresponding to puberty, between 6 and 9 weeks, there is a marked increase in their development and they are again conspicuous during pregnancy and lactation (Fig. 109).

In the adult animal separate units of the gland are indistinguishable, and upon dissection in the lactating animal the gland appears as an almost continuous sheet, just beneath the panniculus carnosus, extending from neck to anus. It is interrupted for a short distance just below the border of the ribs. It is so closely associated with the panniculus carnosus that it is practically impossible to remove this muscle without injury to the gland.

From the ventral cervical region it spreads over the shoulder and as far as the bend of the elbow. From the axillary region a portion runs dorsally toward the shoulder. The abdominal portion spreads out over the lateral wall of the body and is continuous with the relatively thick mass which occupies the entire inguinal region.

### Thoracic Viscera

#### (Figs. 111, 112)

The thoracic viscera need no particular description. Upon removal of the ventral wall of the thorax to display the viscera in situ, the two lobes of the *thymus* appear between the two superior venae cavae and covering the trachea.

There are three lobes of the right *lung*, superior, middle and inferior, readily visible, and a fourth deeper median lobe which lies in contact with the diaphragm and apex of the heart, and is notched to accommodate the inferior vena cava. This is sometimes called the post caval lobe. The left lung has but one lobe.

### Digestive Tract and Associated Glands

(Figs. 111, 113)

In Fig. 111 the abdominal portion of the digestive system is shown in situ while in Fig. 113, with the heart, lungs, and liver removed, and the intestines spread out, the whole length of the digestive tract may be traced from œsophagus to rectum.

The *stomach* (Fig. 114) shows externally its division into two distinct portions. As Flower (1872) describes it, "the left portion is translucent, with a pale whitish mucous lining containing only mucous glands. The right portion is opaque, muscular, reddish gray and vascular with a velvety mucous membrane showing longitudinal folds with peptic glands."

The *spleen* (Figs. 111,113,114) has no peculiarities of relationship or structure but does vary greatly in size.

The *pancreas* (Fig. 114) is a very diffuse low dendritic type as is usual in rodents, extending "from the end of the duodenal fold to the left into the gastro-splenic omentum, where it ramifies, the chief part of the duodenal pancreas following the curve of the gut, but ramifying in its wide mesentery. The pancreatic ducts are many and paired" (Morrell, 1872).

The *liver* (Figs. 111, 115) is divided into four parts; the *median* or *cystic lobe*, which bears a deep fissure for the ligamentum teres hepatis; a *right lobe* partially divided into an anterior and posterior lobule; a large *left lobe*; and the small *caudate lobe* (Spigelian lobe) which fits around the œsophagus. The rat has no gall bladder, and the *ductus choledochus* (Fig. 115) is made up of tributaries from the various lobes of the liver.

Mann (1920) studied the sphincter at the duodenal end of the common bile duct with special reference to species of animals without a gall bladder. He found the duct in each species was surrounded by a bundle of smooth muscle but could observe no difference in animals with a gall bladder as compared with those without one.

McMaster (1922) states that from his findings it is "clear that there resides in the rat ducts no ability to concentrate bile such as resides in the gall bladder of the mouse. The gall bladder, then, is not only absent from the rat in form, but in one at least of its important functions. That its other obvious function, that of a reservoir, cannot be assumed in the rat by the ducts would seem to be indicated, not only by the small size of the channels, but by the recent observation of Mann (1920) that the tonus of the sphincter of Oddi is almost negligible in the rat, in contradistinction to animals which possess a gall bladder."

The *intestine* (Fig. 113) is approximately "seven times the length of the body, i.e., from snout to anus; and of this one-seventh is the large intestine" (Morrell, 1872).

The *caecum* (Fig. 116) is another portion of the intestinal tract worthy of note. Morrell (1872) speaks of the "absence of internal septa, dividing the caecum into cells, so commonly found in Rodents." He also states that the "caecum in the rat is bounded toward the small intestine by a tumid white ring, representing the ileocæcal valve, close to which is the commencement of the large intestine marked by long rugae and absence of villi. Rugae do not extend beyond three inches from the pylorus."

Berry (1901) in his studies on the caecal apex or the vermiform appendix, states that all rodents, being herbivorous, possess a caecum, with the exception of the Myoxidae, and describes it as follows, "In mouse and rat the caecum is slightly constricted about its middle. This constriction subdivides the caecum into two parts, an apical portion and a basal portion. The basal portion in both contains no lymphoid tissue. The apical portion on the other hand, contains a distinct mass of lymphoid tissue in its lateral wall." This constitutes the vermiform appendix.

## UROGENITAL SYSTEM

### MALE

Figs. 117-121 represent a series of dissections, from superficial to deep, showing the organs of the male urogenital system. These drawings should prove more adequate than any description.

The *kidneys* and *suprarenals* (Figs. 120, 122, 123, 132). The relative height of the right and left kidneys, and incidentally of the suprarenals in the rat, is quite the reverse of that found in man. In the rat the right kidney and suprarenal lie more cephalad, and the renal arteries and veins show a corresponding difference.

There is a distinct sexual difference in the *suprarenal* which is discussed briefly in Chapter V on endocrines.

In describing the *bulbo-urethral gland* (Figs. 118, 119, 121), Loewenthal (1897) says, "Its surface is smooth and its structure lobular though this latter fact is not apparent externally. The capsule of the gland is mainly striated muscle, thus for some time gland tissue was not reported."

The *testes* have already descended in a 30-40-day rat though the inguinal canal remains open, as is usual in rodents, and during sexually inactive periods the testis may be withdrawn into the abdominal cavity. Covering the anterior end of the testis (Fig. 133) is the *caput epididymis* which passes into a more constricted mass of tubules, *corpus epididymis*, applied along the surface of the testis as far as the posterior end where it becomes the *cauda epididymis* from which the *ductus deferens* leads back through the inguinal canal and crosses the ureter before joining the urethra. In Fig. 120 the *gland of the ductus deferens* is shown lateral to the ureter. This is due to displacement for purposes of drawing as these glands are well hidden by the bladder and prostates and are often omitted in descriptions of the region. Neither Hunt nor Howell makes any mention of them. Disselhorst (1904) shows them for Mus rattus.

The *seminal vesicles* (Figs. 119-121) are large and lobulated except for the rather smooth tip which is doubled back upon itself. Closely applied along the inner curve of the seminal vesicles, within the same sheath, are the *coagulating glands* (Figs. 119-121). These were taken for accessory or third prostates until Walker (1910) showed that they served to coagulate the seminal fluid and were necessary for impregnation. They are shown to be quite villous upon removal of the sheath.

There are two pairs of *prostate glands* (Figs. 119, 120) which practically surround the proximal end of ductus deferens and are "held by fascia extending over the bladder; they also have a special fascicular capsule" (Walker 1910).

(Figs. 122, 123, 134-137)

Fig. 123 shows the organs of the female urogenital system in an adult animal. The ovaries are therefore shown as a mass of follicles. "Variation of the weight of the ovaries during gestation and lactation, depends largely upon the number and size of the corpora lutea present." (Donaldson '24.) The two horns of the uterus while apparently fused at their lower end, nevertheless maintain separate openings into the vagina. There are, therefore, two ossa uteri.

In the rat the vagina is closed by a membrane or plug, until the beginning of puberty. It opens at about 72 days and ovulation begins at about 77 days. (Long and Evans '22.) Gestation occupies 21-22 days.

## LIST OF FIGURES

### IV. VISCERA

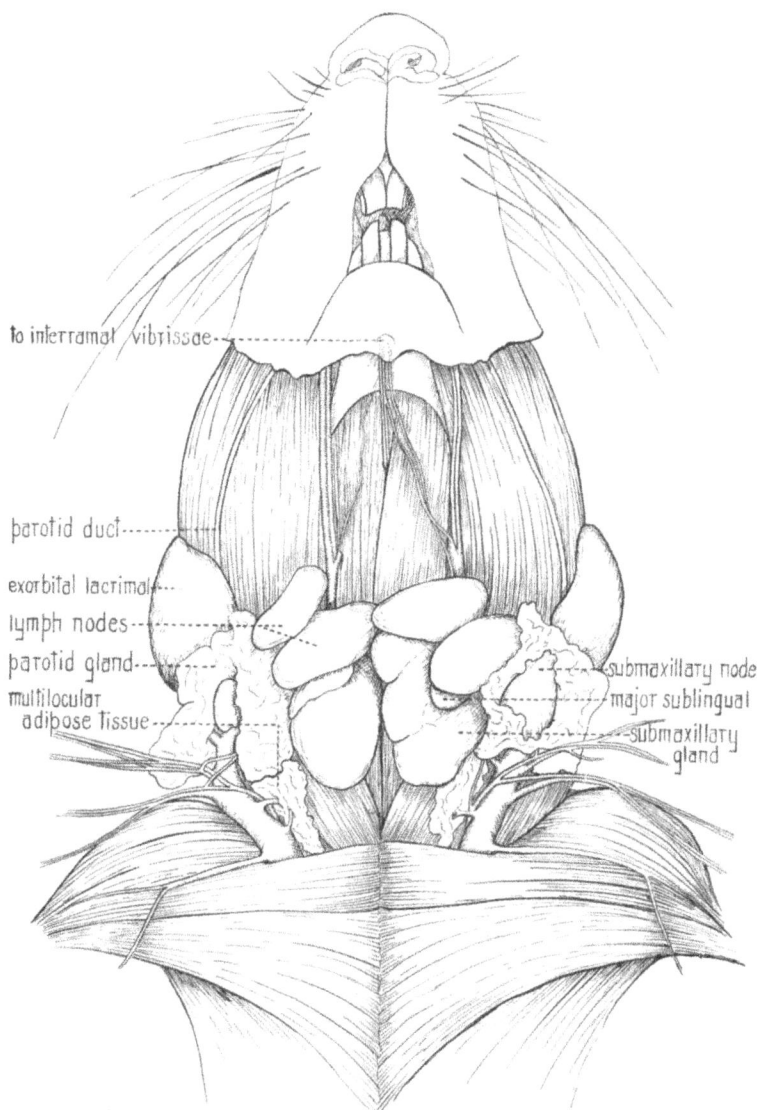

to interramal vibrissae

parotid duct

exorbital lacrimal
lymph nodes
parotid gland
multilocular
   adipose tissue

submaxillary node
major sublingual
submaxillary
   gland

Fig. 103. Superficial Dissection of Neck Viscera

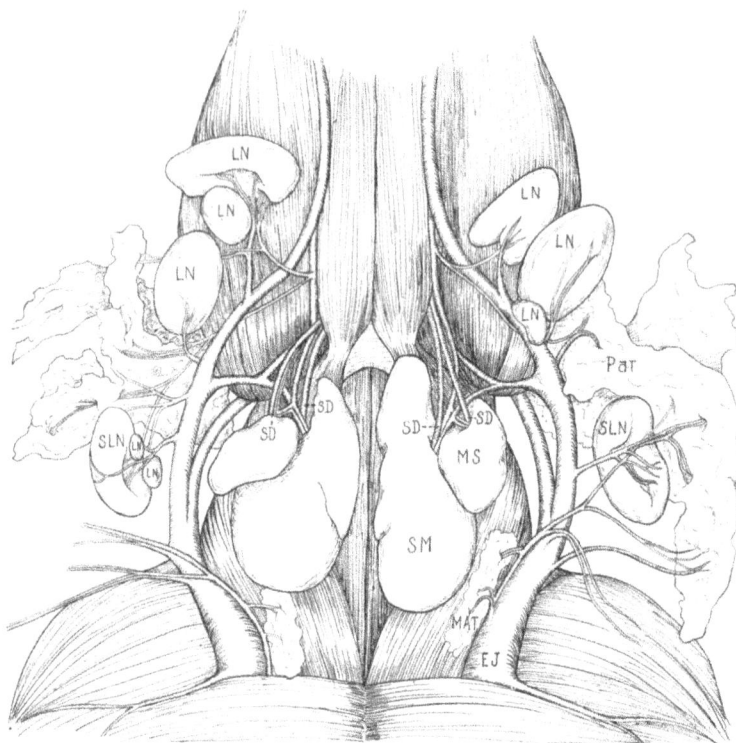

Fig.104. Superficial glands of neck reflected to show relation to underlying structures

LN-lymph node  MAT-multilocular adipose tissue
Par-parotid gland  SD - salivary duct
MS-major sublingual gland  EJ - external jugular vein
SM-submaxillary gland
SLN-submaxillary lymph node

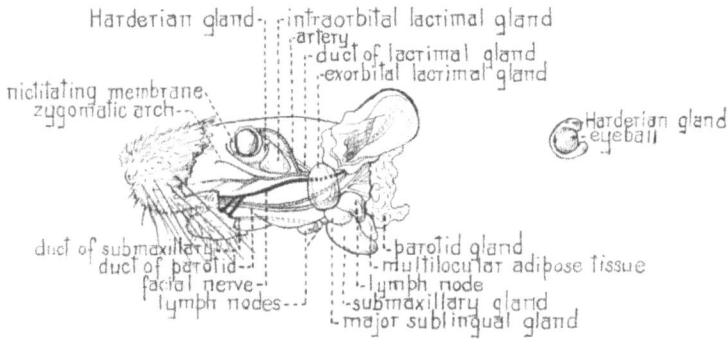

Fig.105. Neck Viscera, and detail showing relation of Harderian gland to eyeball.

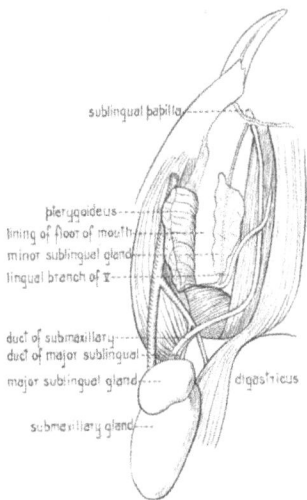

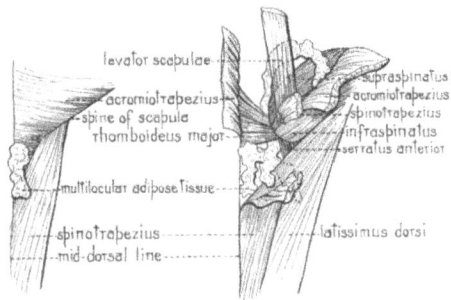

Fig.107. Distribution of multilocular adipose tissue in 90 day rat as shown in dissection of dorsal scapular region.

sternohyoideus
larynx
sternothyroideus
superior thyroid
thyroid isthmus
parathyroid+thyroid

anterior cervical node

common carotid

vagus

posterior
cervical node

internal
jugular

cervical trunk
costocervical
trunk

subclavian
highest intercostal
innominate
vagus

trachea
oesophagus
recurrent
inferior thyroid of
profunda cervicalis
vertebral

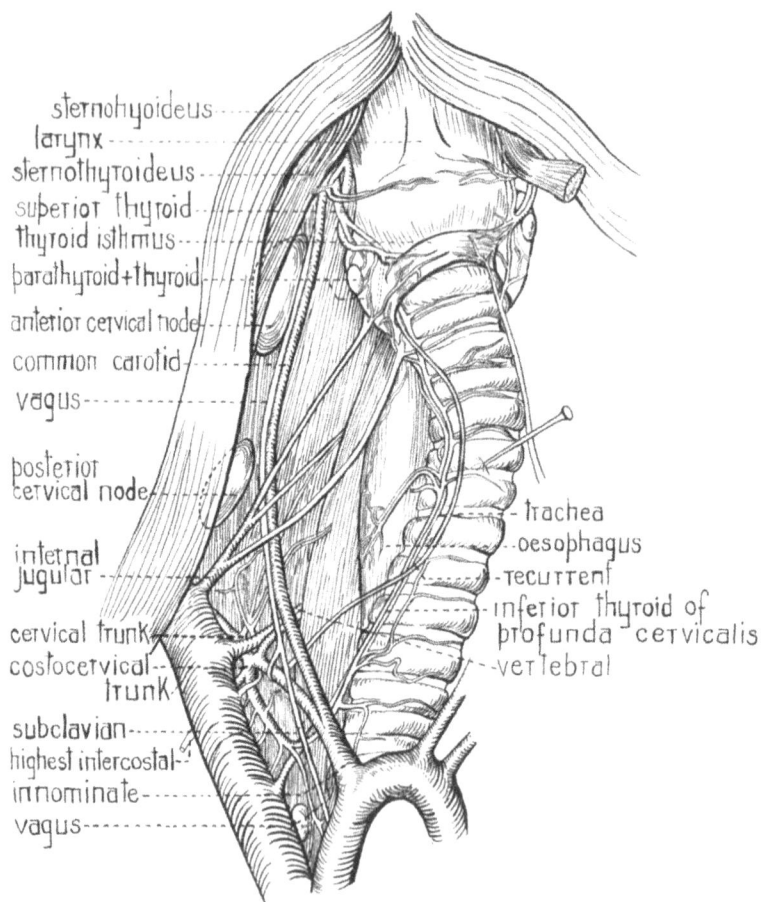

Fig. 108. Deeper Neck Viscera.

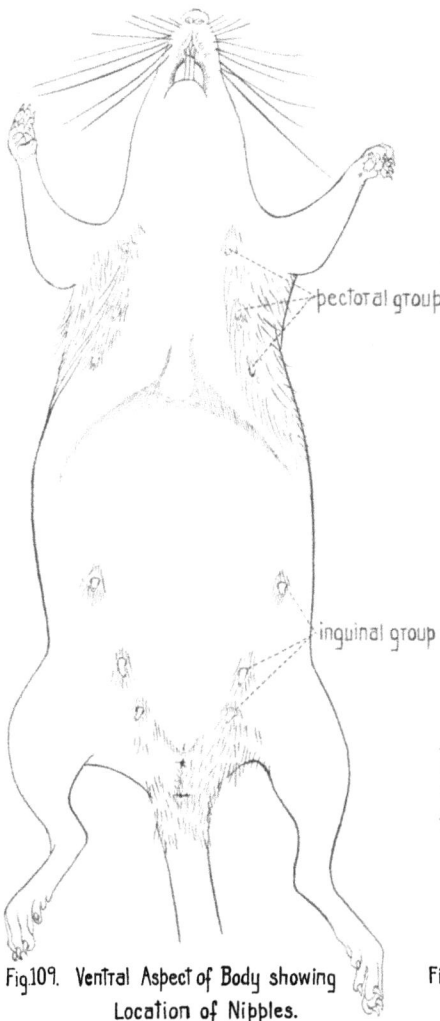

Fig.109. Ventral Aspect of Body showing Location of Nipples.

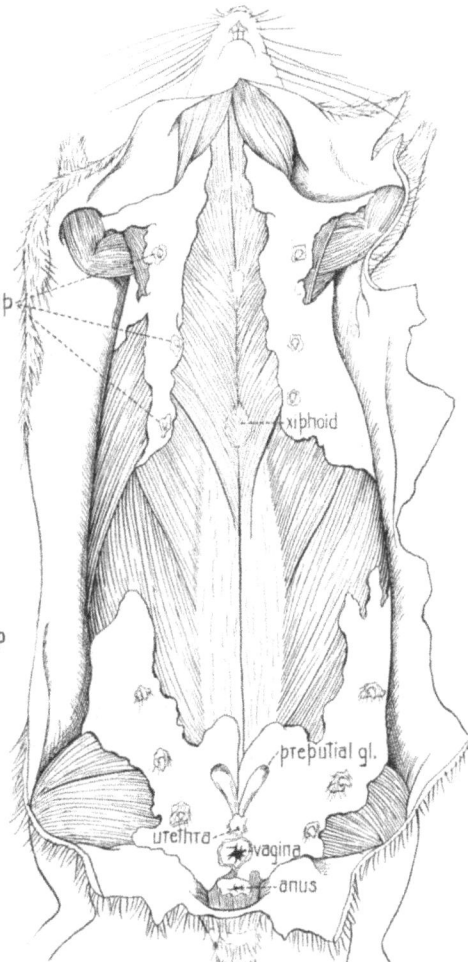

Fig.110. Skin reflected to show Extent of Mammary Gland in Lactating Female.

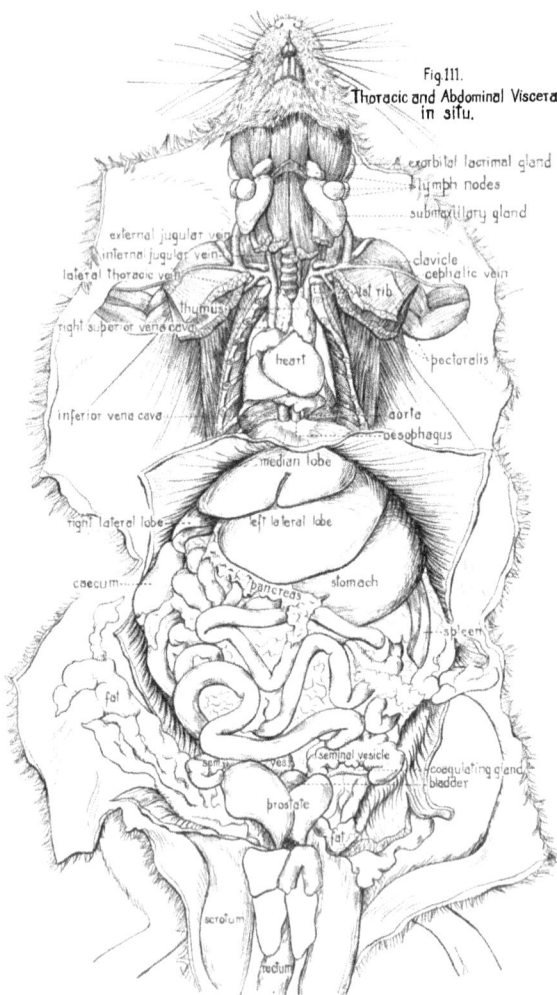

Fig.111.
Thoracic and Abdominal Viscera
in situ.

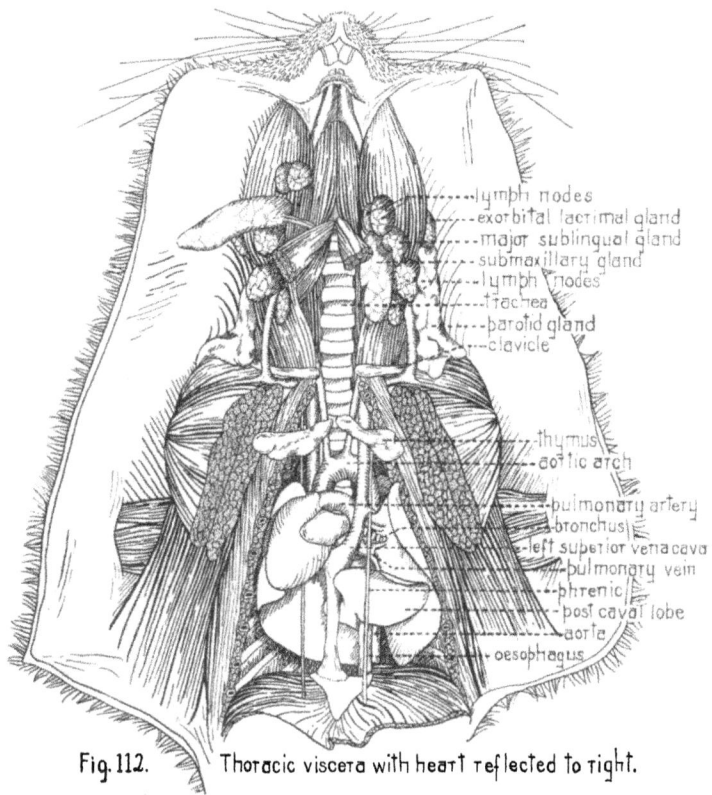

lymph nodes
exorbital lacrimal gland
major sublingual gland
submaxillary gland
lymph nodes
trachea
parotid gland
clavicle

thymus
aortic arch
pulmonary artery
bronchus
left superior vena cava
pulmonary vein
phrenic
post caval lobe
aorta
oesophagus

Fig. 112.    Thoracic viscera with heart reflected to right.

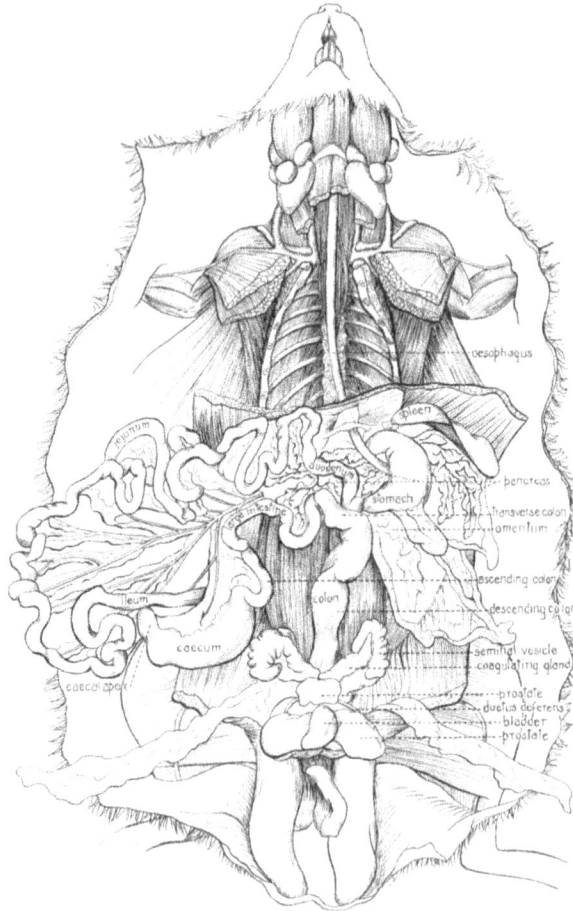

Fig 113.    Thoracic Viscera and Liver Removed to show
Alimentary Tract.

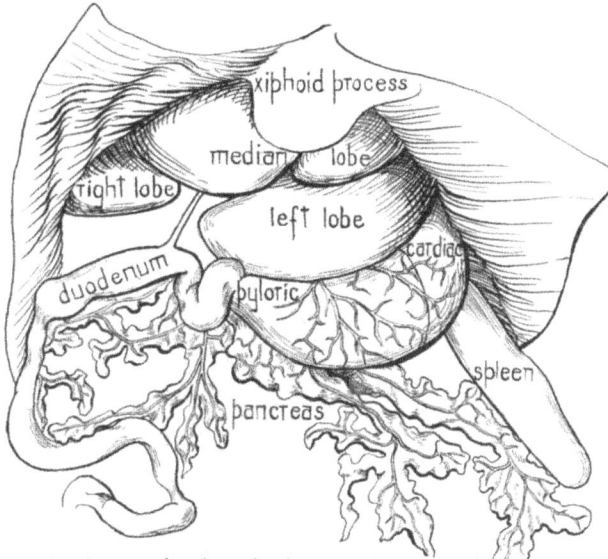

Fig. 114. Cardiac and pyloric portions of the stomach, the pancreas, and liver. Camera lucida.

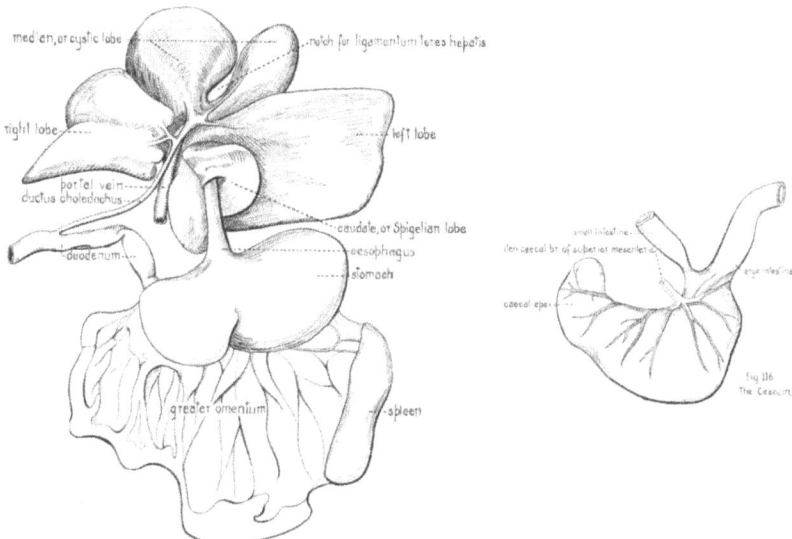

Fig. 115. Ductus choledochus and lobes of the liver, reflected.

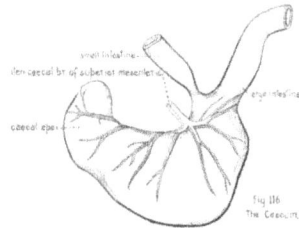

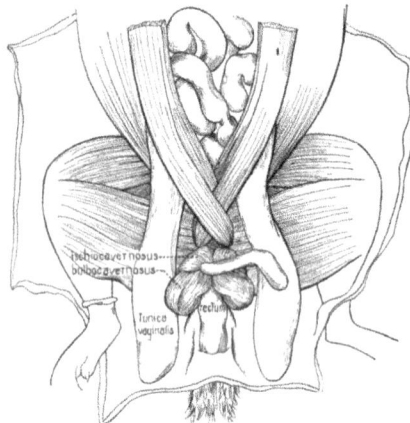

Fig. 117.   Male Urogenital System.   Dissection I

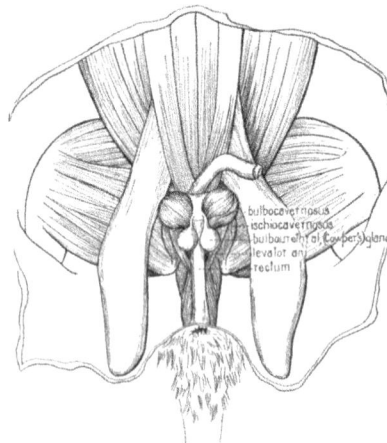

Fig. 118.   Male Urogenital System.   Dissection II.

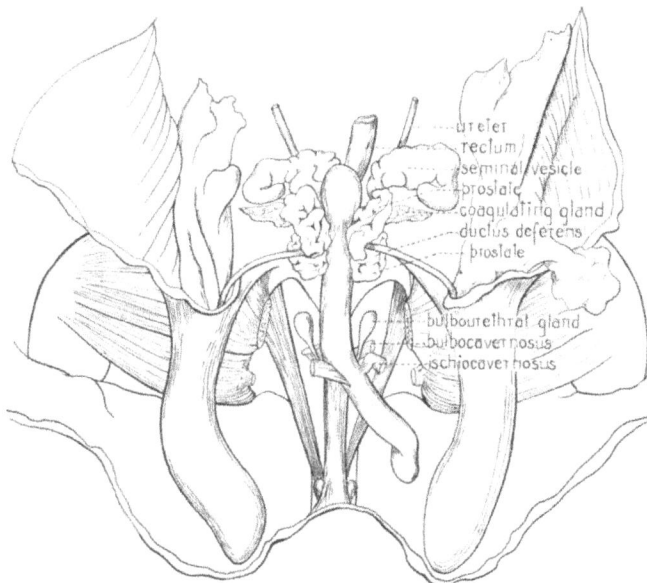

Fig. 119.  Male Urogenital System.  Dissection Ⅲ.

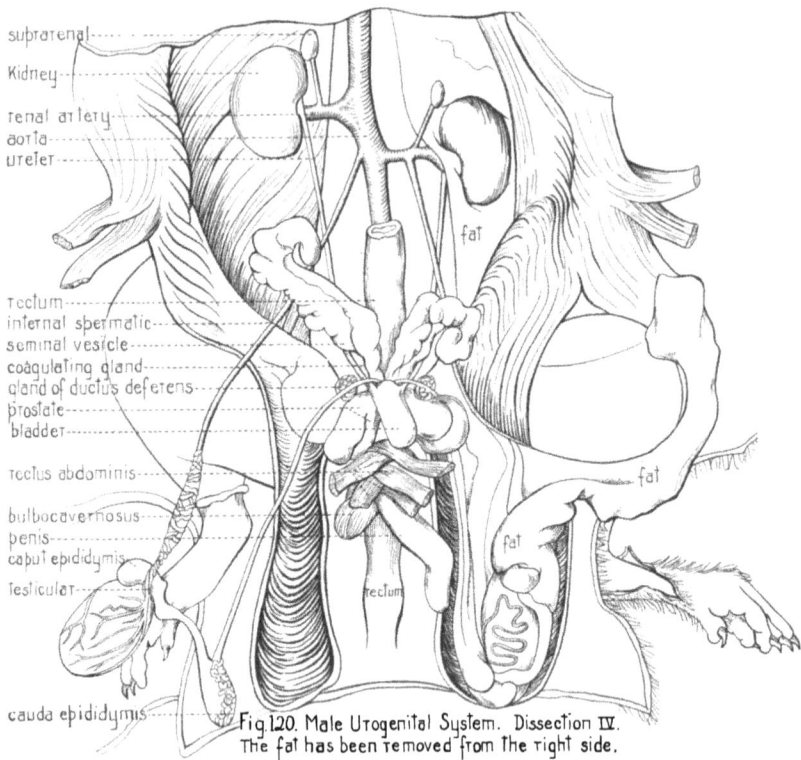

suprarenal
Kidney
renal artery
aorta
ureter
fat

rectum
internal spermatic
seminal vesicle
coagulating gland
gland of ductus deferens
prostate
bladder
rectus abdominis
fat
bulbocavernosus
penis
caput epididymis
fat
testicular
rectum

cauda epididymis

Fig. 120. Male Urogenital System. Dissection IV.
The fat has been removed from the right side.

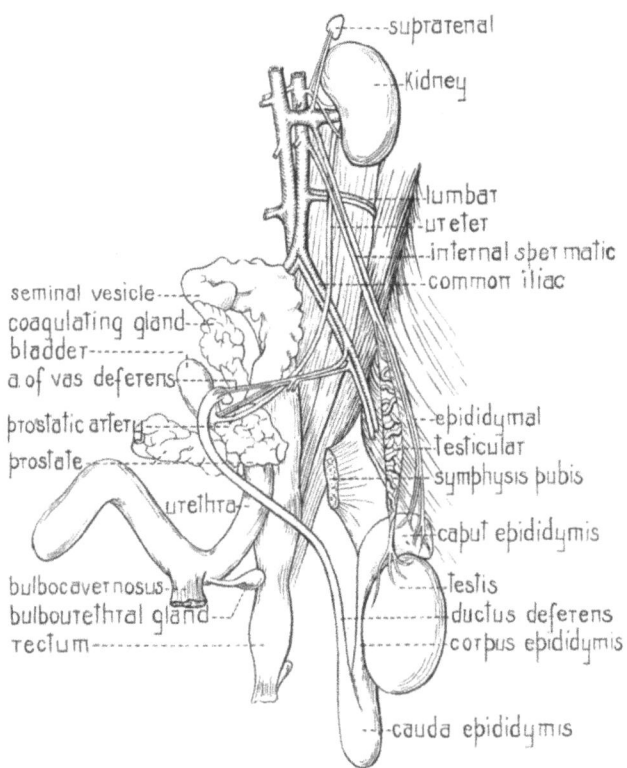

----suprarenal

--Kidney

--lumbar

--ureter

--internal spermatic

--common iliac

seminal vesicle---

coagulating gland---

bladder----------

a. of vas deferens---

prostatic artery---

prostate---

urethra---

--epididymal

--testicular

--symphysis pubis

--caput epididymis

bulbocavernosus---

bulbourethral gland---

rectum----------

--testis

--ductus deferens

--corpus epididymis

--cauda epididymis

Fig.121. Male Urogenital System. Dissection V. Lateral.

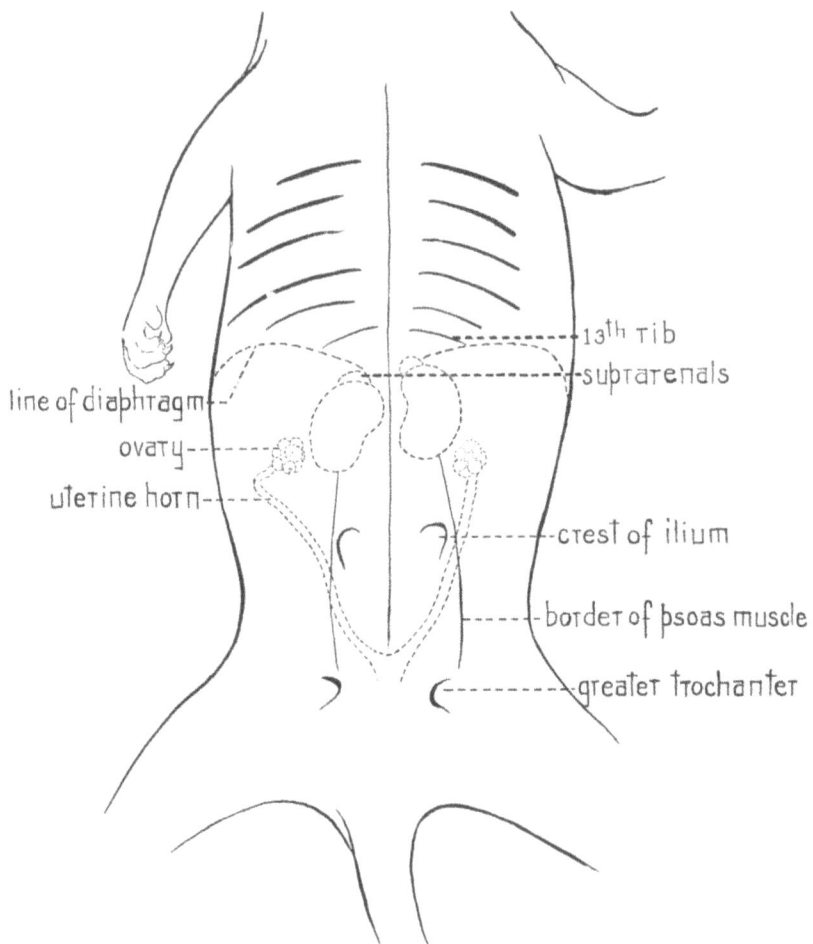

Fig.122. Dorsal View showing in Outline Kidneys, Suprarenals, and Ovaries in Relation to Bony landmarks.

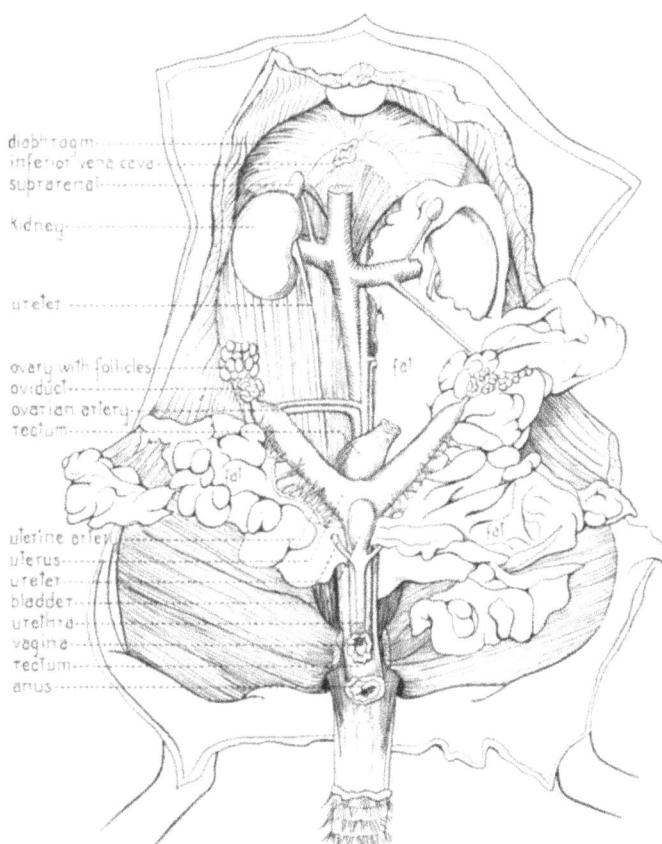

diaphragm
inferior vena cava
suprarenal

kidney

ureter

ovary with follicles
oviduct
ovarian artery
rectum

fat

fat

fat

uterine artery
uterus
ureter
bladder
urethra
vagina
rectum
anus

Fig. 123. Female Urogenital System. The fat has been removed from the right side

# CHAPTER V

## THE ENDOCRINE SYSTEM

(Figs. 124-137)

This system has become so much a subject for very special experimental research that any attempt on the part of the author to describe methods of approach would be presumptuous. Neither is it pertinent to go into the physiology of the endocrine glands to any extent. It seems more appropriate to merely show the organs in relation to surrounding structures. For operative technique the special papers should be consulted.

The organs will be taken up in the following order:

> Pineal body
> Hypophysis
> Carotid gland
> Thyroid and parathyroid
> Thymus
> Suprarenals
> Testis
> Ovaries

The **Pineal Body** (Fig. 124) may be seen, as in all mammals, by spreading apart the cerebral hemispheres and cerebellum.

The **Hypophysis** (Figs. 125-127) is shown from three aspects and in Figs. 140, 141, it appears in its relation to the brain. Stendell 1913, calls it primitive in type with its cavity persisting throughout life. "At about 40-50 days of age there appears a difference in the weight of the hypophysis according to the sex and with advancing age this difference tends to increase. The female has the heavier hypophysis." (Donaldson '24).

The **Carotid Gland** or more properly the **Carotid Body** is situated in the angle between the external and internal carotid arteries, medial to the vagus and posterior to the glossopharyngeal nerve. Since it has at times been looked upon as an endocrine gland it is mentioned here but is more fully discussed in Chapter IV on the viscera.

The **Thyroid** and **Parathyroids** (Figs. 128-130). After spreading apart the superficial cervical gland mass and separating the ventral muscles along the mid-line the lobes of the *thyroid gland* are revealed lying on either side of the trachea just below the larynx, covering from four to five tracheal rings in extent, and connected by an isthmus across the ventral surface of the trachea.

Closely applied to the antero-lateral surface of each lobe of the thyroid is the small *parathyroid gland*. Hoskins, 1925, investigated the question of accessory parathyroids in the rat. As she points out, it has been stated by some writers that accessory parathyroids are common in rats, for example by Erdheim, 1906, and Shapiro and Joffe, 1923. Upon microscopic examination of 94 specimens she found accessory parathyroids in only five cases. Of these five there was never more than one accessory parathyroid. In one case this was imbedded in the thyroid and was evidently the parathyroid in two parts, while in the other four it was cephalad to the thyroid and dorsal to the cartilage.

The **Thymus Gland** (Fig. 131) shows the actual size in a 90-day rat when the gland is at the height of its development. Fig. 111 shows the surrounding structures with some connective tissue removed but with practically no disturbance of the position of the gland.

The **Suprarenal Glands** (Fig. 132) show a distinct sexual difference. They are relatively larger in the female though it is not apparent until about 40-50 days of age according to Donaldson '24.

The **Testis** (Fig. 133) with its adjacent parts, has been described in Chapter IV on viscera.

The **Ovary** (Figs. 134-137) has likewise been included under viscera, Chapter IV.

### LIST OF FIGURES

### V. ENDOCRINES

corpus callosum

pineal body

corpora quadrigemina

hemisphere reflected

Fig.124.
Pineal Body and
Corpora Quadrigemina.

Hypophysis.

stalk

anterior lobe

pars intermedia

groove for post. communicating art.

posterior lobe (pars nervosa)

pontal surface

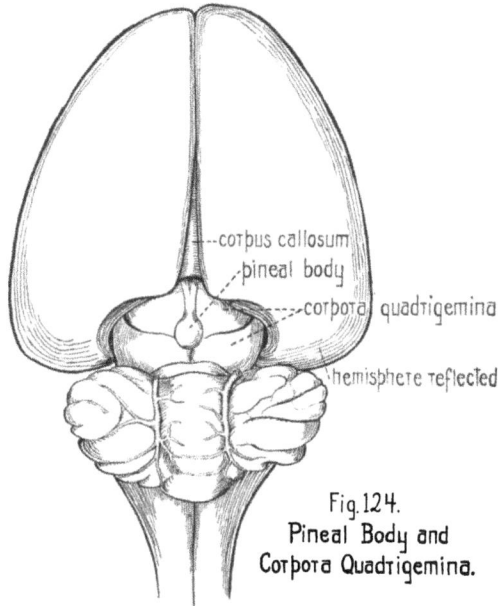

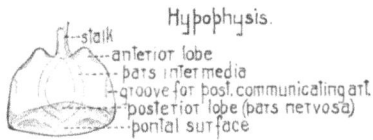

Fig.125.   Dorsal Aspect

surface lies against
floor of cranium

Fig 126.   Ventral Aspect.

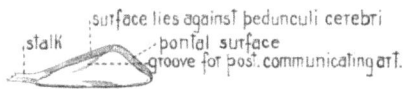

surface lies against pedunculi cerebri

stalk

pontal surface

groove for post. communicating art.

Fig.127.   Lateral Aspect.

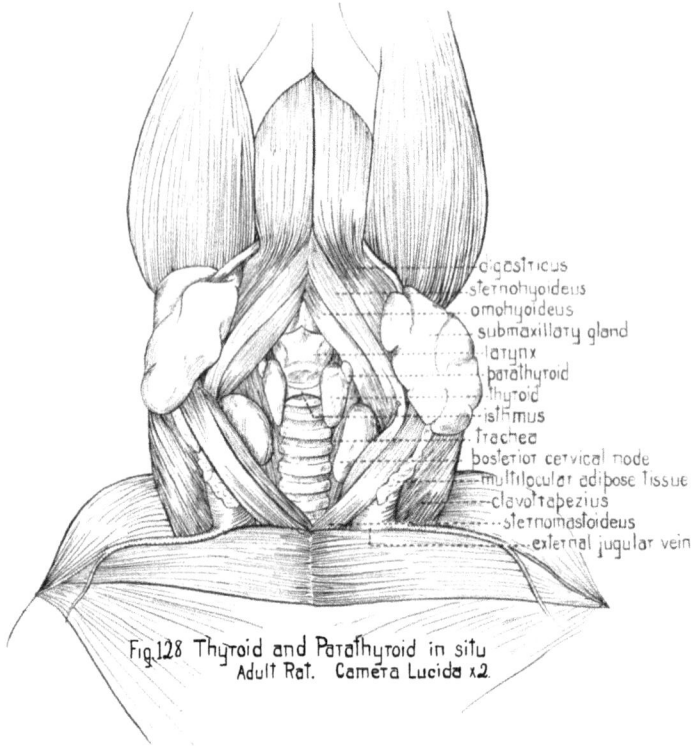

digastricus
sternohyoideus
omohyoideus
submaxillary gland
larynx
parathyroid
thyroid
isthmus
trachea
posterior cervical node
multilocular adipose tissue
clavotrapezius
sternomastoideus
external jugular vein

Fig.128 Thyroid and Parathyroid in situ
Adult Rat.   Camera Lucida x2.

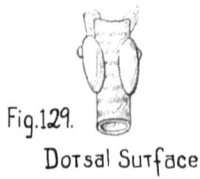

Fig.129.
Dorsal Surface

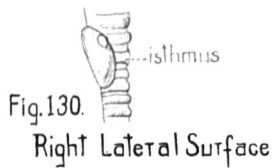

isthmus

Fig.130.
Right Lateral Surface

Thyroid and Parathyroid
Camera Lucida x2.

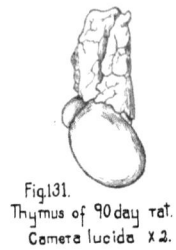

Fig.131.
Thymus of 90 day rat.
Camera lucida x2.

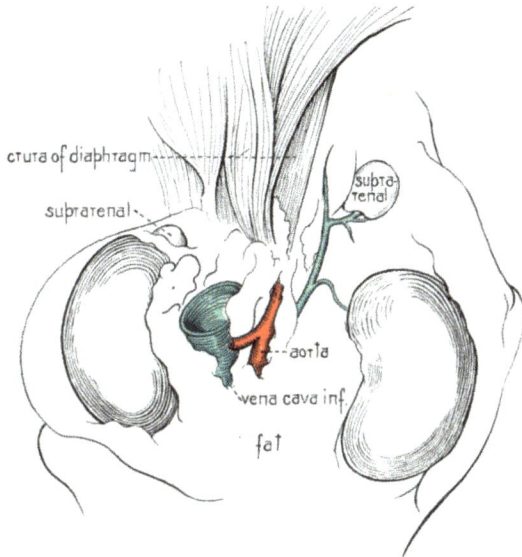

Fig.132. Kidneys and Suprarenals in Situ.

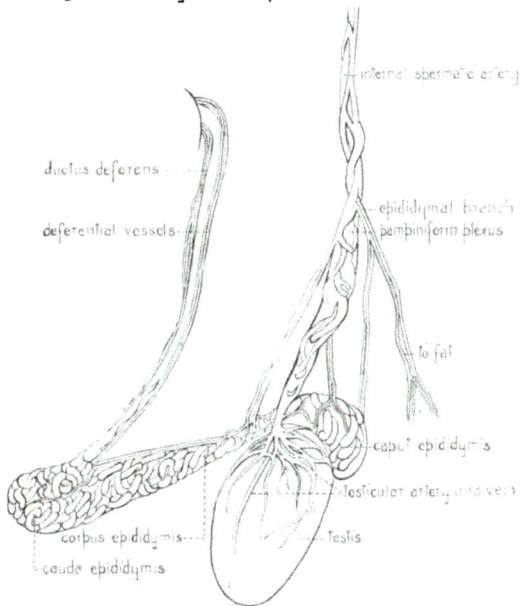

Fig.133  The Testis.

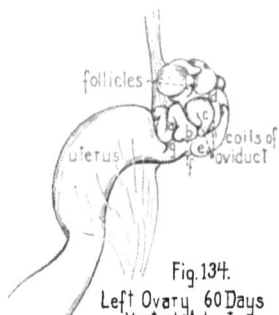

Fig. 134.
Left Ovary 60 Days
Ventral Aspect
Letters represent successive coils.

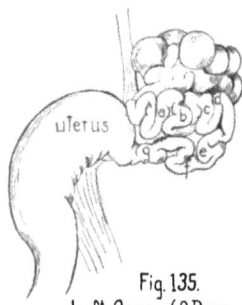

Fig. 135.
Left Ovary 60 Days
Coils of duct slightly separated.

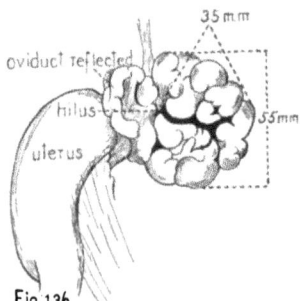

Fig. 136.
Left Ovary 60 Days.

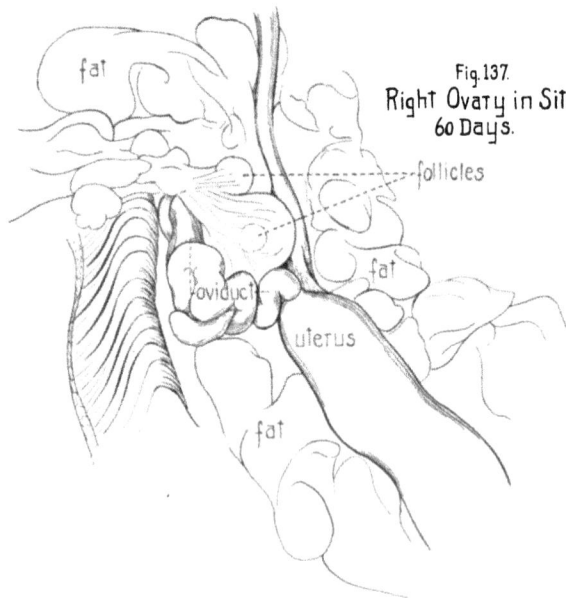

Fig. 137.
Right Ovary in Situ.
60 Days.

# CHAPTER VI

## NERVOUS SYSTEM

(Figs. 138-196)

The nervous system will be described in the following order:

Brain
Cranial nerves
Spinal cord
Spinal nerves
Plexuses
Sympathetic system

### BRAIN

The **Brain** (Figs. 138-141). The *cerebrum* is lissencephalous, with large olfactory bulbs. The *cerebellum* is divided into a median portion, or pyramid, consisting of anterior, central, and posterior lobes; two portions on either side of the median divided into lunate, ansiform, and paramedian lobes; and two conspicuous parafloccular lobes which lie in deep sockets of the periotic capsule of the skull.

### CRANIAL NERVES

The **Cranial Nerves** (Figs. 140, 141). The distribution of the simpler of the cranial nerves is part of the elementary knowledge of every laboratory worker and need not be discussed at length here. All but the olfactory are shown in drawings and the more extensive ones are described.

The **optic** (II), **oculomotor** (III), **trochlear** (IV), and **abducent** (VI), are shown leaving the skull and also distributed to the eyeball muscles (Figs. 142, 143).

The **trigeminal** (V), and **facial** nerves (VII), have been worked out and described below in detail (Figs. 144-153).

#### DISTRIBUTION OF THE TRIGEMINAL NERVE

The **Trigeminal Nerve** (Figs. 140, 141) emerges from the lateral region of the pons by two roots, the *motor* and *sensory* roots, the small motor root lying medial to the *semilunar ganglion* of the large sensory root. After leaving the ganglion the sensory root immediately breaks up into the *ophthalmic, maxillary*, and *mandibular* divisions (Figs. 140, 141), of which the first two are wholly sensory in function, whereas the mandibular portion is joined by the motor root as it leaves the skull.

No attempt has been made to show or describe the four ganglia, (ciliary, sphenopalatine, otic, and submaxillary) associated with the trigeminal nerve, as most of their branches of distribution could not be satisfactorily demonstrated without the use of microscopic sections.

The **Ophthalmic Nerve,** or first division of the trigeminal, gives rise to three branches, *lacrimal, frontal,* and *nasociliary,* all of which enter the orbit through the anterior lacerated foramen (Fig. 143).

The **lacrimal** supplies the intraorbital lacrimal gland and the conjunctiva.

The **frontal** having traversed the orbit, makes its exit in the region of the upper lid, to which it sends branches, then passes on to supply the skin of the forehead.

The **nasociliary** after entering the orbit crosses to the medial wall, passes through the anterior ethmoidal foramen as the **anterior ethmoidal nerve** (Fig. 212), to enter the cranial cavity and run forward to the cribriform plate which it pierces to enter the nasal cavity, where it gives off *internal nasal branches* to the nasal mucous membrane. Emerging from the nasal cavity between the nasal bone and the nasal cartilage as the **external nasal branch** (Fig. 146) it supplies the skin of the wing and tip of the nose.

The **Maxillary Nerve** (Figs. 140, 141) (superior maxillary), or second division of the trigeminal, passes out of the skull with the ophthalmic division, through the anterior lacerated foramen (Fig. 142) (which represents the superior orbital fissure of man plus the foramen rotundum) and crosses the pterygopalatine fossa. This fossa in the rat is wide owing to the lack of a post-orbital plate, and appears as a deep groove at the ventro-posterior limit of the large orbito-temporal fossa. Having thus entered the orbit the nerve traverses the infraorbital groove and canal and passes out through the infraorbital fissure. After entering the infraorbital canal the nerve is sometimes referred to as the infraorbital. Outside the infraorbital fissure, the branches of the nerve spread fan-wise to the side of the nose, the lower and upper lip.

The branches of the maxillary are divided into four groups according to the point from which they leave the nerve. These branches are as follows:

| | |
|---|---|
| In the Cranium.................... | Middle Meningeal |
| In the Pterygopalatine Fossa........ | Zygomatic<br>Sphenopalatine<br>Posterior Superior Alveolar |
| In the Infraorbital Canal........... | Anterior Superior Alveolar<br>The Middle Superior Alveolar of<br>    man is absent in the rat |
| On the Face..................... | Inferior Palpebral<br>External Nasal<br>Superior Labial |

The **middle meningeal** (dural branch), arising close to the origin of the maxillary, supplies the dura mater.

The **zygomatic** (temporomalar, orbital) (Fig. 147) passes through the anterior lacerated foramen to the orbit, where it divides into zygomaticotemporal and zygomaticofacial branches.

The **zygomaticotemporal** branch (temporal) curves dorso-posteriorly, passing beneath the temporalis muscle, and finally pierces the temporal fascia to reach the skin of the side of the head.

The **zygomaticofacial** branch (malar) (Fig. 146) after traversing the lower region of the orbit passes laterally below the eyeball to cross the zygomatic arch and supply the skin of the cheek.

The **nasopalatine** branch with the sphenopalatine artery, runs through the sphenopalatine foramen to the nasal cavity.

The **posterior superior alveolar** branch (posterior superior dental) leaves the nerve just as it enters the infraorbital groove (Fig. 207). Like the corresponding artery it gives off filaments to the gums and the mucous membrane of the cheek, then passes ventrally through minute foramina in the floor of the orbitotemporal fossa to reach the lining of the maxillary sinus and the three molar teeth.

The **middle superior alveolar** branch (middle superior dental) which in man supplies the premolar teeth is not present in the rat since the premolar teeth are also lacking.

The **anterior superior alveolar** branch (anterior superior dental) (Fig. 145) arises from the nerve during its course through the infraorbital canal. It continues through the canal, emerging from the infraorbital fissure only to re-enter the maxillary bone to reach the root of the upper incisor tooth. A minute nasal branch pierces the lateral wall of the nasal cavity to supply the mucous membrane.

The **inferior palpebral** branches, (palpebral) run upward beneath the orbicularis oculi to the lower lid.

The **external nasal** branches (Figs. 144-147), after leaving the infraorbital fissure, extend anteriorly, emerging from the angle between the masseter and levator labii superioris muscles to reach the skin of the nose and muzzle.

The **superior labial** branches (labial) (Fig. 146) accompany the external nasal branches and supply the skin of the upper lip and the mucous membrane of the mouth.

The **Mandibular Nerve** (inferior maxillary) (Figs. 140, 141) the third division of the trigeminal, is composed of a sensory and a motor root. Upon leaving the skull through the foramen ovale, the nerve gives off two branches, *nervus spinosus* and *internal pterygoid* nerve.

The **nervus spinosus** (recurrent branch) re-enters the skull and supplies the dura mater.

The **internal pterygoid** enters the pterygoideus internus muscle. The mandibular nerve then divides into an *anterior* and *posterior trunk*.

1. The **Anterior Trunk** gives off the following branches: *masseteric, deep temporal, buccinator,* and *external pterygoid*.

The **masseteric** (Figs. 144, 147) runs through the mandibular notch to the masseter muscle.

The **deep temporal** (Fig. 147) turns dorsally to reach the temporalis muscle.

The **buccinator** (long buccal) (Figs. 144, 147) passes downward and forward beneath temporalis and the anterior division of masseter, to reach the buccinator muscle.

The **external pterygoid** enters the external pterygoid muscle.

2. The **Posterior Trunk** of the mandibular nerve gives rise to the *auriculotemporal, lingual,* and *inferior alveolar* nerves (Figs. 147, 210).

The **auriculotemporal** immediately turns posteriorly and runs medial to the mandible, turning dorsally between the pinna of the ear and the temporomandibular joint, to cross the root of the zygomatic arch, giving off terminal branches to the skin of the temporal region. In its course the nerve gives off the following branches; the *anterior auricular* branches to the pinna of the ear, the *branches to the external acoustic meatus*; the *articular branches* to the temporomandibular joint; the *parotid branches* to the parotid gland; and the *superficial temporal branches* to the skin of the temporal region.

The **lingual** (Figs. 147, 156, 157) passes downward and forward and enters the tongue, extending to its tip and giving branches to the sublingual gland, mucous membrane of the mouth, gums, and anterior portion of the tongue.

The **inferior alveolar** (inferior dental) (Fig. 147) extends downward along the medial surface of the mandibular ramus, gives off the mylohyoid branch (Fig. 147), enters the mandibular foramen and after extending the length of the mandibular canal from which it gives branches to the teeth, it emerges from the mental foramen just ventral and anterior to the molar teeth (Fig. 147). Its branches are the *mylohyoid, inferior dental, incisor,* and *mental*.

The **mylohyoid** (Figs. 147, 152, 153) leaves the nerve just before it enters the mandibular foramen. It curves ventrally around the border of mylohyoideus, supplies this muscle and gives a branch to the anterior belly of digastricus, and to transversus mandibularis, and ends in the skin of the lower jaw.

The **inferior dental branches,** given off within the mandibular canal, supply the molar teeth.

The **incisor branch** enters the root of the lower incisor tooth.

The **mental branch** (Figs. 146, 147, 153) supplies the skin of the chin and mucous membrane of the lower lip.

Arising from the brain stem at the anterolateral border of the medulla, the **Facial Nerve** (Figs. 140, 141) shows a slight swelling, the geniculate ganglion, as it crosses the root of the trigeminal nerve and enters the facial canal from which it emerges through the stylomastoid foramen (Fig. 154).

The branches of the facial are generally grouped according to the point at which they leave the nerve; (1) in the facial canal, (2) in the neck, (3) in the paratoid gland.

### In the Facial Canal

The **greater superficial petrosal** nerve turns anteriorly, runs along the lateral border of the internal carotid artery and through the middle ear dorsal to the cochlea. In company with the artery it passes through the petrotympanic fissure and comes to lie on the medial aspect of the pterygoid plate. Thence it passes through the anterior lacerated foramen, with the ophthalmic division of the trigeminal nerve, and is joined by the deep petrosal nerve from the sympathetic plexus to form the **nerve of the pterygoid canal** (Vidian nerve) to the sphenopalatine ganglion, from which point it is eventually distributed by orbital and zygomatic branches of the trigeminal to the lacrimal gland and soft palate.

Within the canal the **chorda tympani** also has its origin from the facial nerve. Entering the tympanic cavity, it crosses the tympanic membrane and leaves the cavity through the petrotympanic fissure. Running medial to the branches of the posterior trunk of the mandibular nerve, the chorda tympani becomes incorporated with the lingual nerve (Figs. 210, 217) and with the latter is distributed to the tongue.

### In the Neck

The **posterior auricular** (Figs. 146, 148-151, 156) passes backward and upward between the external ear and the mastoid process to the auricle.

The **digastric** (Figs. 148, 149) arises with the stylohyoid nerve below the posterior auricular and enters the posterior portion of the digastric muscle.

The **stylohyoid** (Figs. 148, 149) arising with the digastric passes forward to the stylohyoid muscle.

### In the Parotid Gland

These branches in turn may be grouped in *temporofacial* and *cervicofacial* divisions.

*Temporofacial.*—The **temporal.** (Figs. 146, 148-151.) Superficial branches pass anteriorly and dorsally over the zygomatic arch to supply the anterior region of the auricle, the exorbital lacrimal gland, the frontalis muscle, and the upper part of the orbicularis oculi. A deep branch, *lower zygomatic* (Figs. 150, 151), which in man is a portion of the buccal, here arises from the temporal. Its branches supply the muscles of the nose, the buccinator, and orbicularis oculi.

The **zygomatic** branch (upper zygomatic, malar) (Figs. 148-151) is small, extending forward and upward to supply the upper portion of the orbicularis oculi.

*Cervicofacial.*—The **buccal** (infra-orbital) branch (Figs. 146, 148-151) crosses the masseter and supplies the superficial muscles of the face, the nose, and the upper lip.

The **ramus marginalis mandibulae** (marginal mandibular branch, mandibular) (Figs. 148-151, 153) runs forward in company with the parotid duct then extends forward to the muscles of the lower lip.

The **cervical** (Figs. 146, 150, 151, 156) goes to the platysma through the parotid gland.

The **Acoustic Nerve** (VIII) is shown only as it leaves the brain stem (Figs. 140, 141).

The origin from the medulla (Figs. 140, 141), the exit from the skull (Figs. 155, 157, 159, 192), of the last four cranial nerves, IX, X, XI, and XII may well be shown together. Their topographical relationships and distribution outside of the skull will be discussed briefly.

The **Glossopharyngeal Nerve** (IX) (Figs. 155, 157, 192) runs antero-medially between the external and internal carotid arteries. It sends a small branch to join a pharyngeal element from the vagus (Fig. 157) and help form the **pharyngeal plexus,** then runs forward to the stylopharyngeus muscle and mucous membrane of the tongue.

The **Vagus Nerve** (X) (Figs. 155, 157, 191, 192, 227) emerges from the posterior lacerated foramen, displays the ganglion nodosum then extends posteriorly in front of the vertebral column through the neck, thorax, and into the abdomen where it ends in the sympathetic plexus.

The **ganglion nodosum** gives off two branches (Figs. 155, 157). The anterior one is the **pharyngeal branch** which joins the pharyngeal branch of the glossopharyngeal nerve to form the pharyngeal plexus. From the posterior end of the ganglion comes the **superior laryngeal** nerve (Figs. 155, 157, 192, shown but not labeled), to the larynx and thymus gland.

**In the neck** the vagus supplies **cardiac branches** (Figs. 191, 193), and the **recurrent nerve** (Figs. 157, 191-193). On the right side the recurrent (laryngeal) nerve makes a loop around the beginning of the subclavian artery (Fig. 192). On the left side, strictly speaking, this nerve arises in the thorax. It runs through the aortic arch (Fig. 157), turns anteriorly along the boundary between œsophagus and trachea, to both of which it supplies branches, and ends as the **inferior laryngeal nerve** to the intrinsic muscles of the larynx.

**In the thorax** the vagus nerves form pulmonary and œsophageal plexuses. These may be seen in dissection but their ramifications are too minute and involved to be adequately shown in any gross anatomical drawing. From the œsophageal plexus two nerves emerge, containing fibers from each vagus, and follow the œsophagus through the diaphragm. The right nerve (Fig. 191) sends branches to the posterior surface of the stomach and to the cœliac ganglion through which it is distributed to the various plexuses of the sympathetic system. The left vagus supplies branches to the lesser curvature and ventral surface of the stomach and joins the sympathetic plexuses by way of the cœliac ganglion (Fig. 194).

The **Spinal Accessory Nerve** (XI) (Figs. 155, 157, 158 detail, 164, 166, 192) consists of two parts, an *internal* and *external ramus.* The **internal ramus** is accessory to the vagus and joins the ganglion nodosum through which it reaches the recurrent branch of the vagus. The **spinal** or **external ramus** appears in the cervical region with the ninth, tenth, and twelfth cranial nerves. As it leaves the posterior lacerated foramen it is separated from the ganglion of the vagus by the hypoglossal nerve which crosses it ventrally. Accompanied by the occipital artery the nerve runs posteriorly between the occipital condyle and the paramastoid process of the skull (Fig. 155).

Appearing from beneath the posterior border of the digastricus the nerve runs downward through the shoulder and joins nerves from the second and third and from the third and fourth cervical nerves. Together they constitute the **sternomastoid** and **subtrapezial plexuses** from which nerves are distributed to sternomastoideus, cleido-mastoideus, clavotrapezius (Fig. 158 detail) and to acromiotrapezius and spinotrapezius muscles (Figs. 157, 166).

The **Hypoglossal Nerve** (XII) (Figs. 155-159, 164, 191, 192) emerges from the hypoglossal canal and runs between the vagus and spinal accessory nerves, dorsal to the former, ventral to the latter. Turning forward it crosses the ventral surface of the occipital artery (Figs. 155, 157) and divides into *ascending* and *descending rami* (Figs. 155, 157, 164, 192). Running posteriorly the **descending ramus of the hypoglossal** joins the **n. descendens cervicalis,** from the first, second, and third cervical nerves, to form a loop, the **ansa hypoglossi,** (Fig. 158) from which nerves are distributed to most of the infrahyoid muscles, the posterior belly of the digastricus, and to the sternohyoideus, sternothyreoideus, and omohyoideus muscles (Figs. 157, 158, 164). The **ascending ramus of the hypoglossal** runs forward beneath the hyoid bone, crosses the ventral surface of the external carotid artery as the latter turns dorsally, and enters the hypoglossal region to supply branches to both extrinsic and intrinsic muscles of the tongue (Figs. 156, 157).

### Spinal Cord

The **Spinal Cord** (Figs. 162, 163), showing typical cervical and lumbar enlargements, ends at the level of the fourth lumbar vertebra, with the filum terminale traceable into the tail beyond the third caudal nerves, while the cauda equina, made up of all the lumbar, sacral, and caudal nerves, conceals the extent of the cord itself.

### Spinal Nerves

The **Spinal Nerves** (Figs. 162, 163), consist of 34 pairs of nerves, 8 cervical, 13 thoracic, 6 lumbar, 4 sacral, and 3 caudal. All the ganglia of the spinal nerves are enclosed within the vertebral canal with the exception of the second cervical, which is exposed in the angle between the atlas and epistropheus.

From the 2nd or 3rd thoracic to the 1st sacral nerves, the pairs of rami communicantes to the sympathetic ganglia may be readily demonstrated (Figs. 159, 184). The rami are considerably longer in the lumbar region, and may be slightly irregular in their distribution. A branch from one ganglion may be distributed to two nerves or two branches may be given to the same nerve.

Typical plexuses are formed in the cervical, brachial, and lumbosacral regions by the anterior (ventral) primary divisions of the nerves. From the posterior (dorsal) primary divisions come the dorsal and lateral cutaneous branches to the integument (Fig. 175). Any lengthy description of the distribution of terminal branches of the plexuses has been omitted but many of these branches are shown in considerable detail in the drawings of the circulatory system of any given region and the more important ones are described.

The **Cervical Nerves** (Fig. 160). The anterior ramus of the *first cervical nerve* emerges from the vertebral canal and passing over the posterior arch of the atlas forms a loop with the second cervical nerve before joining the hypoglossal.

The anterior ramus of the *second cervical nerve* passes between the arches of the first two vertebrae, curves forward around the lateral side of the vertebral artery and divides into an ascending and descending portion. The former joins the first cervical nerve and together they unite with the hypoglossal. The descending portion of the second cervical forms a loop with the *third cervical* and enters into the cervical plexus.

The ganglion of the second cervical nerve is the only ganglion lying outside the vertebral canal in the rat though it is not apparent in a ventral view where it is hidden by the broad transverse process of the epistropheus.

PLEXUSES

CERVICAL PLEXUS

The **Cervical Plexus** (Figs. 157, 159, 160, 164, 166), made up chiefly from the anterior rami of the first four cervical nerves, also receives a component from the fifth, though by far the greater portion of the latter enters into the formation of the brachial plexus. The first cervical also takes very little part, contributing mainly to the hypoglossal. Under cover of the sternocleidomastoid group of muscles, the cervical plexus gives off its branches which may be divided, as in man, into *superficial* (*cutaneous*) *branches* and *deep muscular* and *communicating branches*.

Superficial Cutaneous Branches

(Figs. 157, 164, 175)

**Ascending Branches**, from CIII, CIV. As a rule these arise as one trunk in the rat.

> Lesser occipital
> Great auricular
> Cutaneous colli (transverse superficial cervical)

The common trunk of the above nerves crosses omohyoideus covered by the sternomastoid group of muscles to emerge above the shoulder between clavo- and acromiotrapezius in company with the supraclavicular nerves. Just below the ear it breaks up into three branches, **lesser occipital** to the scalp above and behind the ear and to the skin of the neck; **great auricular** to the ear and the skin behind it, to the parotid gland and to the skin of the lower cheek, and **cutaneous colli** distributed through the platysma to the skin of the ventral region of the neck.

**Descending branches**, from CIII, CIV, CV.

> Anterior supraclavicular (suprasternal)
> Middle supraclavicular (supraclavicular)
> Posterior supraclavicular (supra-acromial)

The two latter arise together. The ascending branches and the anterior supraclavicular generally spring from a loop made up of components from CIII and CIV. These supraclavicular nerves supply the skin of the side of the neck, and of the front of the chest and shoulder.

Deep (Muscular and Communicating Branches)

(Figs. 157, 164, 166)

**Lateral branches**, from CII, CIII, CIV, comprise *muscular* branches to the sternomastoid, cleidomastoid and clavotrapezius, levator claviculae (Figs. 164, 166) levator scapulae (Figs. 164, 166), and scalenes, as well as *communicating* branches to the accessory nerve and thence to trapezius (Figs. 157, 164).

**Medial branches** (Fig. 166) from CI–CV are *muscular* to the prevertebral and infrahyoid muscles, to the diaphragm (phrenic nerve from CIV, CV) or *communicating* branches to X, XII, and to the sympathetic system.

Paterson '87, in his studies on the "Limb Plexuses in Mammals", has worked out in great detail the constitution, arrangement, and distribution of these plexuses in ten different animals including rat. He uses the porcupine as the typical example but where his description is applicable it has been freely quoted, with whatever changes were necessary to make the following account a description of the rat. In several instances it has been necessary to change the nomenclature to conform to the B.N.A. but the older terms, used by Paterson, are also included.

"The terms adopted are as follows: The two primary divisions of the nerve resulting from the division of the original mixed nerve, are called the superior and inferior primary divisions. When these subdivide the chief trunks are called dorsal (posterior, B.N.A.) and ventral (anterior, B.N.A.)."

### BRACHIAL PLEXUS

The **Brachial Plexus** (Figs. 159, 164, 167, 169, 237) is formed by the inferior primary divisions of the last four cervical, the greater part of the first thoracic, and in most cases a small contribution from the second thoracic nerves. The nerves emerge between the scalene muscles, the plexus formation occurring in the axilla (Fig. 167). The plexus is more flattened in the rat than in man and is not divisible into lateral, medial, and posterior cords comparable to those in man.

The *fourth cervical nerve* (Figs. 164, 166), on its appearance, divides into a ventral and dorsal division, both of which enter into the formation of the cervical plexus. From the ventral division, before it joins the plexus, a branch may unite with a portion of the fifth cervical nerve to form the **phrenic nerve** to the diaphragm (Fig. 166).

The *fifth cervical nerve* (Figs. 164, 166) also divides immediately into dorsal and ventral divisions; the former after giving off the **dorsal scapular nerve** (Figs. 164, 174), to levator scapulae and rhomboideus, enters into the formation of the cervical plexus, while the ventral division, after giving a portion to the phrenic, unites with the ventral division of the sixth cervical.

The *sixth cervical nerve* (Figs. 164, 166, 167), first gives off a dorsal (posterior) branch which unites with similar branches from the seventh and eighth cervical to form the **long thoracic nerve** to serratus anterior (Figs. 168, 174). Then immediately upon receiving the major portion of the ventral division of the fifth, the sixth cervical gives off a small **branch to the subclavius** muscle (Fig. 167). The inferior division of the sixth nerve divides into one ventral and three dorsal branches (Figs. 164, 167). Of the *dorsal* the first is the **suprascapular** (Figs. 169, 170, 172, 237) to supraspinatus and infraspinatus; the second, a small nerve, in the **upper (first or short) subscapular** (Figs. 170, 273) to subscapularis; and the third is the first root (head, B.N.A.) of the

**axillary** (circumflex) nerve, described below, (Fig. 164). This joins the second root from the dorsal division of the seventh nerve. The *ventral* branch of the nerve forms a loop with part of the ventral division of the seventh nerve, and gives rise to the *musculocutaneous* (Fig. 164).

The *seventh cervical nerve* (Figs. 164, 167) splits into dorsal and ventral parts. The ventral part subdivides into anterior and posterior branches. The anterior branch forms a loop with the ventral part of the sixth nerve. According to Paterson, before doing this, it gives off two fine nerves to the pectoralis major muscle. So far these have not been demonstrated in the rat. From the loop come, at the one end, the *musculocutaneous nerve*, at the other end the *lateral (external) anterior thoracic*, or these two nerves may be formed from two separate loops.

The **Musculocutaneous Nerve** (Figs. 164, 169-171, 240) crosses the border of teres major and enters the brachium where it runs between coracobrachialis and the short head of biceps brachii giving branches to these muscles (Figs. 170, 240). It then reaches the elbow where it sends a branch curving around to brachialis (Fig. 171) before entering the antibrachium as the **lateral cutaneous nerve of the antibrachium** (Figs. 240, 243, 244). A small **muscular branch** is distributed with the radial branch of the brachial artery to pronator teres and flexor carpi radialis (Fig. 243). Below the elbow the nerve is accompanied by the ramus anastomoticus radialis. After giving off a cutaneous branch the nerve divides into two branches, one of which supplies the skin over the base of the pollex (Fig. 243), while the other communicates with the superficial branch of the radial nerve before reaching the first interdigital space as the **dorsal digital nerve** (Fig. 244).

The **Lateral Anterior Thoracic Nerve** (Figs. 164-167) supplies the pectoralis major, and, piercing the muscle, gives small branches to the skin over it.

The posterior branch of the ventral division of the seventh nerve forms the first root of the *median* (Fig. 164). The *dorsal* division contributes to the long thoracic nerve before splitting into anterior and posterior branches. The anterior branch forms the second root of the *axillary* (circumflex) (Figs. 164, 167), joining part of the dorsal division of the sixth nerve. The posterior branch forms the first root of the *radial* (*musculospiral*) (Figs. 164, 167, 169, 170). From each of these branches a nerve is given off before they join their companion nerves. From the anterior branch a **lower** (**middle**) **subscapular** nerve (Figs. 164, 170) to subscapularis arises, before the formation of the axillary. From the posterior branch the **thoracodorsal** (long subscapular) (Figs. 164, 167) is given off to the latissimus dorsi.

The **Axillary Nerve** made up from CVI and CVII (Figs. 164, 167, 171, 237) divides into an anterior and posterior branch. The *anterior branch* supplies **muscular rami** to the deltoid, an **articular ramus,** to the shoulder joint, and the **lateral cutaneous nerve of the brachium.** The *posterior branch* supplies **muscular rami** to teres major and minor. There seems to be no reason for designating the nerve to teres major as "special subscapular" as given by Paterson. A second short subscapular nerve to subscapularis, mentioned by Paterson, seems to be lacking in the rat.

The **Median nerve** (Figs. 164, 167, 237, 240-243) runs parallel to and in company with the axillary artery and the radial (musculospiral) nerve to the brachium. Here it may be found in one of three relationships to the brachial artery as described in the chapter on the circulatory system (Figs. 237, 241). No branches leave the median in the brachium, but in the hollow of the elbow two branches arise (Fig. 242); one **muscular branch** which passes between the heads of flexor digitorum sublimis and profundus supplying both these muscles and palmaris longus; the other, the **volar interosseous nerve** to the various heads of flexor digitorum profundus and to pronator quadratus.

Continuing through the antibrachium, in company with the corresponding vessels, the median nerve divides just above the transverse carpal ligament into three **common volar digital nerves** (Figs. 240, 243) to the first, second, and third interdigital spaces respectively. Each nerve then divides into two **proper volar digital nerves** to the adjacent sides of the first and second, second and third, and third and fourth digits.

The **Ulnar Nerve** (Figs. 164, 167, 237, 240, 241, 243), smaller than the median and arising behind it, enters the brachium in company with the radial and median nerves and brachial artery (Figs. 167, 169). Here it may be separated from the median by the artery (Fig. 237, 240), or the artery may pass around the ulnar side of the ulnar nerve (Fig. 241), or around the radial side of the median. These variations will be discussed under the brachial artery. Upon reaching the elbow the ulnar nerve crosses the ulnar collateral artery (Figs. 240-242) and passes beneath anconeus to the antibrachium where it divides into a *dorsal* (deep) and *volar* (superficial) branch (Fig. 243) as it runs between flexor carpi ulnaris and flexor digitorum profundus which it supplies.

The **volar branch** (Fig. 243) gives a *cutaneous* branch halfway down the antibrachium, then proceeds with the corresponding branch of the ulnar artery beneath the pisiform bone to the manus where it divides into two branches; a **common volar digital nerve** running through the fourth interdigital space to supply two **proper volar digital nerves** to adjacent sides of digits four and five; and a **proper volar digital nerve** to the ulnar side of **the fifth digit.** Undoubtedly further dissection or microscopic examination would reveal the **deep branch** of this ramus volaris manus supplying the interosseous muscles, part of the lumbricales and parts of the adductor pollicis and flexor pollicis brevis, but this has not been worked out as yet.

The **dorsal branch** (Fig. 244) with the dorsal branch of the ulnar artery, runs between the ulnar head of flexor digitorum profundus and flexor carpi ulnaris to reach the superficial dorsal surface of the antibrachium. It gives rise to a cutaneous branch then divides into two **dorsal digital nerves**, one of which supplies digit five, while the other supplies branches to adjacent sides of digits four and five.

The **eighth cervical nerve** (Figs. 164, 167) divides into dorsal and ventral portions. The *ventral* and larger portion is in two parts, one passing outwards to form the second head of the median, the other passing backwards to join the ventral division of the first thoracic nerve, and being the first root of the **medial anterior thoracic nerve** (Figs. 164, 165, 167, 237) of the thorax. Paterson calls this nerve the lateral cutaneous nerve of the thorax, and adds a note as follows: "This nerve which is fairly constant as a considerable trunk in mammals, being found in seven out of the ten animals dissected, may be looked upon as the homologue of the nerve of Wrisberg (lesser internal cutaneous) of human anatomy. Its origin is the same (from the ventral trunks of the eighth cervical and first thoracic nerves); and it supplies the integument of the fold of the axilla, communicating in its course with the intercostal nerves. In these animals, moreover, there is no other nerve corresponding to the nerve of Wrisberg." The evidence that this nerve is homologous to the nerve of Wrisberg (otherwise known as the lesser internal cutaneous, cutaneous brachii medius), is not convincing as it does not have the same relationship or distribution. The cutaneous brachii medius nerve is accompanied by the basilic vein and is primarily a nerve of the arm, whereas, this nerve trunk never reaches the arm but lies beneath the pectoralis muscles and is distributed in part at least, with the lateral thoracic artery and long thoracic vein to the cutaneous maximus muscle of the thoracic wall and to pectoralis major and minor. *Medial anterior thoracic nerve* would, therefore, seem to be the correct name since it corresponds to the nerve of that name in human anatomy in relationship and distribution if not in size.

The *dorsal* smaller portion of the eighth nerve contributes to the long thoracic nerve (Fig. 168), then forms the second (and larger) root of the *radial* or *musculospiral* (Fig. 164). This nerve is thus derived from the dorsal divisions of the seventh and eighth nerves.

The **Radial (Musculospiral) Nerve** (Figs. 164, 167, 170, 171, 237, 240, 243, 244) leaves the axilla with the ulnar and median nerves and brachial artery, passing between latissimus dorsi and cutaneous maximus to the brachium (Fig. 167), where it gives **muscular branches** (Figs. 170, 171) first to the long and medial heads of triceps brachii, then to the short head of biceps brachii, before curving around the humerus to reach the lateral surface of the brachium where it emerges from between the long head of triceps and brachialis under cover of the lateral head of triceps brachii, in which position it supplies the two latter muscles and coracobrachialis, then gives the **dorsal antibrachial cutaneous** branch (Fig. 244) to the dorsal surface of the forearm before continuing through the antibrachium. In the bend of the elbow the radial nerve divides into a *superficial* and *deep branch*. The **superficial branch** (**superficial radial nerve**) (Fig. 244) accompanies the a. collateralis radialis. Halfway down the antibrachium it divides into two **dorsal digital nerves,** which enter the second and third interdigital spaces to supply adjacent sides of the second and third, and third and fourth digits. The lateral one of these dorsal digitals communicates with the musculocutaneous nerve and the resulting branch forms the dorsal digital nerve of the first interdigital space (Fig. 244). More extended observation would doubtless show a branch from this to the first digit but this was not demonstrated in gross dissection.

The *first thoracic* nerve divides into the following branches:

1. A large branch which is joined by part of the ventral division of the eighth nerve to form the *medial anterior thoracic* nerve (Fig. 164), distributed to cutaneous maximus and the pectoralis muscles. This nerve in the rat apparently represents the first two branches of the first thoracic nerve as described by Paterson.

2. The third head or root of the *median* (Fig. 164). This nerve, formed from the ventral divisions of the seventh and eighth cervical nerves, and from the first thoracic, is the largest in the plexus. Its distribution has been described above.

3. The *ulnar* nerve (Fig. 164), smaller than the median, and arising behind it.

The first thoracic nerve in the rat, as in the porcupine, gives no trunk corresponding to the dorsal trunk of the other nerves engaged in the plexus. In this respect these two animals represent an exception, as, according to Paterson, "In eight out of the ten animals studied the axillary portion of the first thoracic nerve resembled the other nerves joining the plexus, and, like them, divided into dorsal and ventral parts, which combined with the corresponding cords of anterior nerves to complete the plexus."

The *second thoracic* nerve may contribute a small branch to the first thoracic nerve (Figs. 159, 162). Cunningham states that this occurs in one-third or more cases in man, but it is impossible to state at the present time the frequency with which this occurs in the rat.

The **Lumbo-Sacral Plexus** (Figs. 176, 177). For the sake of convenience this plexus will be described as consisting of two parts, lumbar and sacral, although here, as in other cases, there is no line of demarcation between them, and certain nerves enter into the formation of both.

While Paterson's description of the brachial plexus in the porcupine applies to the rat with very few changes, this is not true of the lumbo-sacral plexus where there is considerably more variation, not only within the plexus, but also between the different animals described by Paterson.

In tabulating the number of spinal nerves in the rat, Paterson gives 8 cervical, 11 thoracic, 6 lumbar, 2 sacral, and 3 caudal, making a total of 30. Actually, as has been stated above, the rat has 8 cervical, 13 thoracic, 6 lumbar, 4 sacral, and 3 caudal, totaling 34 (Figs. 159, 161). In comparing the spinal nerves which enter into the lumbo-sacral plexus, he therefore gives the 21st to 25th inclusive for the rat, whereas the 22nd to 28th would be correct and incidentally would be in much closer accord with his findings for the other animals of the series.

Since the description of plexuses given for man by Cunningham is with few exceptions more directly applicable to the rat than Paterson's description it will be extensively quoted with whatever changes are necessary to make it conform to the animal in question.

The lumbo-sacral plexus is formed by the union of the anterior rami of the six lumbar and first sacral nerves. Frequently a communicating branch of the thirteenth thoracic nerve joins the first lumbar nerve (Fig. 177). The nerves to the lower limb are derived from the plexus, and, in addition, nerves arise at its superior limit which are distributed to the trunk above the level of the limb.

### Lumbar Plexus

The **Lumbar Plexus** (Figs. 177, 178, 184) is formed by the first four or five lumbar nerves, and is often joined by a branch from the thirteenth thoracic nerve. It is limited below by the fourth lumbar nerve (n. furcalis) which enters also into the composition of the sacral plexus. The nerves of the lumbar plexus are formed in the loin and supply that region as well as part of the lower limb. They are separated from the nerves of the sacral portion of the plexus by articulation of the os coxa with the sacrum (Fig. 161).

This plexus furnishes another example of the similarity of rat and man. Comparing the normal for man (Cunningham) and rat as follows:

|  | Man | Rat |
|---|---|---|
| Iliohypogastric | T12?, L1 | T13?, L1 |
| Ilio-inguinal | T12?, L1 | L1, L2? |
| Nervus furcalis | L4 | L4 |
| Obturator | L2, 3, 4 | L2, 3, 4 |
| Femoral | L2, 3, 4 | L2, 3, 4 |
| Tibial | L4, 5, S1, 2, 3 | L4, 5, 6 |
| Common peroneal | L4, 5, S1, 2 | L4, 5, 6 |

According to Cunningham the position of the n. furcalis may be taken as a guide to the arrangement of the plexus. This plexus shows considerable variability as compared with the brachial.

The plexus is formed in the substance of the psoas muscle in front of the transverse processes of the lumbar vertebrae. The nerves, on emerging from the intervertebral foramina, are connected with the sympathetic system (Fig. 178) and then divide in the substance of the psoas muscle as follows:

**Muscular branches** to the quadratus lumborum muscle arise independently from the lumbar nerves.

The *first lumbar nerve* (Fig. 178) divides into two branches, a superior one, the *iliohypogastric nerve*, which may be joined by a branch from the thirteenth thoracic, and an inferior one, the *ilio-inguinal nerve*, which is somewhat variable. In some cases these two nerves may be joined in a single trunk for a short distance, or the inferior branch of the first lumbar may unite with a branch from the second lumbar (Figs. 181, 184 right side). In some cases the branch from the second lumbar remains as an independent muscular branch (Figs. 178, 184 left side). Occasionally the iliohypogastric is formed by the thirteenth thoracic nerve and the ilio-inguinal by the first lumbar (Fig. 181).

The **Iliohypogastric Nerve** (Fig. 176) after traversing the psoas muscle obliquely appears at its lateral border on the surface of the quadratus lumborum and behind the kidney. It enters the lateral abdominal wall lying between the transversus and obliquus abdominus internus muscles above the crest of the ilium.

The **Ilio-inguinal Nerve** (Fig. 176) which appears between the psoas major and psoas minor muscles at the level of the anterior end of the left kidney, and at the level of the hilus of the right, runs downward behind the renal vessels and the ureter to reach the iliolumbar vessels which it parallels as they cross the quadratus lumborum. The ilio-inguinal then enters the abdominal wall and extends forward between the transversus and obliquus internus.

The *second lumbar nerve* gives off at its root a small branch which either unites with one from the first lumbar to form the ilio-inguinal (Fig. 184 right) or remains as an independent muscular branch to the abdominal wall (Fig. 184 left). The main portion of the nerve passes backward to form a loop with the third lumbar (Fig. 177). On its way this branch gives off two nerves, the *genitofemoral* and the *lateral cutaneous nerve of the thigh*.

The **Genitofemoral Nerve** (Figs. 176-179, 181) which pierces the psoas major not far from the mid-line (Fig. 176), runs dorsal to the ureter then parallels the internal spermatic vessels to the inguinal region where it divides into two branches, the **lumbo-inguinal branch** (Figs. 176, 177, 273) which pierces the abdominal wall just below the inguinal ligament to reach the skin of the femoral triangle; and the **external spermatic branch** (Figs. 176, 177, 179, 273) which crosses the external iliac vessels and, with the ductus deferens, testicular, and external spermatic vessels, enters the inguinal canal to supply the cremaster muscle and the skin of the scrotum and adjacent part of the thigh. In the female it runs to the labium majus.

The **Lateral Cutaneous Nerve of the Thigh** (lateral femoral cutaneous) (Figs. 176, 177, 181) is generally given off at the point where the branch from the second lumbar nerve joins the third lumbar. In one case it arose as two separate roots, one from the usual source, one from the third lumbar somewhat lower down but before the junction of the third and fourth lumbar. The lateral cutaneous nerve appears on the surface of the psoas under cover of the iliolumbar vessels (Fig. 176). Passing dorsally it runs along the posterior surface of the iliolumbar vessels and accompanies the iliac branch of these vessels to the anterior superior iliac spine where it perforates the muscle wall to reach the skin of the thigh (Figs. 253, 254).

The *third lumbar nerve* (Fig. 177) runs posteriorly across the root of the transverse process of the fourth lumbar vertebra (Fig. 178). After receiving the major portion of the second lumbar nerve it crosses the transverse process of the fifth lumbar vertebra and divides, at the level of emergence of the fifth lumbar nerve, into ventral and dorsal parts.

The *ventral branch*, or division (Figs. 177, 184), forms the first root of the *obturator*, the *dorsal branch* the first root of the *femoral* nerve. From the latter a small nerve is given to iliacus.

The *fourth lumbar nerve* (n. furcalis) crosses the transverse process of the fifth lumbar vertebra in company with the third lumbar nerve (Fig. 178), and divides into two parts. The main portion continues on to join the sacral plexus. The smaller portion also divides into two parts, the *ventral branch* unites with the ventral branch of the third lumbar nerve to form the *obturator*. The *dorsal branch* forms the second root of the *femoral* nerve (Fig. 177).

The **Femoral Nerve** (Figs. 176-184) from the third and fourth lumbar nerves, appears from between psoas minor and iliacus, and runs under the inguinal ligament in company with the external iliac vessels. Before entering the thigh the nerve is divisible into its *anterior* and *posterior* divisions (Figs. 180-183). The former consists of muscular branches to iliacus and pectineus; the latter of a muscular branch to the various parts of quadriceps femoris, and saphenous nerve (long saphenous) which runs superficially down the medial surface of the thigh and lower leg, accompanied by the saphenous artery and large saphenous vein. It supplies cutaneous branches to the medial surface of the lower leg, some of which take the place of the medial sural branch of the tibial. Upon reaching the ankle the nerve divides into several branches (Fig. 179) to the medial surface of the heel, to the tarsus and to the skin in the region of the first metatarsal (Fig. 322). Its extent in the foot depends upon the distribution of the superficial peroneal nerve.

The **Obturator Nerve** (Figs. 177, 178, 182, 183), from the third and fourth lumbar nerves, runs through the psoas major muscle and enters the pelvis, deep to the common iliac vessels, to join the obturator branch of the hypogastric artery (Fig. 178) which it accompanies between abductor caudae externus and internus to reach the obturator membrane through which it enters the thigh. Just before penetrating the membrane, the nerve divides, like the artery, into an *anterior* and *posterior division* or *branch* (Fig. 283). In the thigh the **anterior branch** (Figs. 182, 183) after giving an **articular branch** to the hip joint, emerges under cover of the adductor longus, from between the pectineus and adductor magnus. Extending obliquely caudad it supplies adductor longus, pectineus, gracilis anticus and posticus, and terminates in a **cutaneous branch** to the medial surface of the thigh. The **posterior branch** (Figs. 182, 183) lies somewhat deeper beneath adductor brevis which it supplies, together with adductor magnus, quadratus femoris, and obturator externus.

*Sacral Plexus*

The **Sacral Plexus** (Figs. 163, 177, 184) in the rat, is more limited in the extent of its origin than in man. It is formed by part of the fourth lumbar nerve and by the fifth and part of the sixth (n. bigeminus).

The latter also assists in forming the pudendal plexus. The divisions of the nerves mentioned above unite in a large trunk, the **lumbo-sacral trunk** (Figs. 177, 184-186, 277) which runs parallel with the remainder of the sixth lumbar nerve over the ventral aspect of the sacrum and becomes the *sciatic nerve* in the pelvis minor where it is separated from the pudendal nerve by the superior gluteal artery. Together they run through the deep groove between the dorsal border of the ischium and the root of the tail, as far as the caudal extent of the sciatic notch, where the sciatic nerve enters the thigh. The posterior cutaneous nerve continues in the groove while the pudendal nerve takes a more medial course through the pelvis (Fig. 186).

**Branches.**—The sacral plexus gives rise to two series of nerves of distribution, designated, according to their origin, as *anterior* (ventral) or *posterior* (dorsal), and consolidated in the pelvis as the **sciatic nerve** (**n. ischiadicus**) (Fig. 177, 178, 186, 187, 296, 302) but divisible, upon dissection, into two main *terminal* components, the *tibial* and *common peroneal* nerves, and several smaller *collateral* branches as follows:

| Anterior (ventral) branches | Posterior (dorsal) branches |
|---|---|
| Tibial nerve | Common peroneal nerve |
| Muscular branches | Muscular branches |
| Nerves to hamstring muscles | Nerve to anterior head of bicep |
| Nerves to quadratus femoris | Nerve to piriformis |
| Nerves to gemelli | Superior gluteal nerve |
| Nerve to obturator internus | Inferior gluteal nerve |
| Articular branch to hip joint | Articular branch to knee joint |

Upon removal of the sheath the tibial and common peroneal nerves are separable up to their origin in the plexus; contained in their sheath they appear as a single trunk until they reach the thigh through the sciatic notch. Within the sheath the nerve to the hamstring muscles arises from the tibial nerve, the *nerve to the anterior (short) head of biceps femoris* plus the *inferior gluteal nerve* arises from the peroneal.

The **Superior Gluteal Nerve** (Figs. 177, 184, 187) is the first branch to be given off from the plexus. It runs through the anterior end of the sciatic notch to the hip (Fig. 184), where it is joined by the corresponding artery and divides into a superior and inferior branch. The *superior branch* (Fig. 188) curves immediately around the border of the sciatic notch and between the gluteus medius and minimus muscles, both of which it supplies. The *inferior branch*, running between piriformis and gemellus superior, and beneath gluteus minimus, curves around the inferior border of gluteus medius, which it supplies, to reach the deep surface of tensor fasciae latae and gluteus maximus in the angle between the trunk and the thigh (Figs. 182, 183).

The next branch from the plexus is a small combined **nerve to piriformis, to obturator internus,** and **gemellus superior** (Figs. 177, 186, 188). Branching from the medial aspect of the sciatic, just before it enters the thigh at the level of piriformis is the **inferior gluteal nerve** (Figs. 177, 186-188) and, incorporated with it, the **nerve to the anterior head of biceps femoris** (Figs. 186-188). The inferior gluteal is distributed to gluteus maximus. In the thigh the tibial portion of the sciatic supplies the **nerve to quadratus femoris,** the **nerve to the hamstring muscles** and **to caudofemoralis** (Figs. 183, 187, 315), and an **articular branch** to the hip joint.

After giving the various collateral branches mentioned above, the sciatic divides into its two terminal components (Fig. 187), the *common peroneal* and *tibial nerves*, which in their course through the thigh, cross obturator externus, quadratus femoris and adductor magnus, lying between these muscles and biceps femoris, until they reach the popliteal fossa where they separate. The common peroneal runs down the lateral side of the leg while the tibial takes a more medial course.

At the level of the greater trochanter, the **Common Peroneal Nerve** (external popliteal) (Figs. 187, 188) gives an **articular branch** to the knee joint, and the *sural nerve* to the skin of the lower leg, as described below. The nerve then traverses the popliteal fossa, crosses the lateral surface of the superior muscular branch of the popliteal artery and enters the lower leg between biceps femoris and the lateral head of gastrocnemius (Figs. 187, 311, 320).

The **Sural Nerve** (Figs. 187, 312, 315, 316, 318, 326) branching from the common peroneal nerve in the hip, continues in the sheath with the latter and with the tibial to a variable point in the thigh. It may give a distinct **lateral sural branch** to the skin over the lateral surface of the calf (Fig. 316 shown but not labeled) as it runs through the popliteal fossa. As it continues between biceps femoris and the lateral head of gastrocnemius it is accompanied by the superficial sural artery and vein. Beyond the posterior border of the hamstring muscles, it becomes superficial, sends a cutaneous branch to the skin of the lateral side and back of the distal third of the leg (Fig. 316), a **peroneal anastomotic branch** (Figs. 318, 322) beneath the tendo calcaneous to the lateral plantar nerve, and through it, to the lateral side of the fifth digit (Fig. 326), then runs behind the lateral malleolus and ends in the skin and fascia of the ankle and heel (Fig. 318).

The above origin of the sural nerve which appears to be very definitely the rule in the rat, is not the usual one in man but does occur infrequently. In man the sural nerve is most commonly formed by the medial sural branch of the tibial plus a peroneal anastomotic branch from the common peroneal. Another variation occurring frequently in man is that in which the medial sural forms the sural nerve without any contribution from the peroneal. The more unusual condition in man, whereby the peroneal anastomotic branch forms the sural nerve and the medial sural ends on the heel, is most nearly like the condition obtaining in the rat though we find no medial sural in this animal. When the condition described occurs in man the peroneal anastomotic branch supplies the lateral side of the foot and fifth digit which it does in the rat by way of the lateral plantar nerve. Hence it would seem best to reserve the term peroneal anastomotic for this communicating branch and to designate the long, slender branch from the peroneal, which has the proper distribution to the lower leg, ankle, and heel, and is accompanied by the superficial sural vessels, as the sural nerve.

Below the knee the common peroneal nerve divides (Fig. 321) into a *superficial* and *deep peroneal* branch.

The **Superficial Peroneal Nerve** (musculocutaneous) extends as a long, slender nerve in front of the fibula and beneath peroneus longus to join the peroneal artery and vein. The nerve supplies peroneus longus and brevis before emerging, with the superficial cutaneous vessels, from beneath peroneus brevis and extensor digitorum longus to become superficial in the lower third of the leg (Fig. 327) and reach the dorsum of the foot which it supplies, as well as the medial distal surface of the first digit, and the adjacent sides of digits two and three, three and four, and four and five. Its distribution to the medial side of the first digit depends upon the extent of the saphenous nerve.

The **Deep Peroneal Nerve** (anterior tibial) (Figs. 321, 327) crosses the neck of the fibula beneath the peroneus muscles to join the anterior tibial artery and vein. It supplies branches to tibialis anterior and extensor digitorum longus before accompanying the blood vessels along the lateral surface of the interosseous membrane and thence to the tarsus where together they pass under the transverse crural ligament lying between the tendons of extensor proprius hallucis and tibialis anterior. Crossing the cruciate ligament, the deep peroneal nerve runs deep to the tendons of extensor digitorum longus and brevis of digit II and reaches the second interdigital space. This is contrary to the condition in man where the deep peroneal ends in the first interdigital space from which it sends **dorsal digital nerves** to adjacent sides of digits one and two. Presumably further detailed investigation would reveal this same distribution in the rat, but it is difficult to avoid destruction of these fine terminal nerves to the digits.

The **Tibial Nerve** (internal popliteal) (Figs. 187, 312, 313, 316-318, 320, 323, 325, 326) after giving collateral branches described above, follows its course through the thigh enclosed in the sheath with the common peroneal but, upon dissection, may be separated from the latter as far back as their origin in the plexus. Running obliquely through the popliteal fossa, behind the popliteal vessels, the tibial nerve enters the lower leg, crosses the medial surface of the superior muscular vessels (Fig. 318) and takes a deep course between the two heads of gastrocnemius where it gives off three **muscular branches** (Fig. 320) distributed with the external sural, internal sural, and posterior tibial vessels respectively. The first of these, with the external sural vessels, supplies plantaris, soleus, and the lateral head of the gastrocnemius; the second, with the internal sural vessels, supplies the medial head of gastrocnemius; while the third, in company with the posterior tibial vessels, is distributed to flexor hallucis longus, tibialis posterior and flexor digitorum longus (Fig. 323).

After giving these branches the tibial nerve continues through the calf between plantaris and the medial head of gastrocnemius, and divides just above the ankle into *lateral* and *medial plantar nerves* (Figs. 320, 322, 323, 325, 326).

A **medial sural branch** of the tibial (medial cutaneous nerve of the leg) is apparently lacking, its place being taken by branches of the saphenous nerve.

The **Lateral Plantar Nerve** sends an anastomotic ramus to the medial plantar nerve (Fig. 320), then, in company with the lateral plantar artery and vein (Fig. 322) it crosses the medial aspect of the tuber calcanei to the plantar surface of the foot. The peroneal anastomotic branch of the sural nerve runs under the tendons of soleus and gastrocnemius to reach the lateral plantar nerve just above the heel (Figs. 318, 322). Reaching the flexor surface of the foot, the lateral plantar nerve, under cover of the flexor digitorum brevis, divides into two **common plantar digital nerves,** one to the lateral surface of the fifth digit; one, which runs between the flexor quinti digiti brevis and flexor digitorum brevis to the fourth interdigital space where it divides into two **proper plantar digital nerves** to adjacent sides of digits four and five (Fig. 326).

The **Medial Plantar Nerve** (Figs. 322, 325, 326), accompanied by the deep branch of the medial plantar vessels, passes behind the medial malleolus to the foot where it divides into two branches which in turn divide, giving in all four **common plantar digital nerves.** The first becomes the **proper plantar digital nerve of the first digit.** The three remaining enter the first, second, and third interdigital spaces with the plantar digital arteries, and divide into **proper plantar digital nerves** to the adjacent sides of digits one and two, two and three, and three and four (Fig. 326).

*Pudendal Plexus*

The **Pudendal Plexus** (Figs. 163, 177, 184, 186-190) is not sharply delineated from the sacral plexus. It is formed by the anterior rami of the sixth lumbar, the four sacral, and the first and second caudal nerves. It lies on the posterior wall of the pelvis minor and supplies branches to the perineum and particularly to the tail muscles. It gives rise to the following; *parasympathetic branches, posterior cutaneous nerve of the thigh, pudendal nerve, muscular branches, perineal,* and *cutaneous nerve,* and finally the *inferior caudal trunk,* the latter homologous with the ano-coccygeal nerve of man.

The **n. bigeminus** (Figs. 177, 178, 184, 189) which represents the boundary between sacral and pudendal plexuses is the sixth lumbar in the rat. It divides into two parts, one which enters into the formation of the sciatic nerve, while the other unites with a large branch from the first sacral to form the common origin of the **pudendal** and **posterior cutaneous nerve of the thigh.**

The *first sacral nerve* divides into two parts, the larger of which unites with the branch of the bigeminus to form the common trunk of pudendal and posterior cutaneous nerves. At its origin this trunk lies parallel to the sciatic nerve but becomes separated from it by the superior gluteal and internal pudendal vessels which pass dorsally between the two nerve trunks (Figs. 277, 296). The *pudendal nerve* then accompanies the internal pudendal vessels, while the *posterior cutaneous nerve* emerges on the lateral surface of the thigh in the angle between the two heads of semitendinosus (Figs. 186-190).

The smaller portion of the first sacral joins the next two nerves below it to form the *inferior caudal trunk* (Fig. 163). From this same portion a branch arises and unites with a small branch from the pudendal nerve to form the *perineal nerve* (Figs. 177, 188-190). This in turn gives a *cutaneous branch* (Figs. 189, 190) and the *posterior scrotal branch* (Figs. 189, 190) to the integument of the scrotum.

Branches from the four sacral and first caudal nerves supply a series of small nerves to the caudal flexors (Figs. 184, 185).

The *inferior caudal trunk* (Figs. 177, 178, 184, 189) is formed from the ventral divisions of the four sacral and first caudal nerves or from the second, third, and fourth sacral and the first two caudal nerves, and from it are given off cutaneous branches to the skin of the tail.

SYMPATHETIC SYSTEM

The **Sympathetic System** (Figs. 159, 178, 184, 191-196) in so far as it can be demonstrated in a gross preparation of so small an animal, consists of two *trunks* or chains bearing 24 paired *ganglia;* 3 pairs cervical, 10 thoracic, 6 lumbar, 4 sacral and 1 caudal. (Figs. 159, 184). "About ¼ inch to the right of the midline is a slight long seam, which marks the course of the sympathetic trunk from the second rib to the diaphragm." (Morrell, 1872.)

The two trunks are traceable from the carotid canal of the skull (Fig. 192) into the tail where, as two small filaments they become lost on either side of the middle caudal artery (Fig. 196).

In the cervical region the ganglia are consolidated in 3 pairs, *superior, middle*, and *inferior* cervical ganglia (Fig. 193). The **superior cervical ganglion** (Fig. 192) lies at the level of bifurcation of the common carotid into external and internal carotid arteries and is closely applied to the latter vessel which is accompanied by the sympathetic trunk through the carotid canal. Extending posteriorly the **sympathetic trunk** (chain) lies dorsal to the common carotid artery and the vagus nerve (Figs. 191, 192). Running close to the bodies of the vertebrae it reaches the level of the first rib where it displays the **middle cervical ganglion** (Fig. 193) and close below it, at the level between the second and third ribs, the **inferior cervical ganglion** (Fig. 193). These two ganglia are united by the ansa subclavia (subclavian loop) through which the subclavian artery passes. The middle cervical ganglion may be bilobed with the vertebral artery lying in the groove between the lobes.

From the middle cervical ganglion *gray rami* are given to the fifth, sixth, seventh, and eighth cervical nerves and accompany the vertebral artery to form the vertebral plexus. From the inferior cervical ganglion a *cardiac branch* is given off (Fig. 193).

From the fourth thoracic (or upper thoracic ganglion) to the first sacral pair, the spinal ganglia are connected with the corresponding pairs of spinal nerves through the *gray* and *white rami communicantes* which join the anterior rami of each spinal nerve (Figs. 159, 184, 191).

From the eighth, ninth, and tenth thoracic and first lumbar sympathetic ganglia four branches arise which unite to form the **greater splanchnic nerve** (Fig. 191), which pierces the diaphragm and runs on the right side between the vena cava inferior and the aorta to reach the unpaired **coeliac ganglion** (Figs. 191, 194) with which the splanchnic of the opposite side also connects. This ganglion lies on the ventral surface of the aorta at the level between the coeliac and superior mesenteric arteries. From it one plexus is distributed to the diaphragm, one accompanies the superior mesenteric artery, and other fine plexuses are distributed to the adjacent viscera.

The **lesser** (small) **splanchnic nerve** (Figs. 184, 191) arising at the level of the third lumbar ganglion also joins the coeliac plexus, while the **least** (lowest) **splanchnic** (Figs. 176, 184, 191, 195) from the third or fourth lumbar ganglion becomes closely applied to the abdominal aorta at the level of the renal arteries, curves around the aorta, passing ventral to the iliolumbar vessels and meets the nerve from the opposite side at the bifurcation of the aorta in a ganglion from which a pair of nerves continue with the inferior mesenteric artery to the pelvic viscera. From this pair of fine nerves the **aortic** and **inferior mesenteric plexuses** are formed.

LIST OF FIGURES

VI. NERVOUS SYSTEM

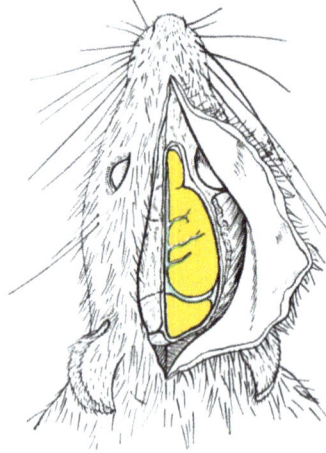

Fig. 138.  Dorsal Aspect of Head showing
Brain in Relation to Exterior.

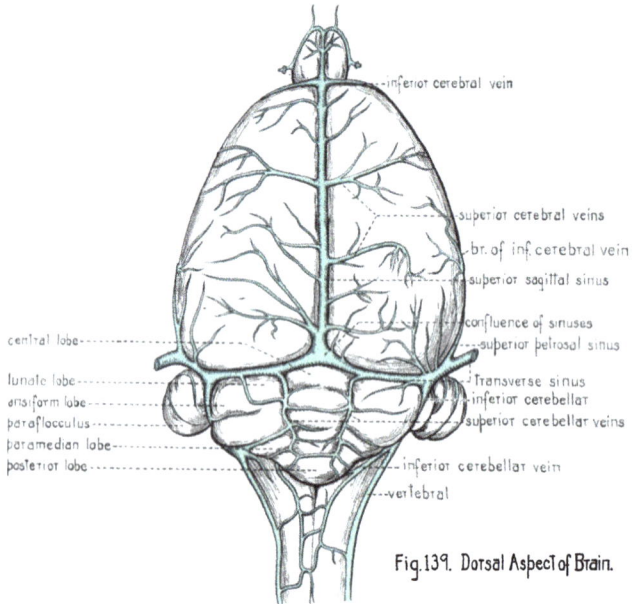

inferior cerebral vein

superior cerebral veins

br. of inf. cerebral vein

superior sagittal sinus

confluence of sinuses

superior petrosal sinus

Transverse sinus

inferior cerebellar

superior cerebellar veins

inferior cerebellar vein

vertebral

central lobe

lunate lobe

ansiform lobe

paraflocculus

paramedian lobe

posterior lobe

Fig. 139. Dorsal Aspect of Brain.

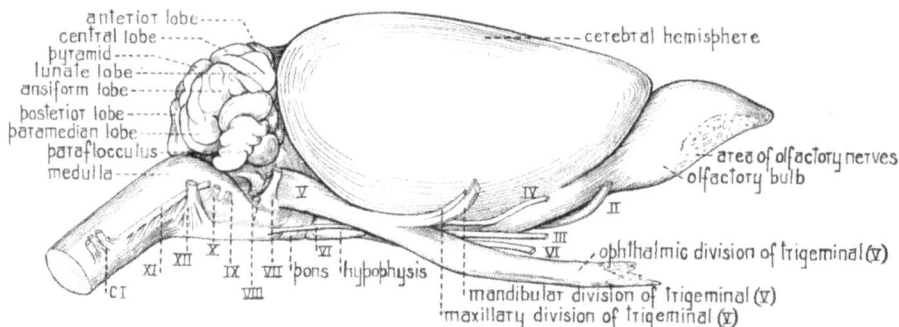

anterior lobe
central lobe
pyramid
lunate lobe
ansiform lobe
posterior lobe
paramedian lobe
paraflocculus
medulla

cerebral hemisphere

area of olfactory nerves
olfactory bulb

V

IV

II

III

VI

ophthalmic division of trigeminal (Ⅴ)

XI   XII   X
     IX   VII   bons   hypophysis
CI          VIII

VI

mandibular division of trigeminal (Ⅴ)
maxillary division of trigeminal (Ⅴ)

Fig. 140.    Lateral Aspect of the Brain and Roots of Cranial Nerves.

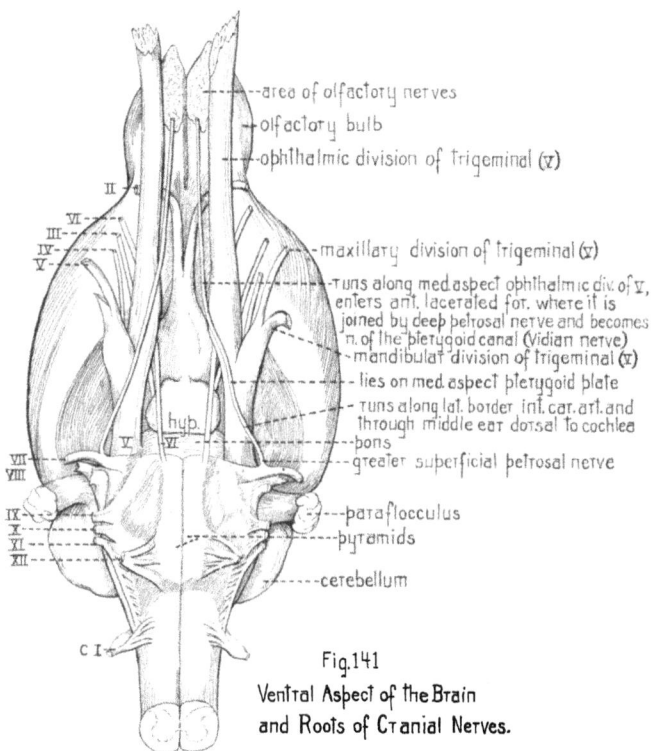

area of olfactory nerves
olfactory bulb
ophthalmic division of trigeminal (Ⅴ)

II

VI
III
IV
V

maxillary division of trigeminal (Ⅴ)
runs along med. aspect ophthalmic div. of Ⅴ,
enters ant. lacerated for. where it is
joined by deep petrosal nerve and becomes
n. of the pterygoid canal (Vidian nerve)
mandibular division of trigeminal (Ⅴ)
lies on med. aspect pterygoid plate
runs along lat. border int. car. art. and
through middle ear dorsal to cochlea
bons
greater superficial petrosal nerve

hyp.

V
VI

VII
VIII

IX
X
XI
XII

paraflocculus
pyramids

cerebellum

C I

Fig. 141

Ventral Aspect of the Brain
and Roots of Cranial Nerves.

Fig. 143. Nerves and Muscles of the Right Eyeball.

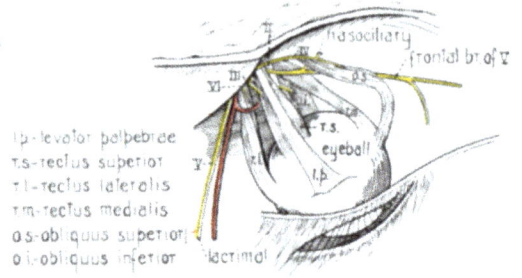

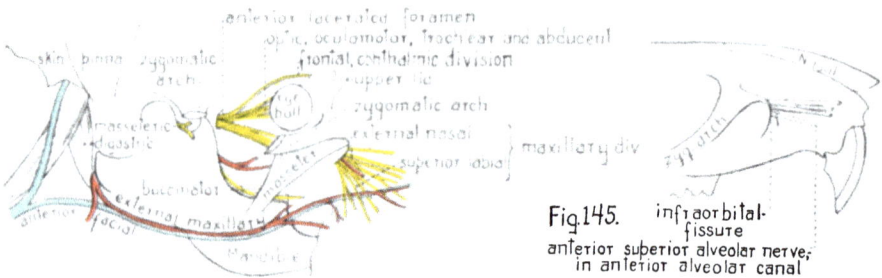

II
III
VI   IV
-optic foramen
-V, ophthalmic + maxillary divisions
-anterior lacerated foramen

-mandibular division of V
-foramen ovale
-stalk of hypophysis
-posterior lobe
-pars intermedia
-anterior lobe

III
II

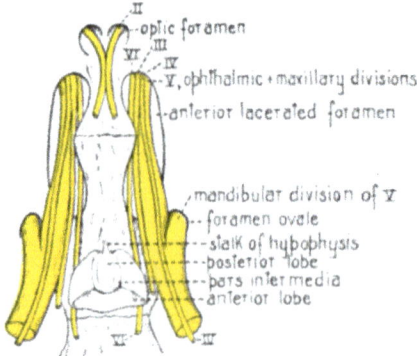

Fig.142. Exit of Orbital Nerves from the Cranial Cavity.

nasociliary
frontal br of V

l.p.-levator palpebrae
r.s.-rectus superior
r.l.-rectus lateralis
r.m.-rectus medialis
o.s.-obliquus superior
o.i.-obliquus inferior

r.s.
eyeball
l.p.

lacrimal

anterior lacerated foramen
optic, oculomotor, trochlear and abducent
frontal, ophthalmic division
upper lid
zygomatic arch
external nasal
superior labial
maxillary div

skin   pinna   zygomatic
arch

masseteric
digastric

buccinator
external maxillary

anterior facial

Mandible

Fig.145.

infraorbital-
fissure
anterior superior alveolar nerve,
in anterior alveolar canal

Fig.144. Ophthalmic and Maxillary Divisions of Trigeminal Nerve

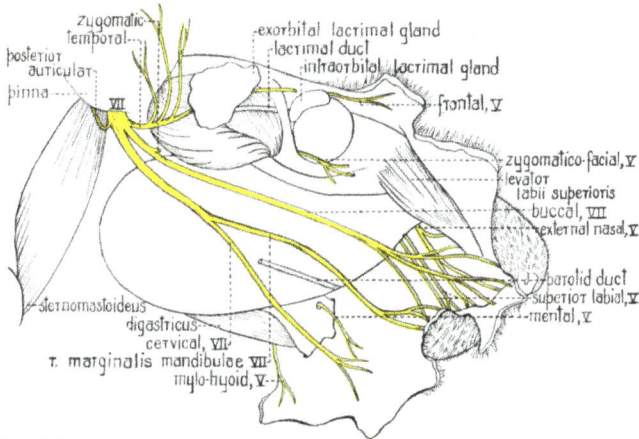

Fig.146. Lateral view of head dissected to show branches of trigeminal and facial nerves

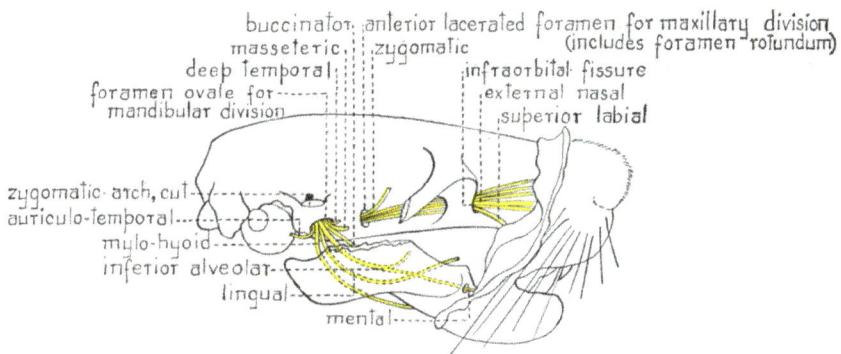

Fig.147. Maxillary and Mandibular Divisions of the Trigeminal Nerve

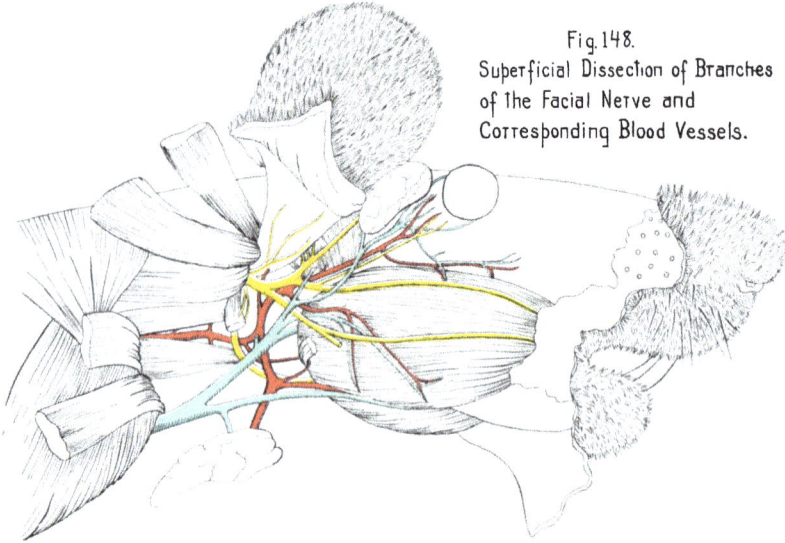

Fig. 148.
Superficial Dissection of Branches
of the Facial Nerve and
Corresponding Blood Vessels.

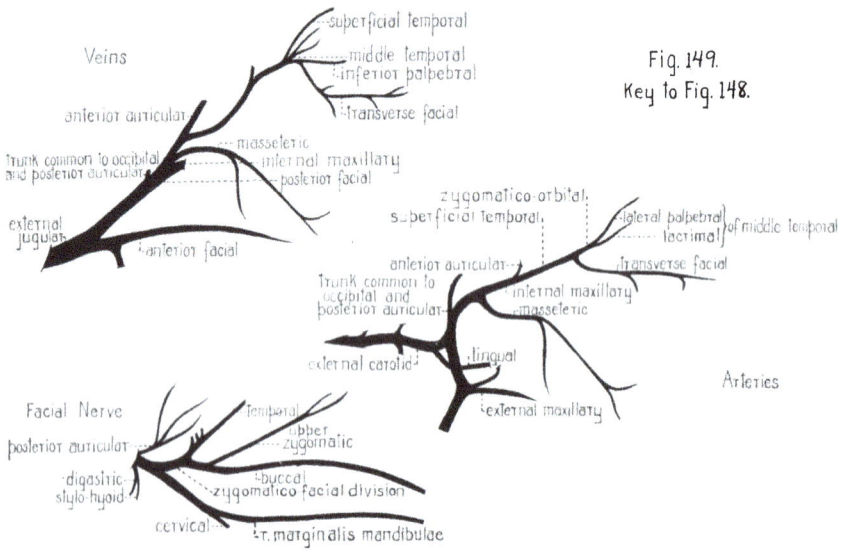

Fig. 149.
Key to Fig. 148.

Veins

superficial temporal
middle temporal
inferior palpebral
transverse facial
anterior auricular
masseteric
Trunk common to occipital
and posterior auricular
internal maxillary
posterior facial
external
jugular
anterior facial

zygomatico-orbital
superficial temporal
lateral palpebral } of middle temporal
lacrimal
transverse facial
anterior auricular
Trunk common to
occipital and
posterior auricular
internal maxillary
masseteric
external carotid
lingual
external maxillary

Arteries

Facial Nerve
posterior auricular
temporal
upper
zygomatic
digastric
stylo-hyoid
buccal
zygomatico facial division
cervical
r. marginalis mandibulae

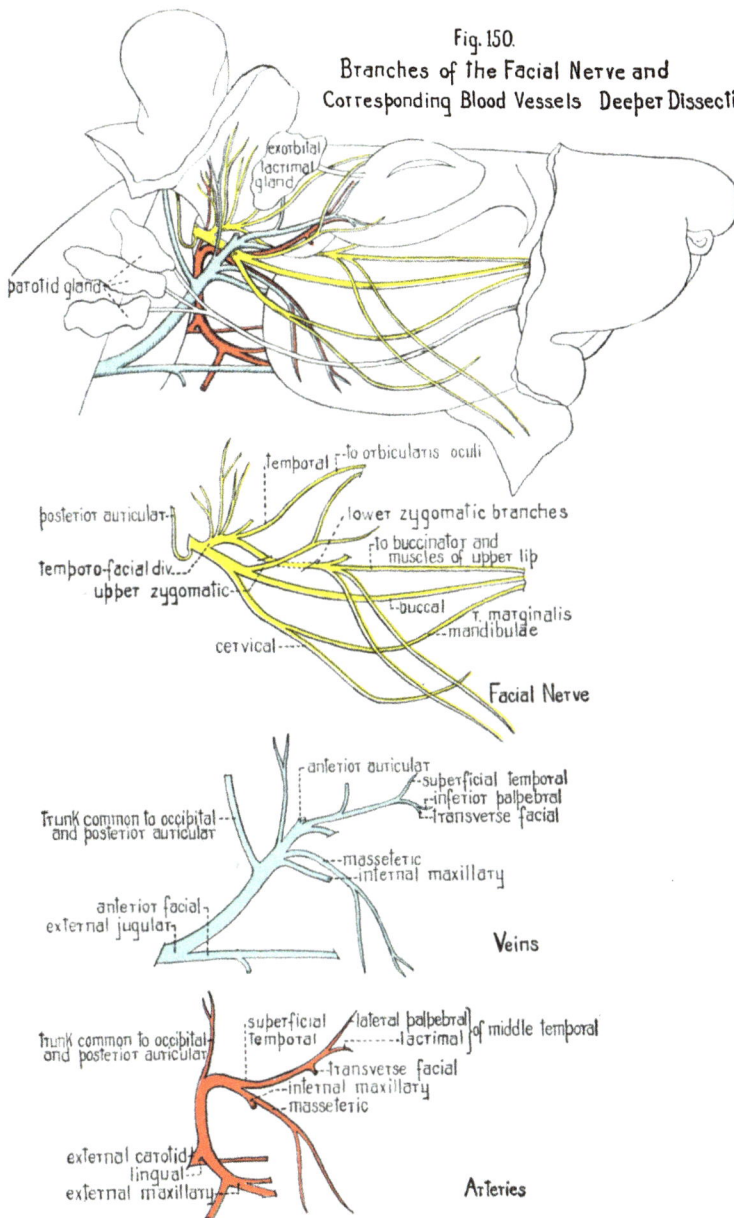

Fig. 150.
Branches of the Facial Nerve and
Corresponding Blood Vessels  Deeper Dissection.

exorbital
lacrimal
gland

parotid gland

temporal — to orbicularis oculi

posterior auricular

lower zygomatic branches

to buccinator and
muscles of upper lip

temporo-facial div.
upper zygomatic

buccal  r. marginalis
mandibulae

cervical

Facial Nerve

anterior auricular
superficial temporal
inferior palpebral
transverse facial

Trunk common to occipital
and posterior auricular

masseteric
internal maxillary

anterior facial
external jugular

Veins

Trunk common to occipital
and posterior auricular

superficial
temporal

lateral palpebral
lacrimal } of middle temporal

transverse facial
internal maxillary
masseteric

external carotid
lingual
external maxillary

Arteries

Fig. 151   Key to Fig. 150.

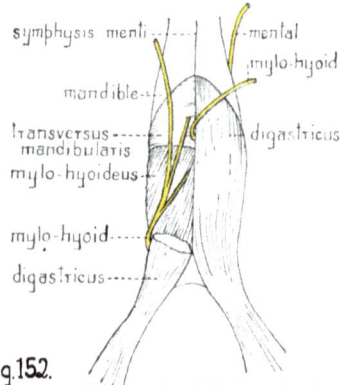

Fig.152.
Mylo-hyoid branch of Trigeminal

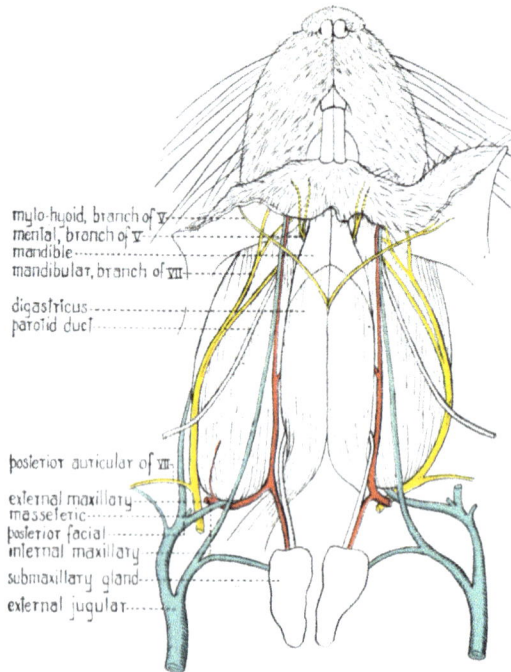

Fig.153. Ventral view of head dissected to show branches of trigeminal
and facial nerves.

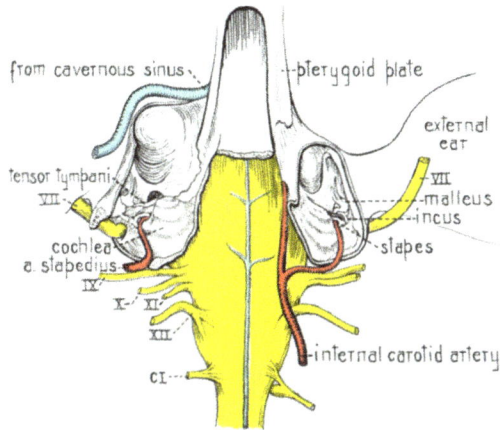

Fig 154  Middle Ear and Cranial Nerves VII, IX, X, XI, XII
in relation to the Medulla.

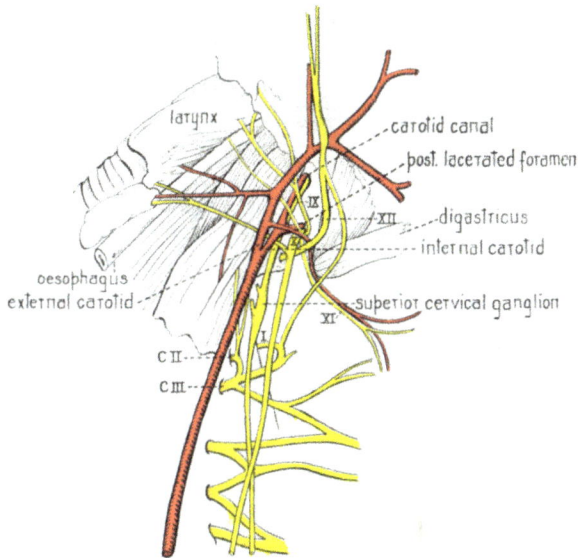

Fig.155.  Exit of Cranial Nerves IX, X, XI, XII, and
Cervical Nerves I, II, III,

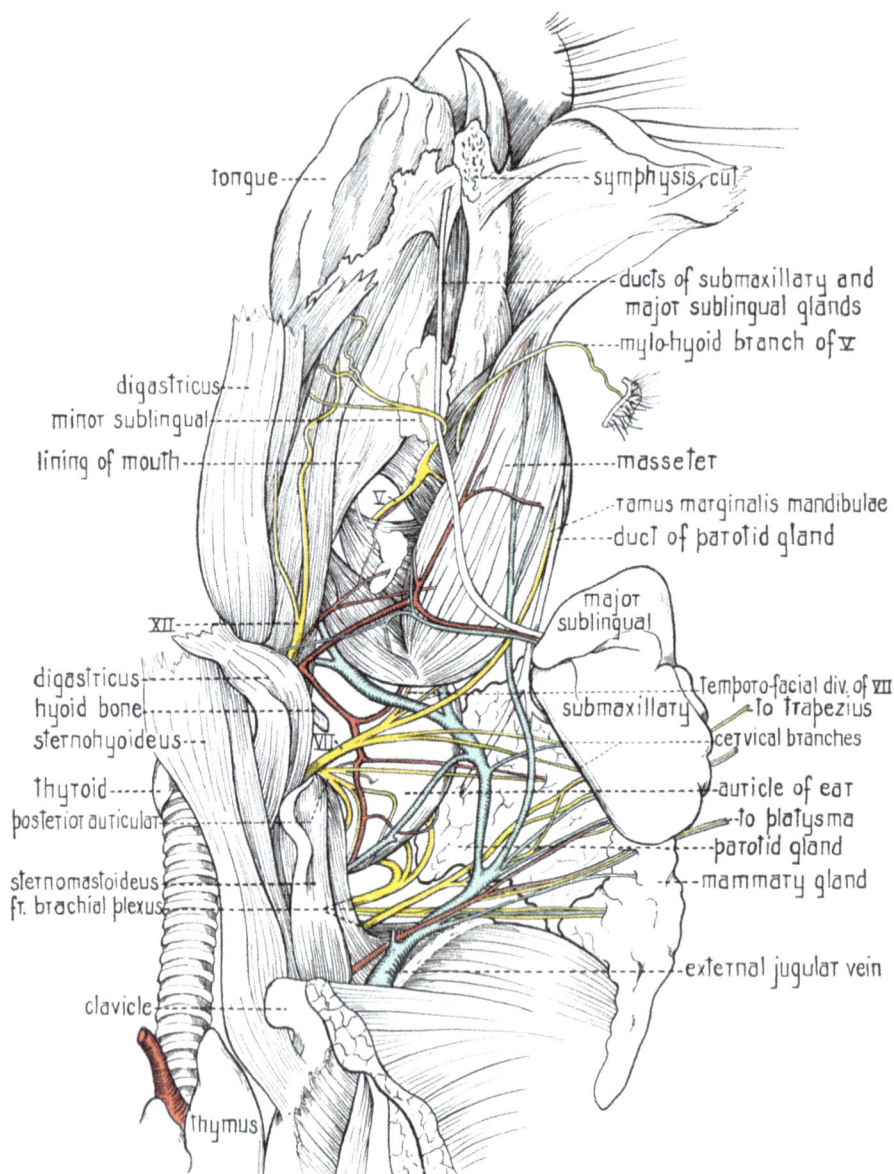

Fig. 156. Superficial Dissection of Ventral Region of Head and Neck showing Distribution of V, VII, and XII.

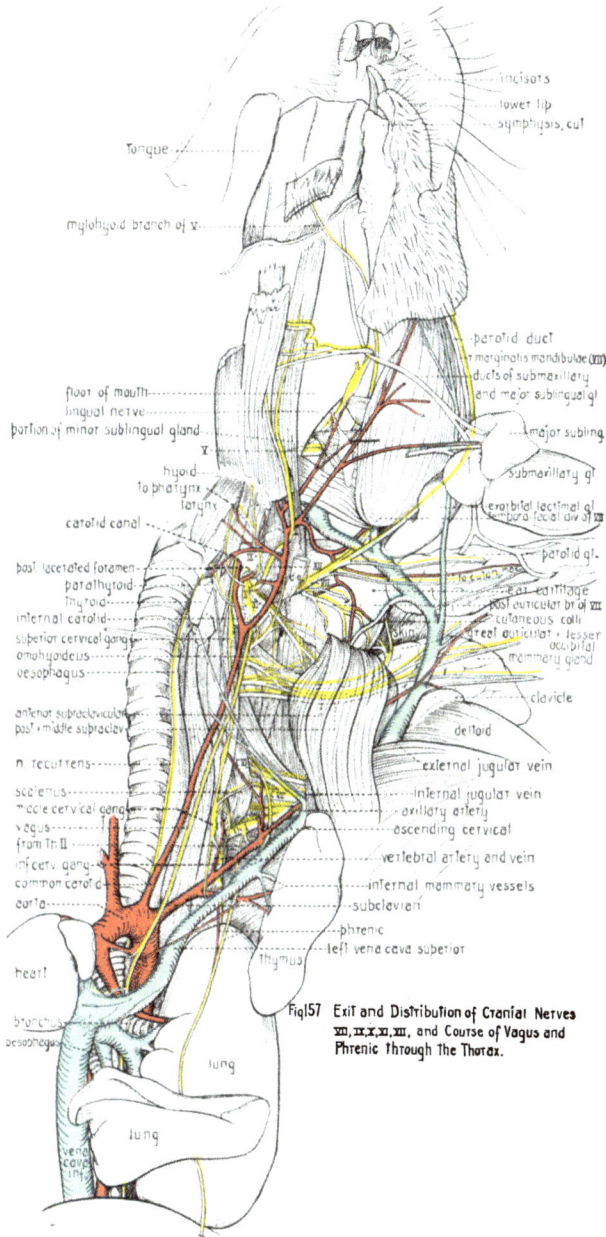

Fig.157 Exit and Distribution of Cranial Nerves VII, IX, X, XI, XII, and Course of Vagus and Phrenic through the Thorax.

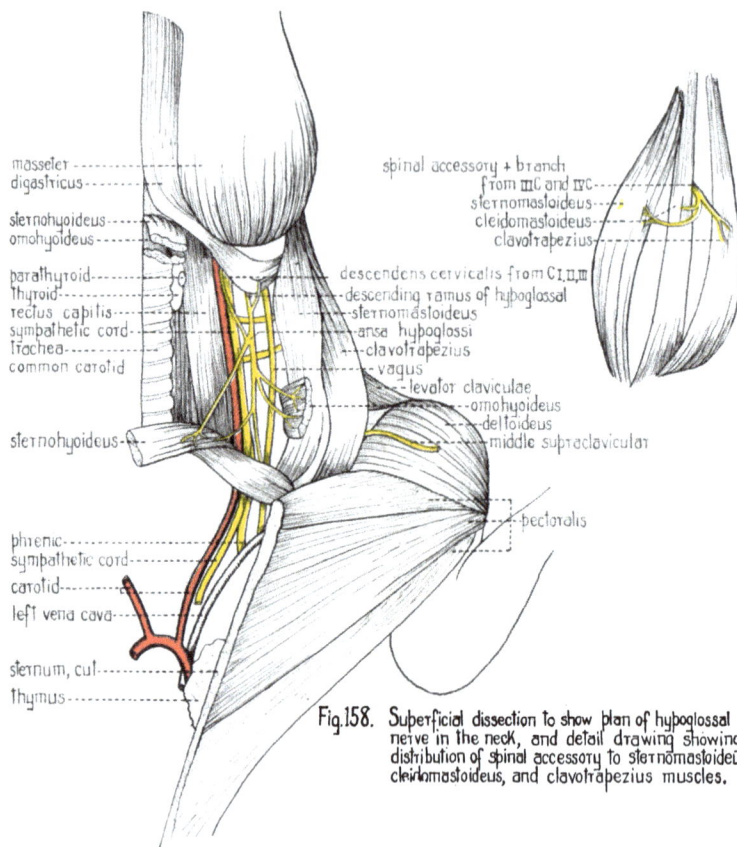

masseter
digastricus

sternohyoideus
omohyoideus

parathyroid
thyroid
rectus capitis
sympathetic cord
trachea
common carotid

sternohyoideus

phrenic
sympathetic cord
carotid
left vena cava

sternum, cut
thymus

spinal accessory + branch
from IIIC and IVC
sternomastoideus
cleidomastoideus
clavotrapezius

descendens cervicalis from C I, II, III
descending ramus of hypoglossal
sternomastoideus
ansa hypoglossi
clavotrapezius
vagus
levator claviculae
omohyoideus
deltoideus
middle supraclavicular

pectoralis

Fig. 158. Superficial dissection to show plan of hypoglossal nerve in the neck, and detail drawing showing distribution of spinal accessory to sternomastoideus, cleidomastoideus, and clavotrapezius muscles.

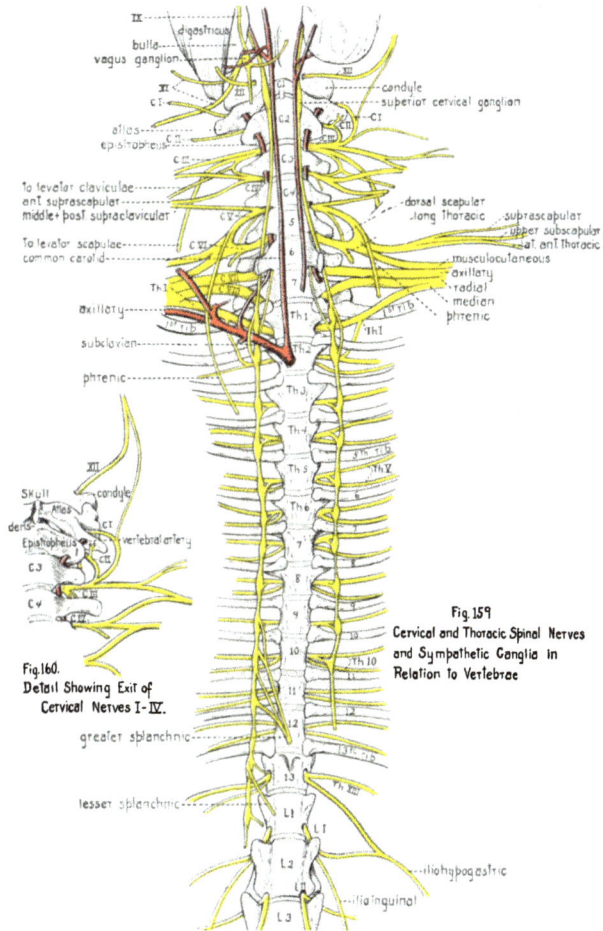

IX
digastricus
bulla
vagus ganglion
XI
CI
atlas
epistropheus
CII
CIII
to levator claviculae
ant. suprascapular
middle+post. supraclavicular
CV
to levator scapulae
common carotid
CVI
Th1
axillary
1st rib
subclavian
phrenic
XII
condyle
superior cervical ganglion
CI
CII
CIII
C2
C3
C4
5
6
7
dorsal scapular
long thoracic
suprascapular
upper subscapular
lat. ant. thoracic
musculocutaneous
axillary
radial
median
phrenic
1st rib
Th1
Th2
Th3
Th4
Th5
Th6
5th rib
Th V
XII
Skull
condyle
Atlas
dens
CI
Epistropheus
vertebral artery
CII
C3
C4
CIII
CIV
10
11
12
13
Th 10
13th rib
greater splanchnic
lesser splanchnic
Th XIII
L1
L2
L3
iliohypogastric
ilioinguinal

Fig. 159
Cervical and Thoracic Spinal Nerves and Sympathetic Ganglia in Relation to Vertebrae

Fig. 160.
Detail Showing Exit of Cervical Nerves I-IV.

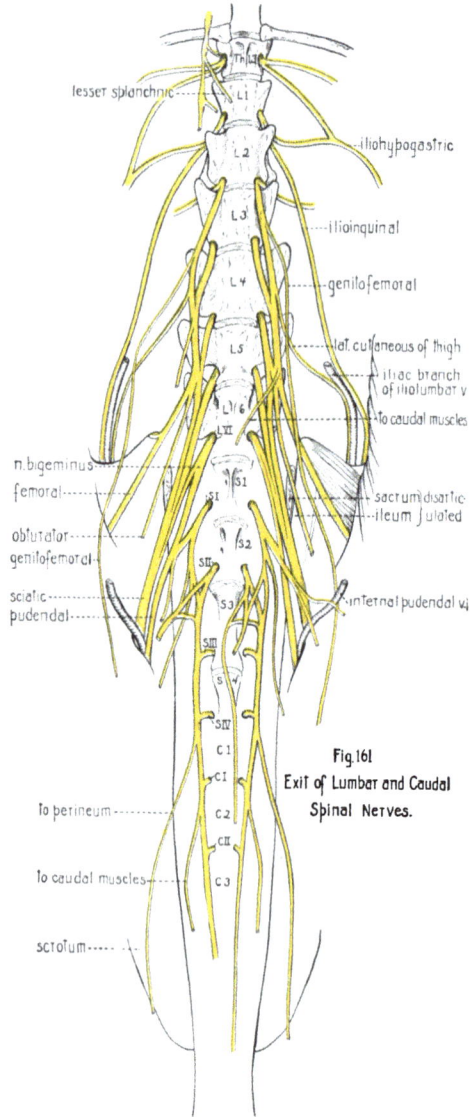

Fig. 161
Exit of Lumbar and Caudal
Spinal Nerves.

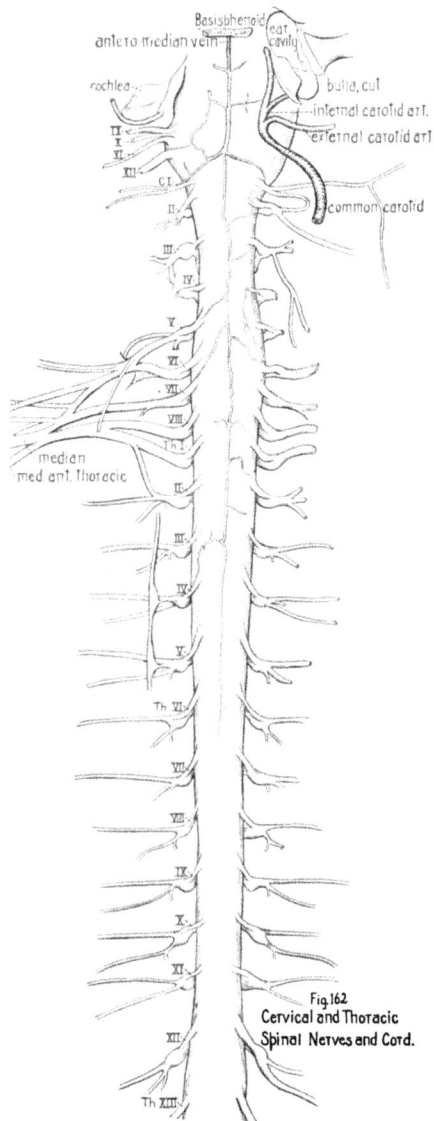

Fig.162
Cervical and Thoracic
Spinal Nerves and Cord.

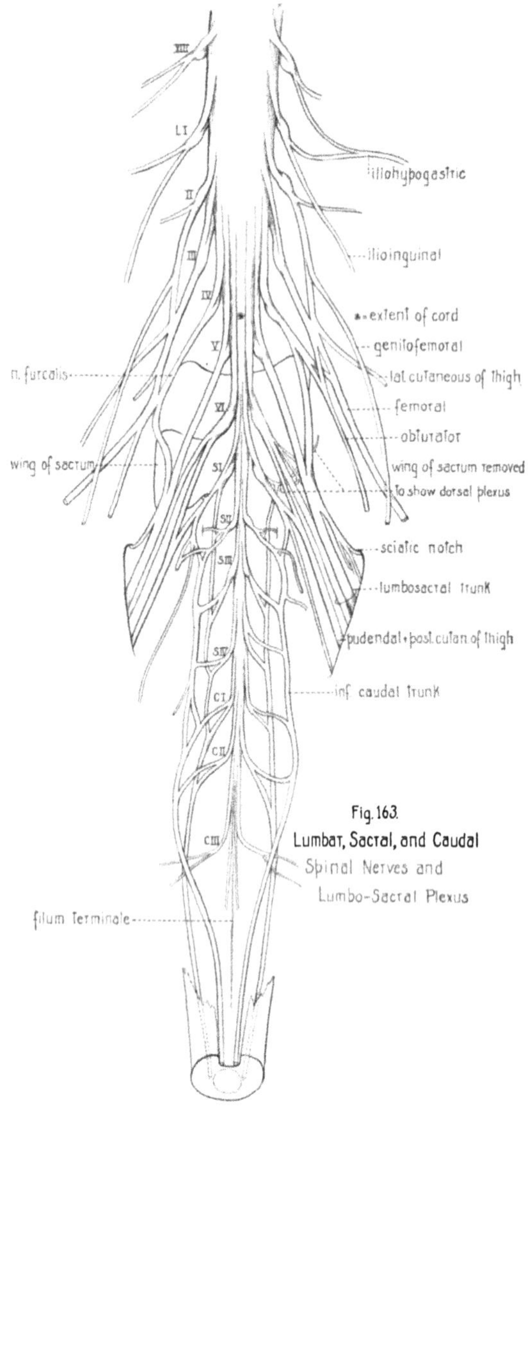

XIII.

L I

II

III

IV

V

VI

S I

S II

S III

S IV

C I

C II

C III

n. furcalis

wing of sacrum

filum terminale

iliohypogastric

ilioinguinal

extent of cord

genitofemoral

lat cutaneous of thigh

femoral

obturator

wing of sacrum removed
to show dorsal plexus

sciatic notch

lumbosacral trunk

pudendal+post cutan of thigh

inf caudal trunk

Fig. 163.
Lumbar, Sacral, and Caudal
Spinal Nerves and
Lumbo-Sacral Plexus

XI

XII ---- To sternohyoid, omohyoid and digastric

C I ----

C II --

C III -- XI

cutaneous colli

great auricular

lesser occipital

To levator claviculae -

subtrapezial plexus To sternomastoid, cleidomastoid and clavotrapezius

C IV --

anterior supraclavicular

posterior supraclavicular

middle supraclavicular

C V -- dorsal division ---- To levator scapulae

vent. div.

dorsal scapular -
To levator scapulae and rhomboideus

phrenic

suprascapular

upper (short) subscapular

To subclavius

C VI -- dors.

vent. lateral anterior thoracic To pectoralis major

lower (middle) subscapular
To subscapularis

long thoracic To ---- serratus anterior

dors. div.

musculocutaneous
ant.

mus. br. To deltoid, plus artic. br. To shoulder joint, plus lateral cutan. of arm

ant. branch 1st root axillary (circumflex)

post.

C VII vent. div.
dors. div. ant. br. = 2d root post. branch post. br. 1st root thoracodorsal To teres major To teres minor

To latis. dorsi

To pectoralis major and minor

2d root musculospiral (radial)

dors. div. 1st root

vent. div. 2d root median

C VIII -- 3d root

ulnar

medial anterior thoracic

Th I --

from Th II. not constant To cutaneous maximus To pectoralis minor

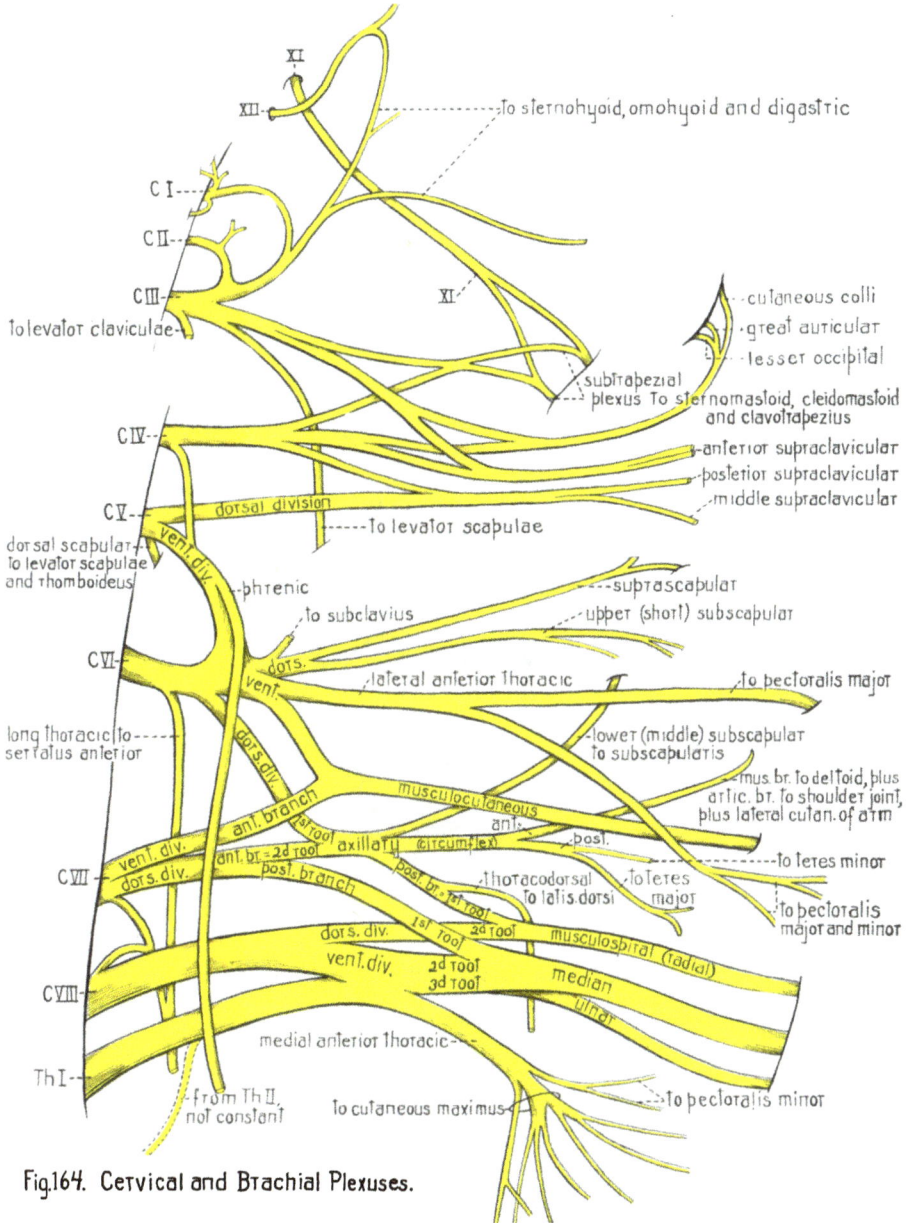

Fig.164. Cervical and Brachial Plexuses.

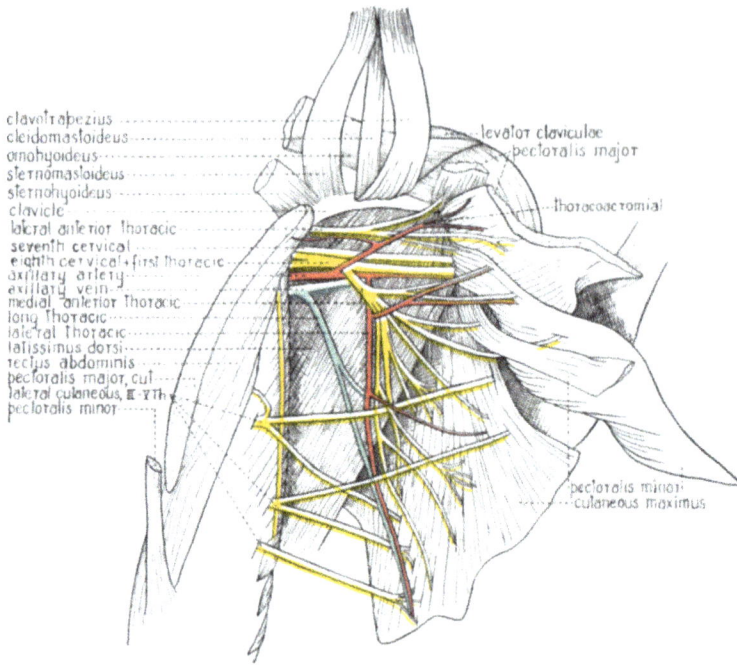

clavotrapezius
cleidomastoideus
omohyoideus
sternomastoideus
sternohyoideus
clavicle
lateral anterior thoracic
seventh cervical
eighth cervical+first thoracic
axillary artery
axillary vein
medial anterior thoracic
long thoracic
lateral thoracic
latissimus dorsi
rectus abdominis
pectoralis major, cut
lateral cutaneous, II · VTh
pectoralis minor

levator claviculae
pectoralis major

thoracoacromial

pectoralis minor
cutaneous maximus

Fig.165. Pectoralis cut and left arm reflected to show course of nerves through axilla.

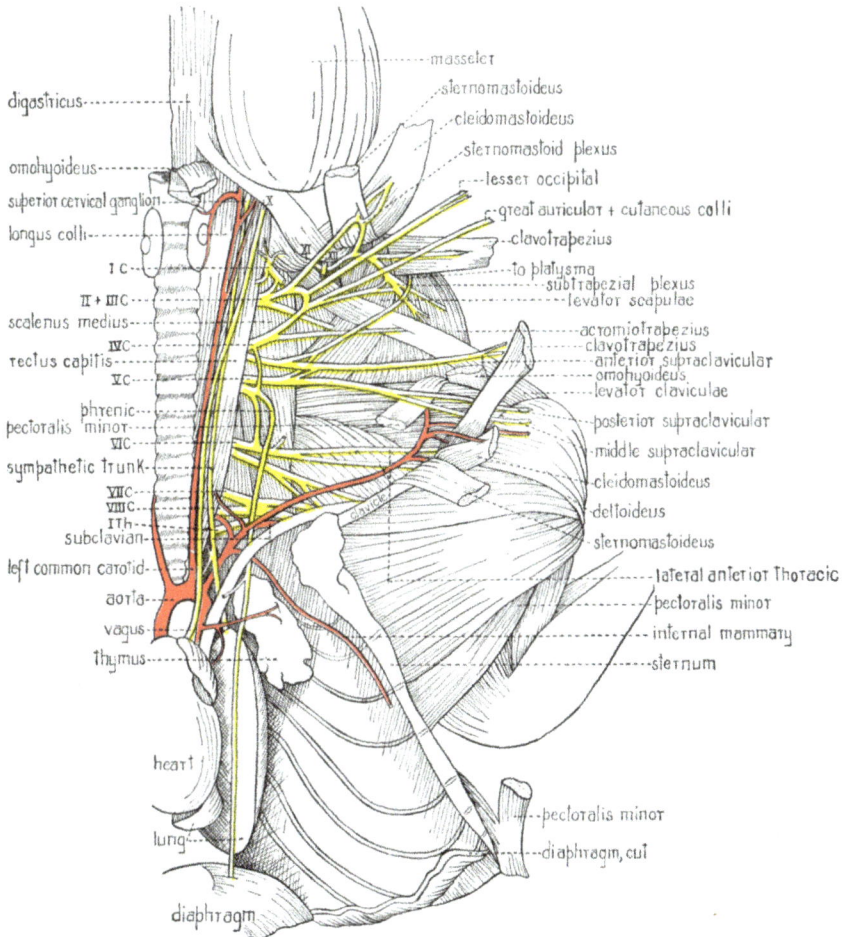

Fig.166. Left Cervical Plexus

digastricus

omohyoideus
superior cervical ganglion
longus colli
I C
II + III C
scalenus medius
IV C
rectus capitis
V C
phrenic
pectoralis minor
VI C
sympathetic trunk
VII C
VIII C
I Th
subclavian
left common carotid
aorta
vagus
thymus

heart

lung

diaphragm

masseter
sternomastoideus
cleidomastoideus
sternomastoid plexus
lesser occipital
great auricular + cutaneous colli
clavotrapezius
to platysma
subtrapezial plexus
levator scapulae
acromiotrapezius
clavotrapezius
anterior supraclavicular
omohyoideus
levator claviculae
posterior supraclavicular
middle supraclavicular
cleidomastoideus
deltoideus
sternomastoideus
lateral anterior thoracic
pectoralis minor
internal mammary
sternum

pectoralis minor
diaphragm, cut

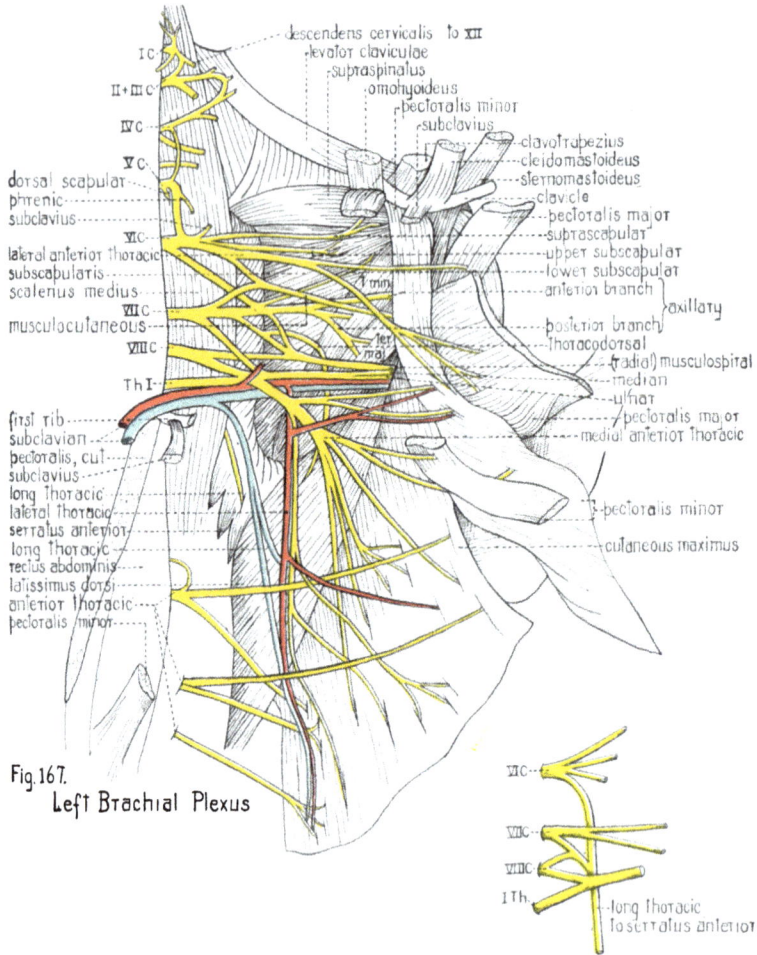

descendens cervicalis  to XII
levator claviculae
supraspinatus
omohyoideus
pectoralis minor
subclavius
clavotrapezius
cleidomastoideus
sternomastoideus
clavicle
pectoralis major
suprascapular
upper subscapular
lower subscapular
anterior branch
posterior branch
thoracodorsal
(radial) musculospiral
median
ulnar
pectoralis major
medial anterior thoracic
axillary

I C
II + III C
IV C
V C

dorsal scapular
phrenic
subclavius
VI C
lateral anterior thoracic
subscapularis
scalenus medius
VII C
musculocutaneous
VIII C
Th I

first rib
subclavian
pectoralis, cut
subclavius
long thoracic
lateral thoracic
serratus anterior
long thoracic
rectus abdominis
latissimus dorsi
anterior thoracic
pectoralis minor

pectoralis minor
cutaneous maximus

Fig. 167.
Left Brachial Plexus

VI C
VII C
VIII C
I Th.
long thoracic
to serratus anterior

Fig. 168.   Deep dissection through scalenes to show
detail of origin of long thoracic nerve.

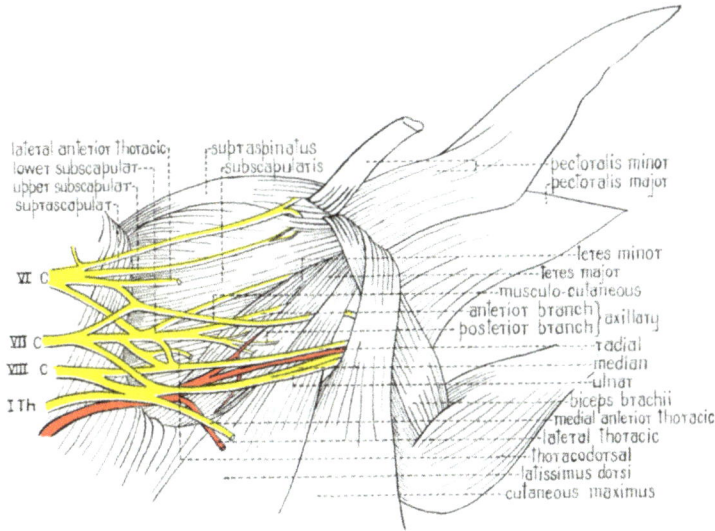

Fig.169. Left shoulder reflected to show branches of brachial plexus

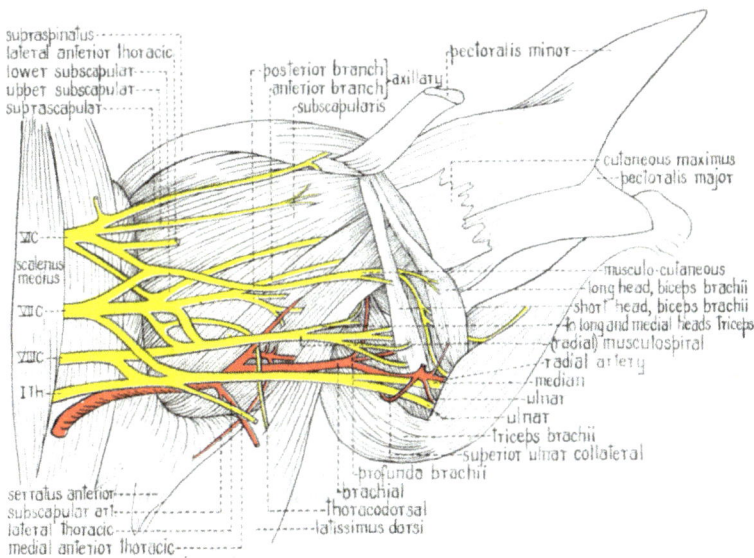

Fig.170. Left shoulder reflected to show nerves of upper arm

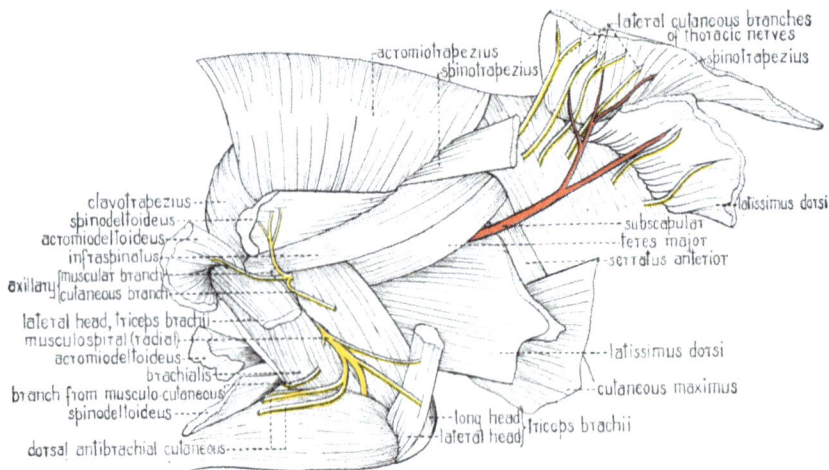

Fig.171. Lateral view of left shoulder, dissected to show distribution of nerves to upper arm

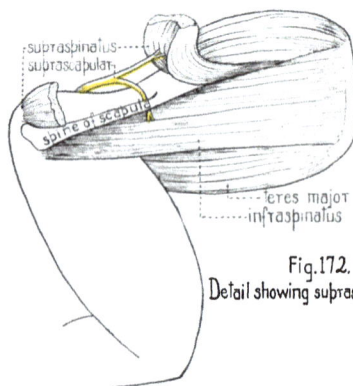

Fig.172.
Detail showing subrascapular nerve.

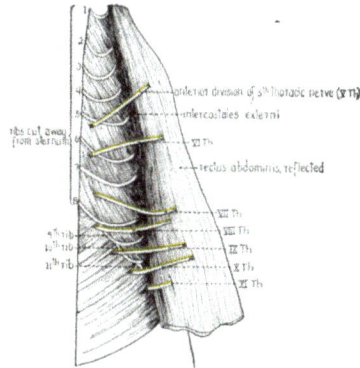

Fig.173. Left half of thorax dissected to show innervation
of rectus abdominis

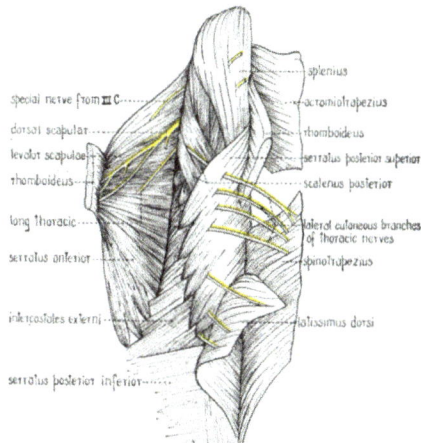

fig.174. Left dorsal aspect of thorax with rhomboideus cut and scapula drawn
laterally, to show nerves to levator scapulae and serratus anterior.

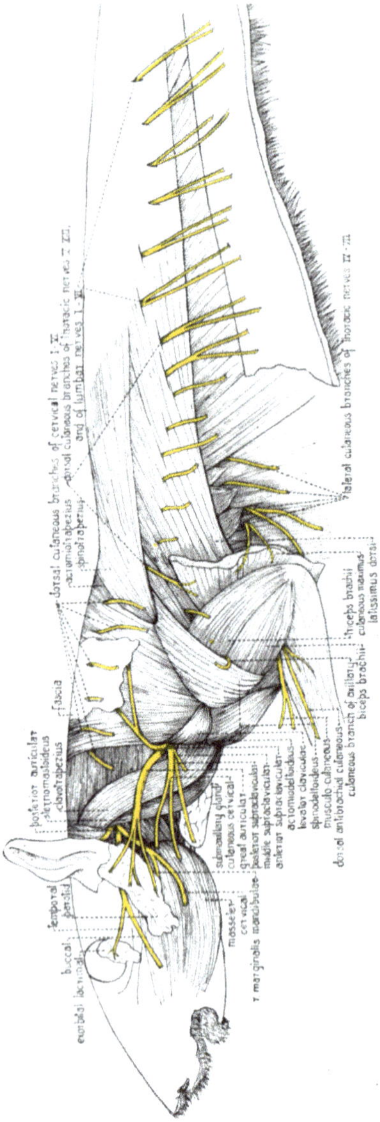

Fig. 175. Superficial dissection of lateral surface to show cutaneous branches of cervical, thoracic, and lumbar nerves.

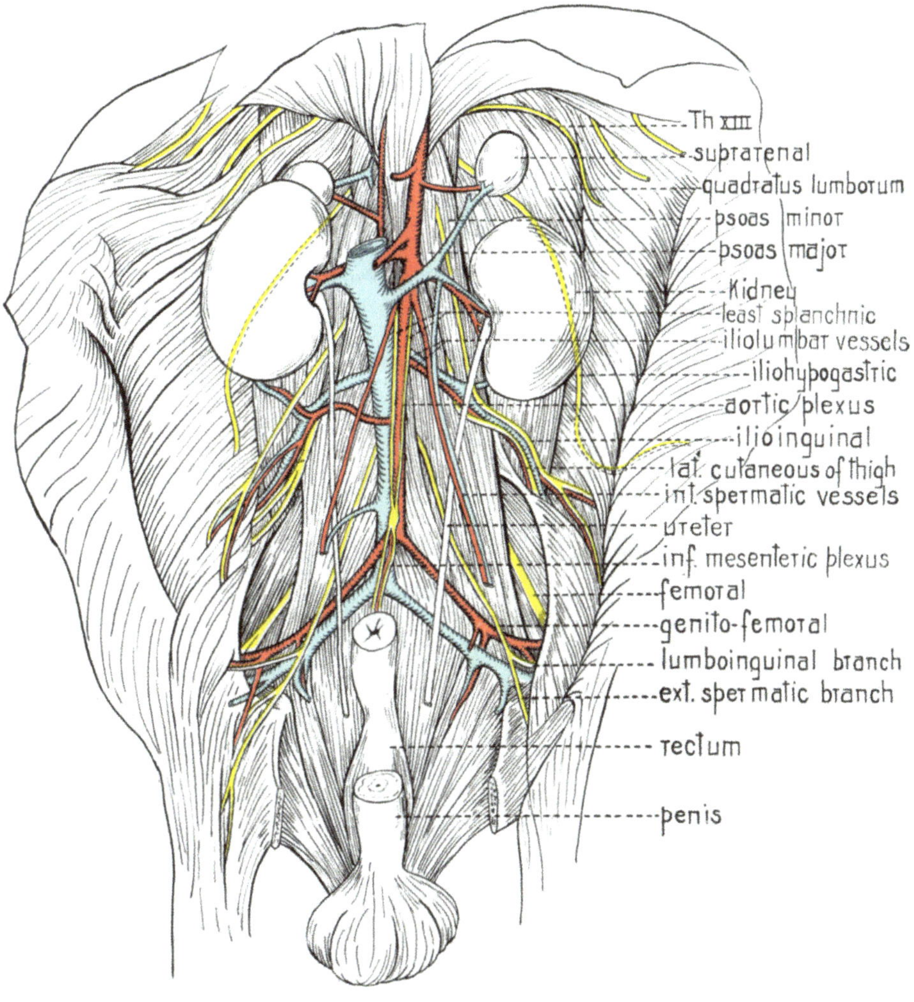
Th XIII
suprarenal
quadratus lumborum
psoas minor
psoas major
Kidney
least splanchnic
iliolumbar vessels
iliohypogastric
aortic plexus
ilioinguinal
lat. cutaneous of thigh
int. spermatic vessels
ureter
inf. mesenteric plexus
femoral
genito-femoral
lumboinguinal branch
ext. spermatic branch
rectum
penis

Fig. 176. Superficial Dissection of Lumbar Plexus.

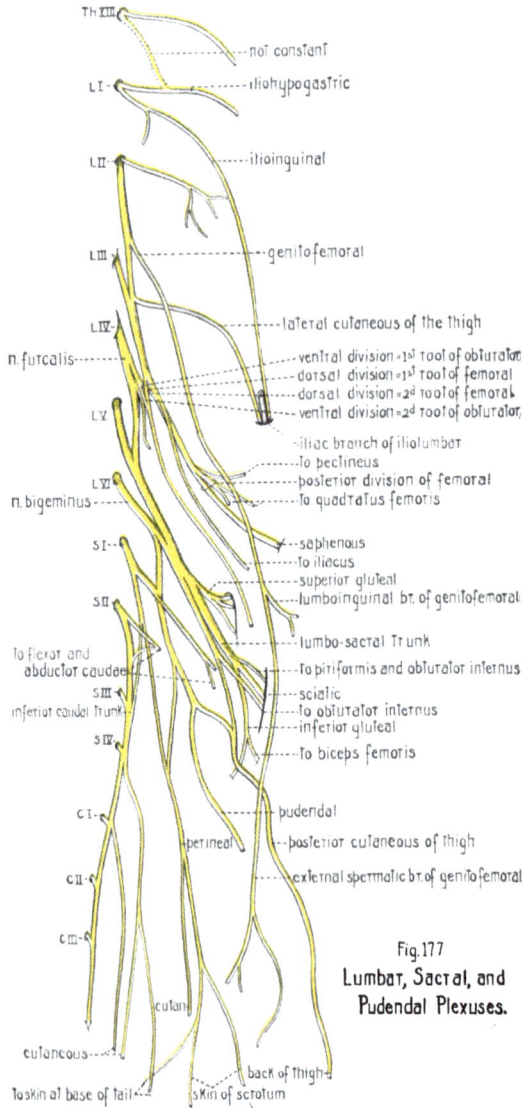

Fig. 177
Lumbar, Sacral, and
Pudendal Plexuses.

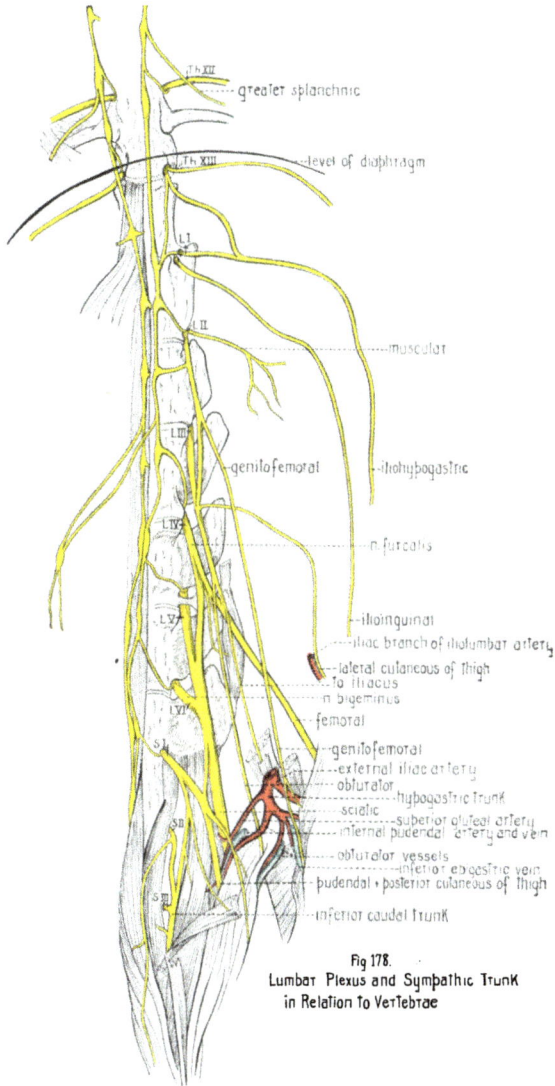

Fig 178.
Lumbar Plexus and Sympathic Trunk
in Relation to Vertebrae

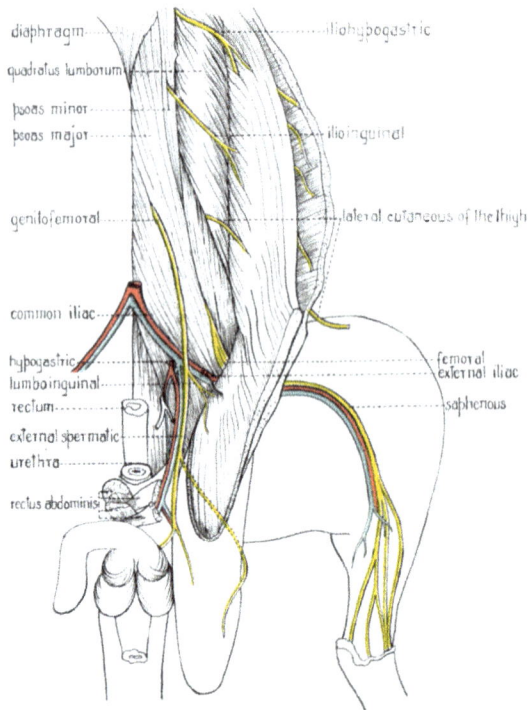

diaphragm
quadratus lumborum
psoas minor
psoas major
genitofemoral
common iliac
hypogastric
lumboinguinal
rectum
external spermatic
urethra
rectus abdominis

iliohypogastric
ilioinguinal
lateral cutaneous of the thigh
femoral
external iliac
saphenous

Fig.179. Superficial dissection of branches of left lumbar plexus
The External Spermatic.,

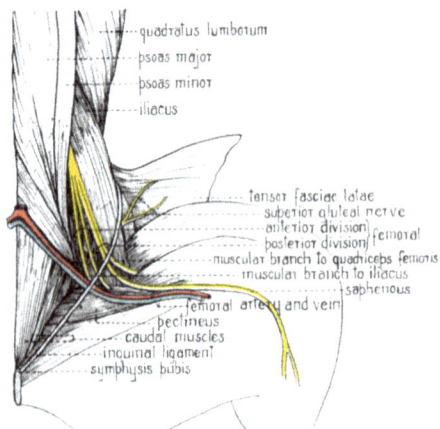

quadratus lumborum
psoas major
psoas minor
iliacus

tensor fasciae latae
superior gluteal nerve
anterior division
posterior division } femoral
muscular branch to quadriceps femoris
muscular branch to iliacus
saphenous
femoral artery and vein
pectineus
caudal muscles
inguinal ligament
symphysis pubis

Fig.180. Superficial dissection to show divisions of left femoral nerve.

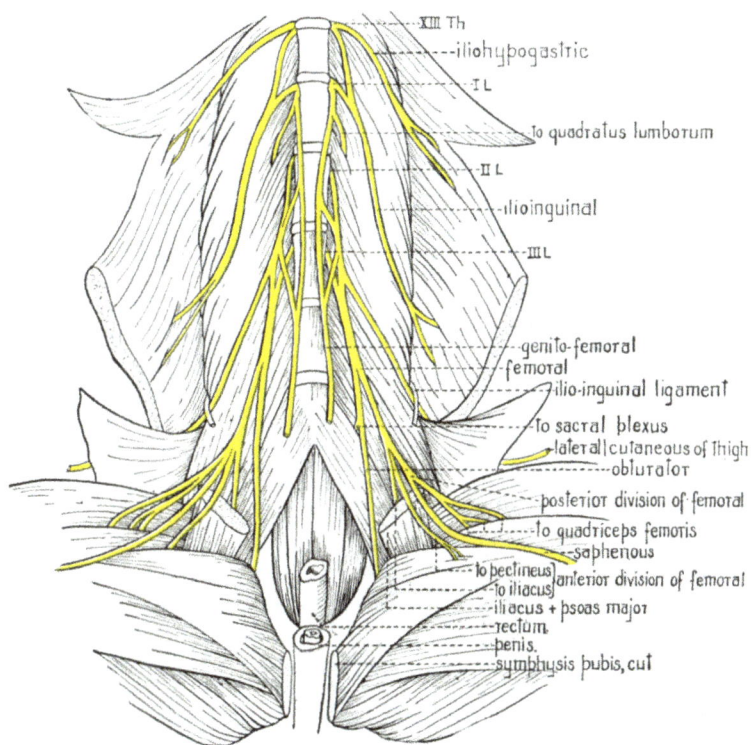

XIII Th
iliohypogastric
I L
to quadratus lumborum
II L
ilioinguinal
III L
genito-femoral
femoral
ilio-inguinal ligament
to sacral plexus
lateral cutaneous of thigh
obturator
posterior division of femoral
to quadriceps femoris
saphenous
to pectineus) anterior division of femoral
to iliacus)
iliacus + psoas major
rectum
penis
symphysis pubis, cut

Fig. 181. Divisions of the Femoral Nerve.

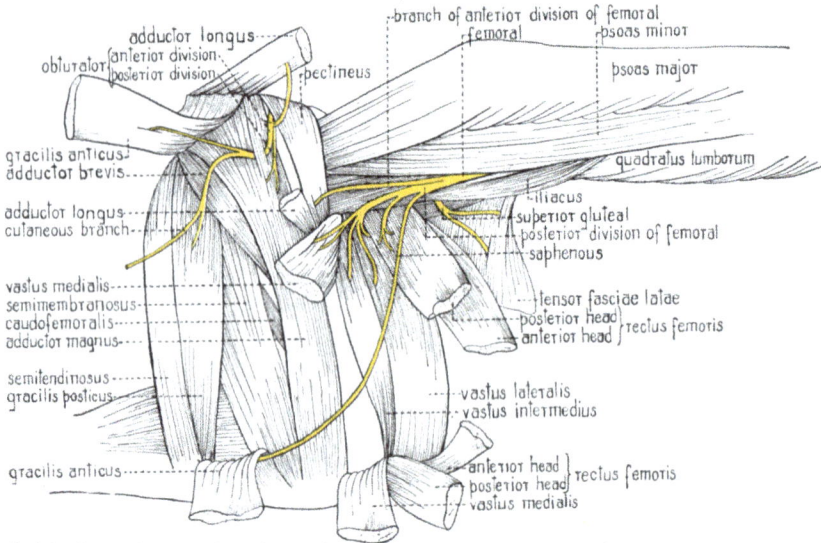

Fig.182. Medial Aspect of Left Thigh showing Branches of Femoral and Obturator Nerves.

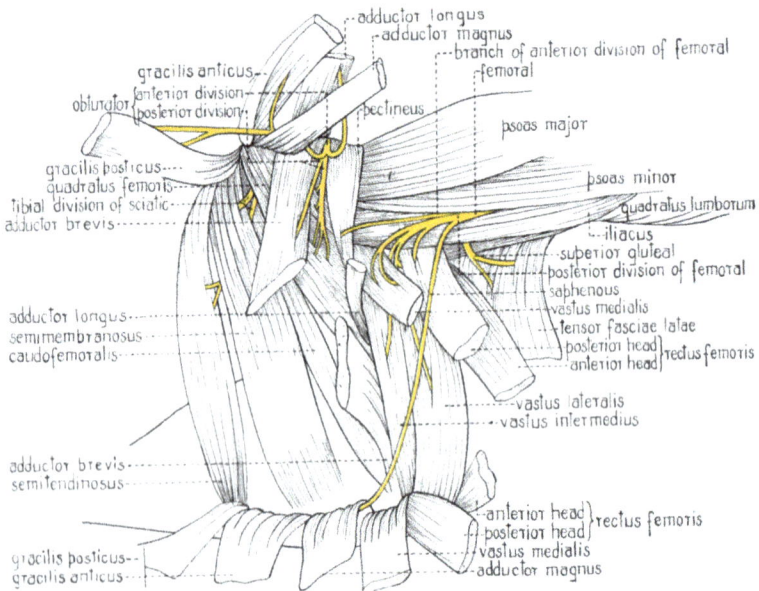

Fig.183. Medial aspect of left thigh, showing branches of femoral, obturator, and sciatic nerves.

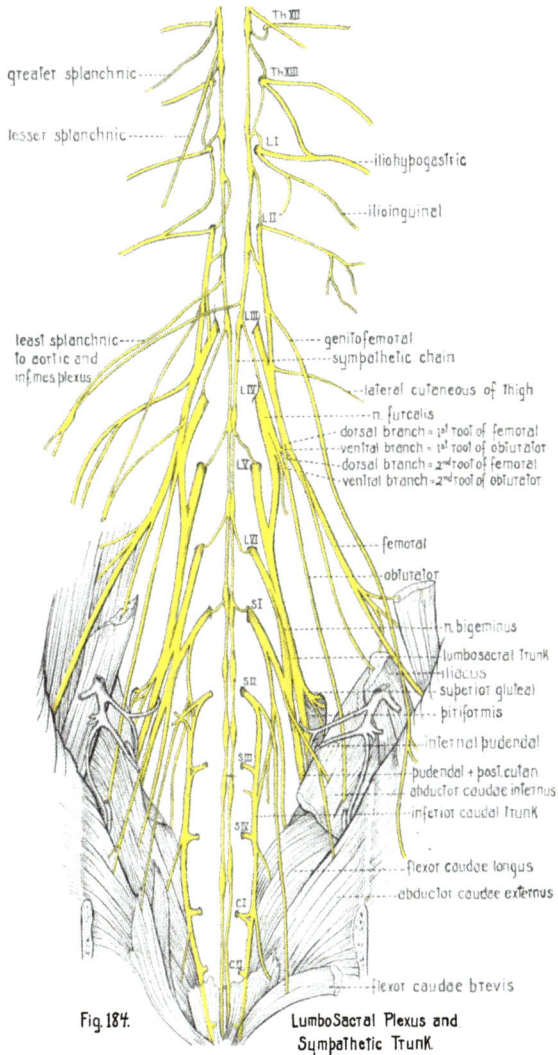

greater splanchnic

lesser splanchnic

Th XII

Th XIII

L I ---------iliohypogastric

L II ---------ilioinguinal

least splanchnic
to aortic and
inf.mes.plexus

L III ----------genitofemoral
------------sympathetic chain

L IV ----------lateral cutaneous of thigh
-----------n. furcalis
-----------dorsal branch = 1st root of femoral
-----------ventral branch = 1st root of obturator
-----------dorsal branch = 2nd root of femoral
-----------ventral branch = 2nd root of obturator

L V

L VI ----------femoral

-----------obturator

S I

-----------n. bigeminus
-----------lumbosacral trunk
-----------iliacus
S II -----------superior gluteal
-----------piriformis

-----------internal pudendal

S III -----------pudendal + post.cutan.
-----------abductor caudae internus
-----------inferior caudal trunk

S IV

-----------flexor caudae longus
-----------abductor caudae externus

C I

C II

-----------flexor caudae brevis

Fig. 184.                    LumboSacral Plexus and
                              Sympathetic Trunk

femoral

flexor caudae brevis

VIL·IS
II S

lumbo-sacral trunk

obturator

pudendal

IIIS

flexor caudae longus

to caudal muscles
abductor caudae internus
abductor caudae externus

IVS

symphysis pubis

caudal muscles

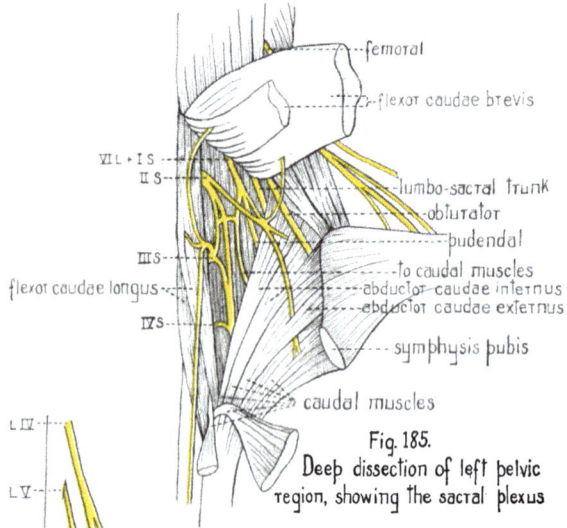

Fig. 185.
Deep dissection of left pelvic
region, showing the sacral plexus

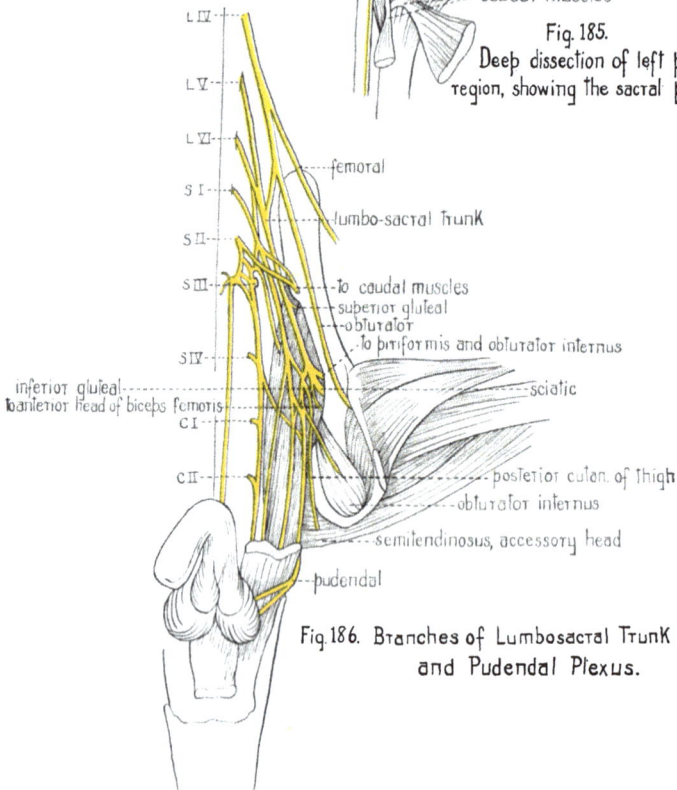

L IV

L V

L VI

femoral

S I

lumbo-sacral trunk

S II

S III

to caudal muscles
superior gluteal
obturator
to piriformis and obturator internus

S IV

inferior gluteal
anterior head of biceps femoris

sciatic

C I

C II

posterior cutan. of thigh
obturator internus
semitendinosus, accessory head

pudendal

Fig. 186. Branches of Lumbosacral Trunk
and Pudendal Plexus.

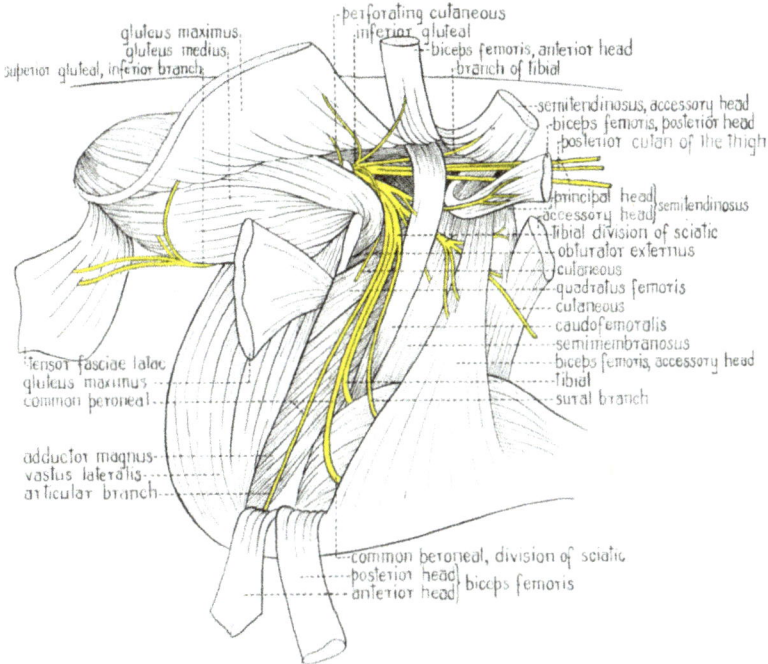

Fig. 187. Lateral aspect of the left thigh dissected to show divisions of the sciatic nerve.

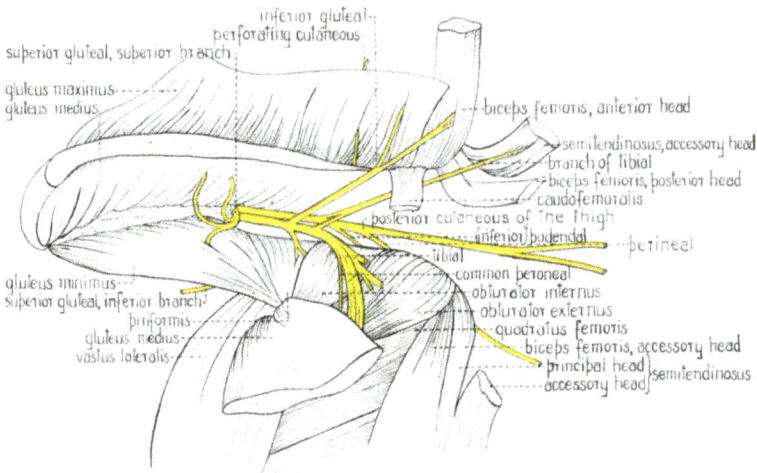

Fig. 188. Deep dissection of left lateral hip region, showing nerves from the sacral plexus

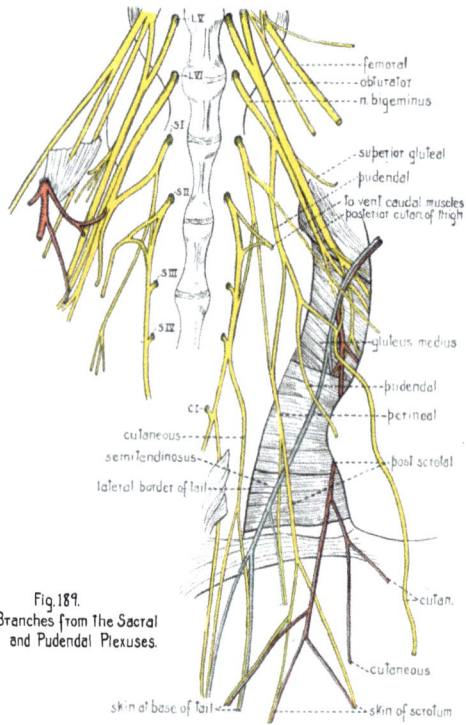

Fig. 189.
Branches from the Sacral
and Pudendal Plexuses.

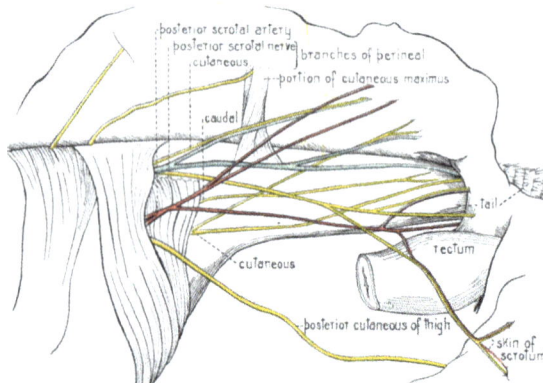

Fig. 190.    Distribution of Branches from Sacral and Pudendal Plexuses.

posterior lacerated foramen
hypoglossal (XII)
vagus ganglion (nodosum)
spinal accessory (XI)
sternomastoid
common carotid
phrenic

ansa subclavia

pectoralis
superior thoracic ganglion
sternum
rami communicantes
internal mammary

diaphragm
intercostals
greater splanchnic
suprarenal gland

lesser splanchnic
renal artery
coeliac ganglion
least splanchnic

carotid canal
glossopharyngeal (IX)
larynx
superior cervical ganglion
n. recurrens
rectus capitis
longus colli
sympathetic trunk
right vagus (X)
middle cervical ganglion
superior cardiac branch
subclavian artery
inferior cervical ganglion
superior vena cava
right auricle
right ventricle
inferior cardiac branch
right lung
aorta
oesophagus
intercostal arteries
left lung

vena cava inferior
phrenic
diaphragm

liver
inferior mesenteric
stomach
coeliac axis
colon

renal artery
internal spermatic art
iliolumbar artery

Th VII
Th VIII
Th IX
Th X
Th XI
XII
XIII
L II
L III
L IV

seminal vesicle
bladder
prostate

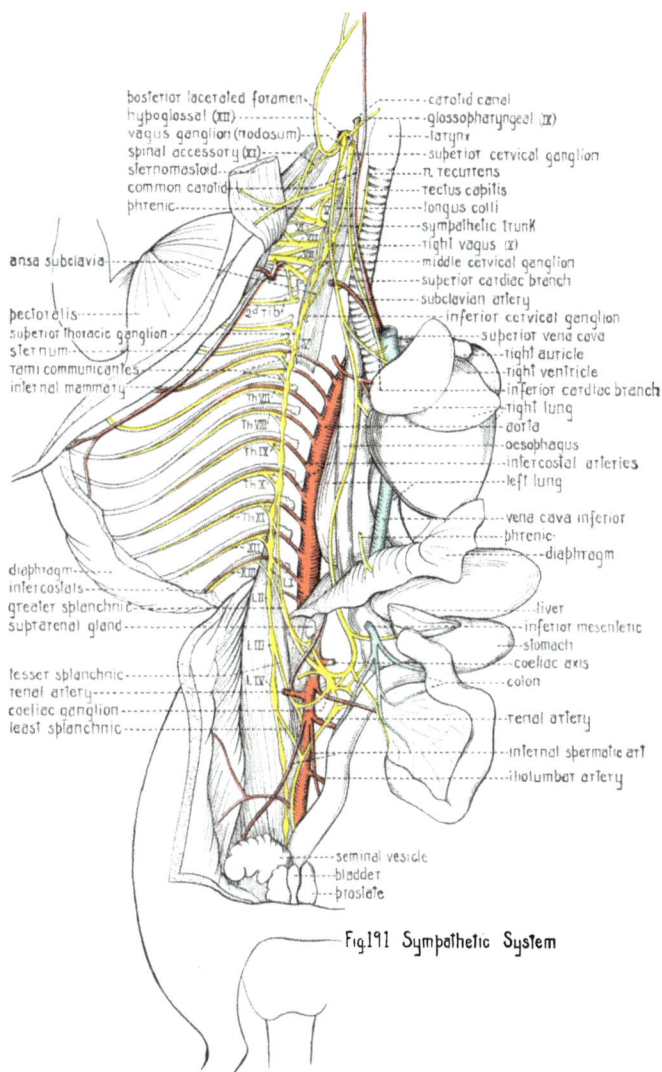

Fig.191 Sympathetic System

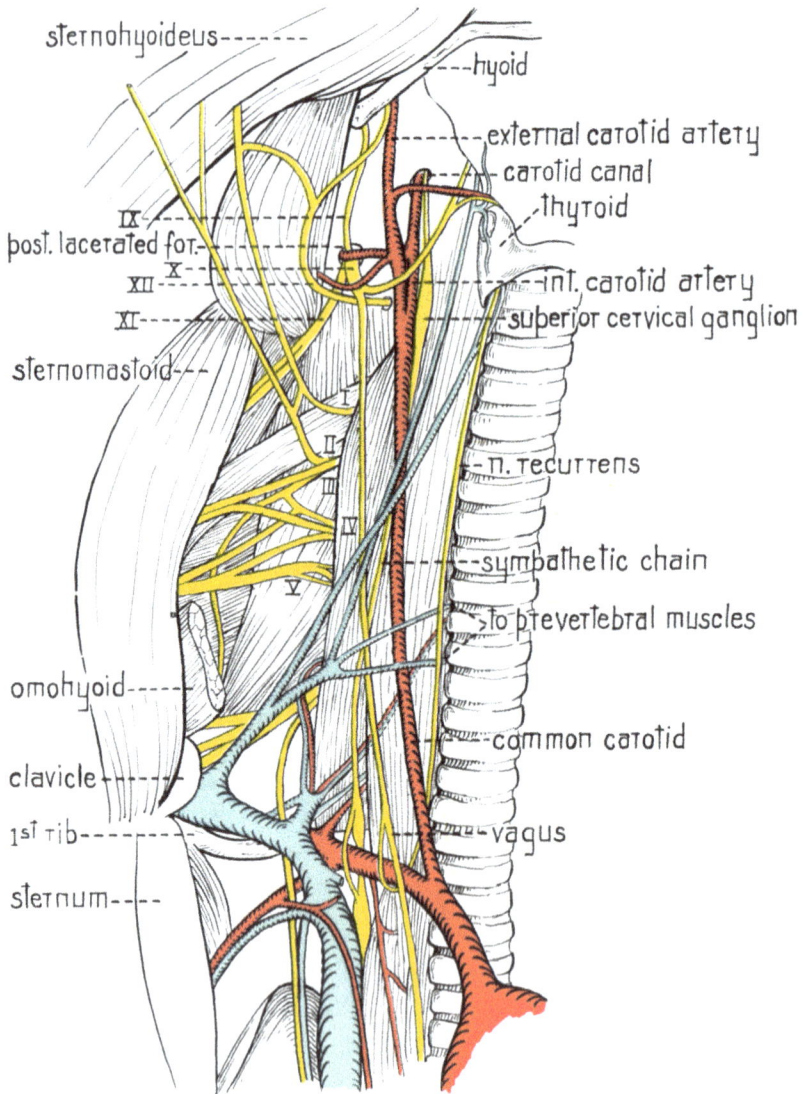

Fig. 19.2. Sympathetic Trunk emerging from Skull.

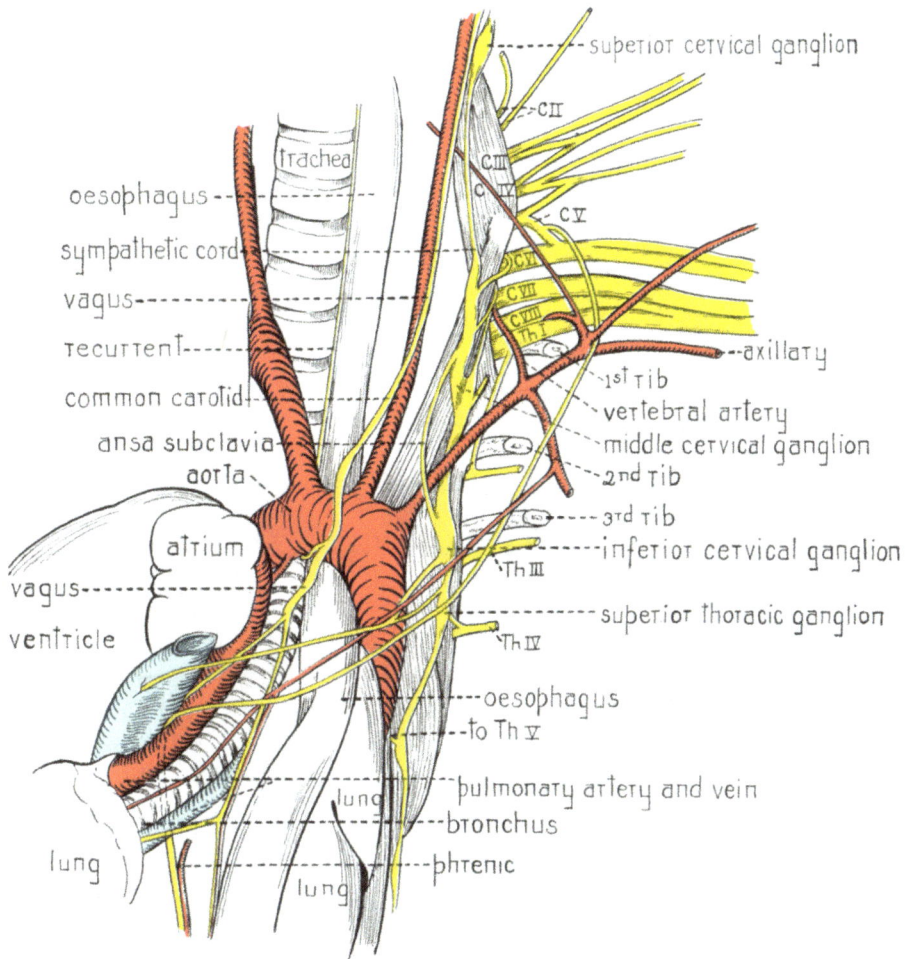

Fig.193. Cardiac and Pulmonary Branches of Sympathetic Cord and Vagus

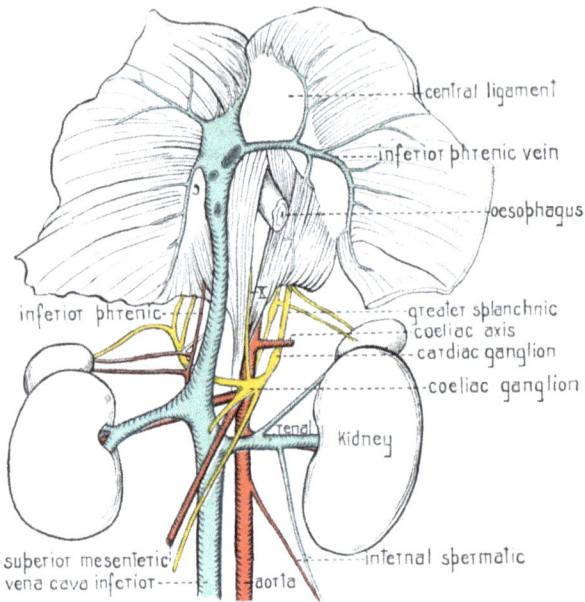

Fig.194 Greater Splanchnic Nerves and Coeliac Ganglion.

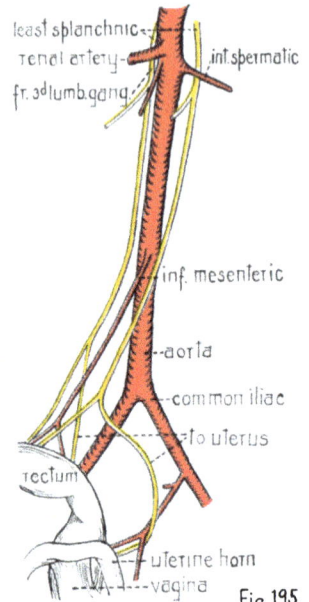

Fig.195.
Detail of Inferior Mesenteric Plexus

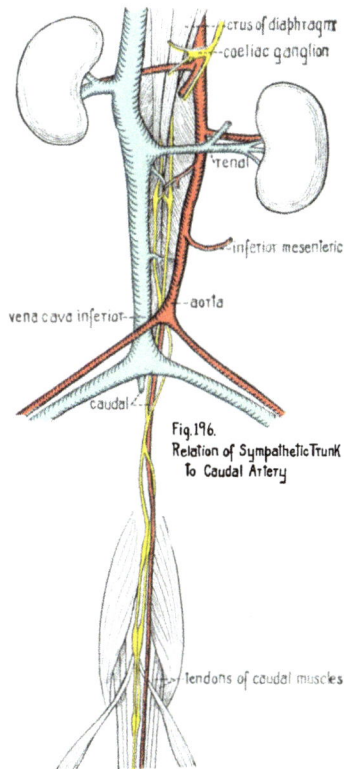

crus of diaphragm

coeliac ganglion

renal

inferior mesenteric

aorta

vena cava inferior

caudal

Fig. 196.
Relation of Sympathetic Trunk
To Caudal Artery

tendons of caudal muscles

# CHAPTER VII

## CIRCULATORY SYSTEM
### (Figs. 197—339)

The circulatory system will be described in the following order:

Arteries
  Pulmonary
  Systemic
    Aorta
    Arteries of the head and neck
      Innominate
      Common carotid
        External carotid
        Internal carotid
    Arteries of the upper extremity
      Subclavian
      Axillary
      Brachial
      Ulnar
      Median
    Arteries of the trunk
      Descending aorta
      Common iliac
        External iliac
        Hypogastric
    Arteries of the lower extremity
      Femoral
      Popliteal
      Anterior tibial
      Posterior tibial

Veins
  Pulmonary
  Systemic
    Veins of the heart
    Veins of the head and neck
      Veins of the exterior of the head
        and face
      Veins of the neck
      Superficial veins of the brain
    Veins of the upper extremity
      Ulnar
      Radial
      Basilic
      Axillary
      Subclavian
  Veins of the thorax
    Venae cavae superiores
  Veins of the abdomen and pelvis
    Vena cava inferior
    Portal veins
    Common iliac
    External iliac
  Veins of the lower extremity
    Veins of the foot and lower leg
    Veins of the thigh
  Lymphatics

## ARTERIES

### PULMONARY

The **Pulmonary Artery** (Fig. 112) shows the typical relationships. It arises from the conus arteriosus of the right ventricle, and after crossing the aorta curves dorsoposteriorly, dividing into the right and left pulmonary arteries to the two lungs. The right pulmonary passes dorsal to the aorta and right vena cava superior and ventral to the bronchus to enter the lung. The left pulmonary crosses ventral to the aorta and bronchus and dorsal to the left vena cava superior to enter the left lung. The azygos, which in the rat lies on the left of the vertebral column, crosses the left pulmonary artery to enter the left vena cava superior.

## SYSTEMIC

### Aorta

The **Ascending Aorta** which is that portion within the pericardium, gives rise to only two branches, namely, the right and left *coronary arteries* which leave the aorta close to its origin and run downward over the heart supplying its walls.

The **Arch of the Aorta** (Fig. 112) crosses ventral to the trachea then turns dorsally and posteriorly to the left of the trachea, continuing downward as the descending aorta.

**Branches.**—From the arch the following branches arise; the *innominate*, the *left common carotid*, and the *left subclavian* (Fig. 197).

### Arteries of the Head and Neck

The **Innominate Artery** (**Brachiocephalic**) (Fig. 197) extends anteriorly and slightly to the right as far as the level of the sternoclavicular joint where it divides into the right common carotid and right subclavian arteries.

The *right* and *left common carotids* supply the head and neck, each common carotid dividing into an *external* and *internal carotid*.

The right **Common Carotid Artery** (Fig. 197) is a continuation of the innominate. The left, arising from the arch of the aorta, follows a similar course through the thoracic region. After leaving the thorax each artery extends anteriorly, practically parallel with the trachea, to the level of the thyroid gland, where it divides into the external and internal carotid arteries.

The **External Carotid Artery** (Figs. 197-203, 206), beginning at the level of the thyroid gland as a direct continuation of the common carotid, gives off five branches before reaching the angle of the jaw where it turns dorsally at almost a right angle, runs for a short distance, and just below the ear gives off three more branches as it again makes a right angle, turning forward and dividing into superficial temporal and internal maxillary arteries.

**Branches.**—As the external carotid artery ascends it gives off the following branches:

| | |
|---|---|
| Occipital | External Maxillary |
| Superior Thyroid | Posterior Auricular |
| Ascending Pharyngeal | Masseteric |
| Lingual | Anterior Auricular |
| Ascending Palatine | Superficial Temporal |
| Internal Maxillary | |

The **Occipital Artery** (Figs. 197, 198, 331) takes its origin from the external carotid close above the bifurcation of the common carotid. It crosses the internal carotid ventrally and, covered by sternomastoideus, divides into a muscular branch to the digastric and stylohyoid muscles; a cleidomastoid branch; an auricular branch to the ear; a posterior meningeal branch, which enters the posterior, lacerated foramen and supplies the dura mater; and a descending branch which supplies splenius, trapezius, and longus colli muscles (Fig. 198).

The **Superior Thyroid Artery** (Figs. 198, 227, 331) leaves the external carotid artery opposite the larynx. It makes a sharp loop on the surface of the larynx, gives a branch to that organ, then runs posteriorly to ramify over the surface of the thyroid. It also supplies the parathyroid and sends a branch beneath the thyroid isthmus to anastomose with the inferior thyroid artery which is distributed to the lower portion of the gland. The superior thyroid and the ascending pharyngeal frequently arise together (Fig. 202). Several muscular branches which in man arise from the superior thyroid are here taken over by the ascending pharyngeal, and the sternocleidomastoid branch is replaced by a branch of the cervical trunk.

The **Ascending Pharyngeal Artery** (Figs. 198, 199) arises from the external carotid artery just above the origin of the superior thyroid, or frequently in common with the superior thyroid (Fig. 202). When the two are separate the ascending pharyngeal gives rise to several of the muscular branches which in man arise from the superior thyroid. This condition plus the fact that the ascending palatine furnishes the palatine, inferior tympanic, and posterior meningeal branches, which in man usually belong to the ascending pharyngeal, leaves the latter mainly muscular in its distribution. In addition to the pharynx it supplies the sternohyoid, sternothyroid, omohyoid, posterior belly of digastric, thyreohyoid, styloglossus, stylohyoid, rectus capitis and longus colli muscles (Fig. 199).

A small **muscular artery** arising close to the bifurcation of the external carotid and external maxillary supplies the posterior belly of digastricus (Fig. 201).

The **Lingual Artery** (Figs. 200, 203, 208) takes its origin from the external carotid covered by the posterior belly of digastricus and the hyoid bone (Fig. 200). It runs forward parallel with and medial to the external maxillary artery to reach the base of the tongue. It supplies branches to that organ and continues forward along the lower surface to the tip as the *profunda linguae* (Fig. 208). The ascending palatine branch of the external maxillary frequently arises with the lingual artery (Fig. 202).

The **Ascending Palatine Artery** (Figs. 197, 199, 202, 213, 215, 216) may arise from the external carotid artery with the lingual (Fig. 202) or as a separate branch in the short interval between the lingual and the external maxillary (Fig. 197). Its origin is covered by the hyoid bone. Running laterally across the ventral surface of the tympanic bulla it supplies twigs to the styloglossus muscle, to the walls and constrictors of the pharynx, and to external and internal pterygoid muscles (Figs. 202, 215). Finally, it turns dorsally between the tympanic bulla and the external pterygoid process and divides into three terminal branches (Fig. 215); the *palatine branch*, which runs forward along the margin of pterygoideus externus to the soft palate; the *tympanic branch* to the tympanum; and a *posterior meningeal branch* (Fig. 216) which enters the skull through the petrotympanic fissure. In man these three branches usually arise from the ascending pharyngeal artery.

The **External Maxillary Artery** (facial artery) (Figs. 197, 205, 207, 208, 214) leaves the external carotid just as the latter reaches the ramus of the mandible. At this point the external carotid artery turns dorsally at a sharp angle, extending for a short distance along the ramus, while the external maxillary artery turns ventrally and runs forward along the boundary between the masseter and digastric muscles (Figs. 197, 205). At the anterior end of the digastric muscle the artery crosses the mandible and appears on the lateral surface of the facial region following the ventral margin of the masseter. It runs upward past the corner of the mouth and after giving off the superior labial is known as the *angular artery* (Fig. 205).

**Branches.**—The branches may be listed in two groups, namely those given off in the cervical region (Figs. 204, 207, 208), and those from the facial portion of the artery (Figs. 205, 207).

| Cervical branches | Facial branches |
|---|---|
| Glandular | Cutaneous |
| Tonsillar | Inferior labial |
| Submental | Muscular |
| Masseteric | Superior labial and lateral nasal |
| | Angular |

The **glandular branches** (Figs. 197, 205, 330), usually two large vessels, are given off just above the anterior margin of the posterior belly of digastricus. The larger one turns posteriorly and supplies the submaxillary and major sublingual glands. The second glandular branch, arising just anterior to the submaxillary branch, also turns posteriorly and supplies twigs to the lymph glands of the region (Fig. 330). This branch may arise from the submental artery.

The **tonsillar branch** (Figs. 203, 204, 207, 208) of the external maxillary arises between the two glandular branches. Though the tonsil in the rat is represented merely by scattered follicles, the name tonsillar branch has been retained for this artery since its course and distribution correspond to those of the same artery in man. Passing medial to the mandible, it supplies styloglossus then ramifies extensively through the internal pterygoid muscle and the superior constrictor of the pharynx before reaching its termination in the lining of the pharynx and the root of the tongue (Fig. 203).

The **submental artery** (Figs. 197, 198, 204, 207) leaves the external maxillary about the middle of the anterior belly of digastricus running between this muscle and mylohyoid (Fig. 207) supplying both, and continues forward to reach transversus mandibularis, the platysma and integument of the submental region (Fig. 198). Sometimes one of the glandular branches arises from the submental instead of from the external maxillary artery.

The **masseteric artery** (Fig. 207) arising just anterior to the submental, ramifies extensively through the masseteric muscle, and anastomoses with the masseteric and buccinator branch of the internal maxillary.

The **cutaneous branches** (Fig. 205) of the *facial portion* of the external maxillary may be several in number. The first one, which is quite constant, arises as the external maxillary crosses the mandible, and supplies the platysma and skin of the mid-ventral region. Another cutaneous branch arises between the inferior and superior labial branches and supplies the corner of the mouth.

The **inferior labial artery** (Figs. 205, 207) is the second branch from the facial portion of the external maxillary, arising as the latter crosses the upper margin of the mandible. After sending a small branch into the mental foramen of the mandible to anastomose with the inferior alveolar branch of the internal maxillary, and a branch to the lining of the mouth (Figs. 208, 214), it runs forward along the mandible to the mucous membrane and muscle of the lower lip. The inferior labial may arise with the muscular branch.

Two rather large **muscular** or **anastomotic branches** (Figs. 205, 207) are given off between the inferior and superior labial arteries. The lower one of these, which may arise with the inferior labial, is directed dorsoposteriorly, and takes a deep course through the masseter muscle which it supplies together with the buccinator (Fig. 205), crosses the inferior alveolar ridge just in front of the ramus of the mandible, and after supplying the lining of the mouth, anastomoses with the posterior superior alveolar artery (Figs. 207, 214).

The second muscular branch, arising just below the superior labial artery at the level of the roof of the mouth, is also directed dorsoposteriorly. It gives off a branch which turns anteriorly to supply the upper corner of the mouth, then continuing under cover of the masseter muscle which it supplies, it reaches the roof of the mouth to which it gives a branch before anastomosing with the posterior superior alveolar artery (Figs. 205, 207, 214).

The **superior labial artery** (Fig. 205) leaving the external maxillary just above the angle of the mouth runs forward to the upper lip giving off the *lateral nasal artery* (Fig. 205) which is directed upward to the nose. In man the lateral nasal artery is a separate branch from the external maxillary.

The **angular artery** (Fig. 205) is the terminal portion of the external maxillary which continues dorsally over the anterior end of the zygomatic arch to the corner of the eye. In its course it gives twigs to the levator labii superioris, and mystacial pad, and to the superficial portion of masseter.

The **Posterior Auricular Artery** (Figs. 197, 201, 205, 206) arises from the external carotid as it crosses the origin of sternomastoideus in the space between the tympanic bulla and the ramus of the mandible. The external carotid turns forward at a right angle while the posterior auricular, covered at first by the parotid gland, continues dorsally behind the ear, supplying the parotid gland, the levator auris longus muscle, the auricle and the integument of the occipital region above and behind the ear.

The **Masseteric Artery** (Fig. 205) springs from the angle of the external carotid as the latter turns forward below the external acoustic meatus. It runs forward and downward for a short distance over the lateral surface of the masseter before dipping into the substance of the muscle, which it supplies.

The **Anterior Auricular Artery** (Figs. 197, 202, 205, 206) which in man is a branch of the superficial temporal, here arises from the external carotid between the temporomandibular joint and the external acoustic meatus, and, covered by the parotid gland, crosses the posterior root of the zygomatic arch dividing into two branches, one of which supplies the exorbital portion of the lacrimal gland, while the other continues dorsally to the anterior portion of the auricle (Fig. 205).

The **Superficial Temporal Artery** (Figs. 197, 205–207) is one of the terminal branches of the external carotid. In the interval between the tympanic bulla and the neck of the mandible, the external carotid turns forward at right angles and, after giving off the *masseteric* and *anterior auricular branches*, divides into the *superficial temporal* and the *internal maxillary arteries* (Fig. 205). The internal maxillary takes a deep course while the superficial temporal runs over the lateral surface of the masseter, crossing the posterior root of the zygomatic arch. It gives a **masseteric branch** (Fig. 205) to the masseter muscle; the **middle temporal branch** (Fig. 205) to the temporalis muscle; the **transverse facial** (Fig. 205) which follows the lower margin of the zygomatic arch, supplying the masseter muscle and the integument of the region; the **zygomatico-orbital** which divides almost at once into *palpebral* and *lacrimal branches* (Fig. 205) to the upper lid and the intra-orbital lacrimal gland respectively.

The **Internal Maxillary Artery** (Figs. 205–208, 214), one of the terminal branches of the external carotid, begins just in front of the temporomandibular joint. It takes a deep course beneath the masseter muscle. In man this artery has three portions, the *mandibular, pterygoid* and *pterygopalatine*. In the rat the branches of the last portion are entirely taken over by the internal carotid artery, while the first and second portions are by no means so distinct as in man. Nevertheless, these terms have been retained for purposes of comparison.

**Branches of the First or Mandibular Portion.**—In the rat the principal branch of the mandibular portion is the *inferior alveolar*.

The **inferior alveolar artery** (r. mandibularis, Tandler) (Figs. 205–208, 214) leaves the internal maxillary artery as the latter crosses the lateral surface of the neck of the mandible, and runs medial to the mandible entering the mandibular canal through the mandibular foramen in company with the corresponding nerve (Fig. 210). Within the canal it gives off an *incisor branch* (Fig. 208) which curves posteriorly and enters the root of the lower incisor tooth. It continues as the *mental branch* (Fig. 214) through the canal, anastomosing with the mental branch of the inferior labial artery which enters the canal through the mental foramen. The pterygoid branch of the second portion of internal maxillary frequently arises with the inferior alveolar.

In her studies of the embryology of the arteries in the rat, Mrs. Leitch ('34) has shown that the inferior alveolar artery is at first associated with the pterygopalatine branch of the internal carotid but between fourteen and fifteen days an anastomosis (the ramus anastomoticus 2, of Tandler) occurs between the inferior alveolar and external carotid arteries by means of which the inferior alveolar is transferred to the internal maxillary branch of the external carotid artery.

**Branches of the Second or Pterygoid Portion** (Figs. 205–208).—From the second portion of the internal maxillary arise: first, the *pterygoid branch* to the pterygoid muscles (Fig. 206), which may, however, arise from the inferior alveolar; the *posterior deep temporal artery* (Fig. 206) which runs upward to the temporalis muscle; the *anterior deep temporal artery* (Figs. 205–208, 214) which crosses the neck of the mandible beneath the posterior root of the zygomatic arch and gives an articular branch to the temporo-mandibular joint before entering the orbital fossa where it supplies temporalis; and finally the *masseteric* plus the *buccinator artery* (Fig. 207), the terminal branches of the internal maxillary which follow a deep course over the lateral surface of the mandible, passing just above the alveolus of the lower incisor, and supplying the deep masseter and buccinator muscles, and anastomosing with the corresponding branch of the external maxillary.

The **Internal Carotid Artery** (Figs. 197, 209–213, 217–220, 331) arising from the common carotid opposite the lower end of the thyroid gland follows the same general direction through the neck as the external carotid but at a much deeper level along the base of the skull. Upon reaching the tympanic bulla it gives off a large *pterygopalatine branch* (Fig. 209) representing what is in man a portion of the internal maxillary branch of the external carotid. Continuing along the medial surface of the bulla for a short distance the internal carotid enters the carotid canal between the bulla and the basal plate of the occipital bone to reach the interior of the skull where it gives branches to the brain.

**Branches.**—The internal carotid gives off the following branches:

> From the cervical portion—
> Pterygopalatine
> From the cerebral portion—
> Posterior communicating
> Chorioid
> Anterior cerebral
> Middle cerebral

The **Pterygopalatine Artery** (a. stapedius, Tandler) (Figs. 209, 210, 212, 213) corresponds to the pterygopalatine portion of the internal maxillary branch of the external carotid in man. It arises from the cervical portion of the internal carotid in the rat, enters the posterior lacerated foramen of the skull with the internal jugular vein, traverses the tympanic bulla and, emerging through the petrotympanic fissure with the chorda tympani (Fig. 210) just medial to the external pterygoid process, divides into a *pterygoid* (Fig. 209) and *palatine portion* (Figs. 209, 211–213). The pterygoid portion turns medially and enters the pterygoid canal through the middle lacerated foramen as described below (Fig. 209). The palatine portion (r. infraorbitalis, Tandler) pierces the external pterygoid process through the foramen ovale (Fig. 209), enters the alisphenoid canal and finally makes its exit through the anterior lacerated foramen into the orbit in company with the ophthalmic division of the trigeminal nerve (Figs. 211, 212), where it divides into the *ophthalmic* and *posterior superior alveolar arteries* (Fig. 213).

Branches.—The branches of the pterygopalatine artery are:

> From the pterygoid portion—
> Pharyngeal
> Artery of the pterygoid canal (Vidian artery)
> From the palatine portion—
> Ophthalmic
> Posterior superior alveolar
> Descending palatine
> Sphenopalatine
> Infraorbital

The **pharyngeal artery** (Fig. 209) arises close to the bifurcation of the pterygopalatine artery, from its pterygoid portion, and supplies the soft palate.

The **artery of the pterygoid canal** (Vidian artery) (Fig. 209) is a continuation of the pterygoid portion of the pterygopalatine artery. In the pterygoid fossa it runs medially and enters the pterygoid canal through the middle lacerated foramen (Figs. 209, 210, 211, 213) on the lateral surface of the internal pterygoid process. Extending anteriorly through this canal it emerges to the surface of the basisphenoid bone through the pterygopalatine foramen (Fig. 209), crosses the suture between the basisphenoid and presphenoid bones and enters the posterior nares to supply the nasal pharynx and the interior of the nasal cavity.

The **ophthalmic artery** (Figs. 209, 211, 213) which in man originates within the skull from the cavernous portion of the internal carotid, and enters the orbit through the optic foramen is, in the rat, a branch of the palatine portion of the pterygopalatine artery, arising in the alisphenoid canal and entering the orbit through the anterior lacerated foramen. It divides into *superior* and *inferior ophthalmic* portions (Fig. 213) which supply the eyeball and its muscles, the lacrimal and Harderian glands, and the eyelids. The superior ophthalmic branch (Fig. 213) sends an **anterior ethmoidal artery** with the nasociliary branch of the trigeminal nerve through the anterior ethmoidal foramen in the frontal bone on the medial wall of the orbital fossa, to reach the ethmoidal region of the nasal cavity (Fig. 212).

It should be noted here that the ophthalmic artery of the adult is not the ophthalmic artery of the embryo, as has been shown by Mrs. Leitch, '34. She states that "in the interval from sixteen to seventeen days and beyond, the embryonic ophthalmic artery from the anteromedial cerebral trunk progressively diminishes in size and an 'ophthalmic' artery develops from the palatine portion of the pterygopalatine branch, and growing into the orbit, replaces the embryonic ophthalmic artery."

The **posterior superior alveolar artery** (Figs. 207, 209, 211, 213, 214) consists of a large branch which springs from the palatine portion of the pterygopalatine artery as it enters the orbit through the anterior lacerated foramen. This portion immediately leaves the orbital fossa. Curving outward it follows the lateral surface of the alveolar ridge, supplies the gums, the buccinator and masseteric muscles, and anastomoses with corresponding branches of the external maxillary (Fig. 207). Medial to the alveolar ridge, as the palatine portion of the artery continues along the infraorbital groove, small alveolar branches are distributed to the olmar teeth. These represent the remainder of the posterior superior alveolar artery of man.

The **descending palatine artery** (Figs. 209, 210, 212, 213) arises as the second branch of the palatine portion of the pterygopalatine artery as it runs forward in the infra-orbital groove. The descending palatine artery passes with the corresponding nerve through the posterior palatine foramen (Fig. 212) to the hard palate where it runs forward in a groove just medial to the alveolar ridge. As it reaches the first molar tooth it sends an anastomotic branch around the anterior surface of the tooth to meet the posterior superior alveolar artery (Fig. 213). Its terminal branch continues to the incisive foramen which it enters and anastomoses with the sphenopalatine artery.

The **sphenopalatine artery** (Figs. 209, 212, 213) is the third branch of the pterygo-palatine artery arising within the orbit. Upon reaching the sphenopalatine foramen it penetrates, with the nasal nerves, to the nasal cavity (Fig. 212) where it supplies the mucous membrane of the septum and conchae. One of its branches anastomoses with the descending palatine by way of the incisive canal.

The **infraorbital artery** (Figs. 205, 209, 211–213) continues the course of the palatine portion of the pterygopalatine artery as its terminal branch. In addition to small twigs to the inferior rectus and inferior oblique muscles it sends a branch obliquely through the orbit. From this the **anterior superior alveolar branch** (Figs. 211, 213), runs to the upper incisor, and two small branches, one passing anterior and one posterior to the root of the zygomatic arch, anastomose with branches of the external maxillary (Fig. 214). A **nasal branch** (Fig. 213) passes through a small foramen at the anterior end of the infraorbital groove, accompanied by the corresponding nerve (Fig. 212), while the terminal branches of the infraorbital, with the infraorbital nerves (Fig. 211), pass through the infraorbital fissure (Figs. 205, 211–213) to the side of the nose, where they supply the mystacial vibrissae and their pad.

The following branches arise from the internal carotid within the skull:

The **Posterior Communicating Artery** (Figs. 217–219) arises from the internal carotid on the basal surface of the brain lateral to the stalk of the infundibulum and curves posteriorly around the hypophysis, lying in a groove of the hypophysis (Figs. 125, 127), to reach the median basilar artery formed by the union of the vertebral arteries of either side (Fig. 219). The posterior communicating arteries of the two sides thus form the posterior portion of the arterial circle of Willis of the brain (Fig. 219), while the anterior cerebral arteries form the anterior portion.

The **Posterior Cerebral Artery** (Figs. 217–219), is a branch of the posterior communicating artery. Its origin is discussed under the Circle of Willis. Curving around the peduncle of the cerebrum it runs upward along the fold of the tentorium to supply the tentorial surface of the hemisphere, and the medial and lateral surfaces of the occipital lobe.

The **Circle of Willis** (Figs. 217–219), formed by the *anterior cerebral* and *posterior communicating arteries*, has been variously interpreted by different investigators.

Adachi '28, who has made one of the most exhaustive studies of the variations of the circle in man, says, in effect, that, "by the a. communicans posterior is understood the connection between the internal carotid and the basilar arteries, so that the posterior cerebral does not, in the ordinary sense, spring from the basilar, but from the posterior communicating artery." The anterior, carotid part, of the posterior communicating may be stronger or the posterior, basilar part, may show the greater caliber. Adachi goes on to show 83 individual cases displaying all types of variation within this generalized scheme.

The three drawings shown in the present work of the circle of Willis in the rat (Figs. 217–219), are random samples, since there has been no attempt to select them or to make any exhaustive study of this region. The fact that all three of these specimens show a marked difference in the branching and relative size of the various vessels would lead one to suppose that we may very well expect to find as extensive variation in the adult rat as Adachi finds in man. The specimen shown in Fig. 217 is a case where the posterior cerebral gets its main supply from the basilar artery. Fig. 218 is a case where the carotid and basilar parts of the posterior communicating are approximately equal in caliber, while Fig. 219 shows a case where the internal carotid is the chief source of supply for the posterior cerebral.

It is interesting to trace the development of these vessels in the rat embryo as described by Mrs. Leitch (unpublished manuscript). She says, "In the early stages of embryos, the primitive internal carotid is considered by many writers as passing forward into the head where in the region of the mesencephalon the right and left internal carotids connect with the primitive basilar artery (or its antecedent, the longitudinal neural vessels).

"Then at a later time, when a branch appears (cranial to the optic stalk) this branch is designated by some authors as an *anterior ramus* of the internal carotid, while the portion of the original primitive internal carotid lying between the telecephalon and the junction with the basilar artery is now referred to as the *posterior ramus* of the internal carotid."

The **Chorioid Artery** (Fig. 218) leaves the internal carotid close to the origin of the posterior communicating artery. Running for a short distance in the groove between the temporal lobe and the cerebral peduncle it curves dorsally around the latter and pushes into the lateral ventricle along the chorioid fissure carrying the thin membranous medial wall of the ventricle with it, in such a way that the plexus formed by the artery is entirely surrounded by this membrane and never actually penetrates into the cavity of the ventricle. In addition to this chorioid plexus of the lateral ventricle the chorioid artery enters into the formation of the tela chorioidea of the third ventricle in a similar fashion.

The **Anterior Cerebral Artery** (Figs. 217-219), one of the terminal branches of the internal carotid, arises within the skull at the base of the brain just lateral to the optic chiasma. Crossing the olfactory tract it joins the corresponding artery of the opposite side to form an azygos vessel which passes into the longitudinal fissure (Fig. 218) where it curves upward over the genu of the corpus callosum and extends the entire length of the upper surface of the latter, finally reaching the parietal region where it anastomoses with branches of the posterior cerebral artery. As the anterior cerebral arteries of the two sides enter the longitudinal fissure they may be connected by a rather simple plexus from which the azygos anterior cerebral vessel arises. Adachi, '28, shows all manner of variation in this respect in man, from the usual simple anterior communicating artery connecting two anterior cerebral vessels, through a plexiform connection of the anterior cerebrals, to a single case of an azygos vessel as seen in the rat. There is, however, as a rule, no **anterior communicating artery** as such in the rat, but it is evident that the circle of Willis is complete.

The **Middle Cerebral Artery** (Figs. 217-220), the larger of the two terminal branches of the internal carotid, arising at the base of the brain lateral to the infundibulum, runs upward over the lateral surface of the hemisphere. On the upper surface it is, directed somewhat posteriorly, ramifying extensively over the entire dorsal surface of the hemisphere (Fig. 220).

### Arteries of the Upper Extremity

A single **arterial trunk** (Fig. 224) supplies each upper extremity. Arising from the innominate artery on the right, and from the aortic arch on the left (Fig. 197), this trunk turns laterally and leaves the thorax between the first rib and the clavicle (Fig. 234). It then traverses the axillary region and becoming more superficial as it emerges from beneath the pectoral muscles extends along the medial surface of the brachium and along the volar aspect of the antibrachium.

From its origin to the lateral margin of the first rib the trunk bears the name *subclavian;* from the rib to the point where it passes between latissimus dorsi and cutaneous maximus and leaves the axilla, it is known as the *axillary;* in the brachium it becomes the *brachial* and in the antibrachium the *median.*

The right **Subclavian Artery** (Figs. 224, 226-228, 230) is quite short since it is only that portion from the innominate to the border of the first rib (Fig. 228). The left subclavian (Fig. 237) is longer, but like the right, gives branches only as it approaches the first rib. These branches correspond on the two sides.

**Branches.**—The following branches leave the subclavian:

> Costocervical trunk
> Internal mammary
> Vertebral
> Cervical trunk

The **Costocervical Trunk** (Figs. 224-227) is short and thick, running downward on the inner surface of the dorsal wall of the thoracic cavity, in a groove along the lateral border of longus colli. The trunk gives rise to three branches; the **cervicalis profunda** (Figs. 224, 226); the **highest intercostal** artery (Figs. 224, 226, 227), and a small **branch to the vena cava superior** (Figs. 224–226). Occasionally the highest intercostal artery arises independently. In addition to these branches, the costocervical trunk of the right side, as a rule, gives rise to the bronchial arteries (Figs. 225, 226). Where this is not the case, they arise in common with the anterior mediastinal and pericardial branches of the internal mammary, a condition which is the general rule on the left side (Fig. 225). In a few instances one or both sides may show the **pericardiacophrenic** branch as a continuation of the costocervical trunk (Figs. 225, 226). More frequently this artery is a branch of the internal mammary (Fig. 228).

The **cervicalis profunda** (Figs. 224, 226) arises close to the origin of the trunk and supplies longus colli, longus capitis and adjacent muscles of the deep cervical region.

In addition to these muscular branches, there is a fairly large branch which turns anteriorly along the trachea. This is the **inferior thyroid artery** (Figs. 225, 227) which in man arises with the more superficial cervical vessels, forming with them, the thyrocervical trunk. The inferior thyroid passes forward along the lateral surface of the trachea in company with the nervus recurrens and inferior thyroid vein (Fig. 227). It gives off *tracheal*, *œsophageal*, and *inferior laryngeal* branches before terminating in the thyroid gland, where it anastomoses with the superior thyroid (Fig. 227).

The **highest intercostal artery** (Figs. 226, 230) supplies in general the upper intercostal spaces but varies somewhat on the two sides, and in the number of spaces supplied. As a rule the artery gives a series of branches to the first, second, and third intercostal spaces. This occurs more constantly on the right. On the left it frequently supplies only the first and second spaces, the third being supplied by an aortic intercostal artery.

The small **branch to the vena cava superior** (Figs. 225, 226) usually supplies an adjacent lymph node before reaching the vein close to its junction with the heart.

The **Internal Mammary Artery** (Figs. 224, 226, 228) takes its origin from the subclavian at practically the same point as the costocervical trunk, passing downward along the inner thoracic wall while the subclavian continues outward across the first rib.

**Branches.—**

| | |
|---|---|
| Pericardiacophrenic | Intercostals |
| Pericardiacomediastinal | Perforating |
| Sternal | Musculophrenic |

Superior Epigastric

The **pericardiacophrenic artery** (Figs. 224–228) displays considerable variation in its point of origin. On the right side, in the majority of cases, it arises from the internal mammary just as this artery passes under the first rib. Almost invariably it gives a small branch to one or more lymph nodes located along its course (Fig. 228), then accompanies the phrenic nerve to the diaphragm. In rare instances its origin may be common with that of the pericardial and anterior mediastinal arteries (Fig. 225

right). It is fairly common for the pericardiacophrenic to arise from the costocervical trunk (Fig. 227). This seldom occurs simultaneously on both sides. On the left side it arises about equally in any one of the three following situations: independently from the internal mammary; from the internal mammary in conjunction with anterior mediastinal and pericardial branches (Fig. 225); or as a continuation of the costocervical trunk.

The **pericardiacomediastinal artery** (Fig. 225) corresponding to the pericardial and anterior mediastinal arteries of man, arises from the internal mammary. The relationship of these two components is constant enough to justify the use of the term pericardiacomediastinal for this artery in the rat. It first gives off a branch to the thymus (Fig. 228) (in rare instances this branch may arise independently from the internal mammary) (Fig. 226), and to associated lymph nodes, then crossing the vena cava superior to which it sends small branches, it supplies the fat and connective tissue of the anterior mediastinum and the pericardium in the region of the pulmonary artery and atrium.

Rarely on the right side, but more frequently on the left, the pericardiacomediastinal artery gives rise to the pericardiacophrenic (Fig. 225 left), in which case the point of branching is variable. Small **bronchial branches** to the lower half of the trachea, to the œsophagus, lungs, bronchi, and lymph glands of the vicinity, may arise from the pericardiacomediastinal on the right side, and, so far as observed, invariably do so on the left (Fig. 225). Those on the right more often are branches of the costocervical trunk (Fig. 226). They seem to correspond to certain aortic branches in man.

The **sternal branches** (Fig. 226) penetrate the posterior surface of the sternum.

The **intercostal branches** (anterior intercostals) (Figs. 226, 228) are distributed to the upper six intercostal spaces. The vessels of each side are arranged in pairs following the upper and lower margins of each rib as far as, and including the sixth. They supply both external and internal intercostal muscles and send small branches through these to the pectoral muscles.

The **perforating branches** (Fig. 229) leave the internal mammary segmentally and penetrate the muscles of the first six intercostal spaces, while the last and largest branch passes through the angle formed by the xiphoid process and the cartilages of the seventh and eighth ribs. These branches supply the intercostals, pectoralis major and minor, rectus abdominis and cutaneous branches to the integument.

The **musculophrenic artery** (Fig. 228), one of the terminal branches of the internal mammary, turns laterally between the sixth and seventh ribs and gives branches which run parallel to the seventh, eighth, and ninth ribs, supplying the corresponding intercostal spaces in the same fashion as the intercostals of the internal mammary. In addition the musculophrenic gives branches to the diaphragm, lower pericardium, and terminal ones to the abdominal muscles.

The **superior epigastric artery** (Figs. 228, 229) continues the course of the internal mammary as the other one of its terminal branches. After passing from beneath the costal border through the diaphragm, it becomes closely associated with the rectus abdominis, supplying this muscle, and the muscles and integument of the ventral abdominal wall, and anastomosing with the inferior epigastric artery.

The **Vertebral Artery** (Figs. 219, 224, 227, 228, 235), arising from the anterior surface of the subclavian, crosses the roots of the brachial plexus (Fig. 193), passes under the carotid tubercle of the sixth cervical vertebra (Figs. 193, 197) and between the rectus capitis and longus colli muscles to extend upward through the transverse foramina of the upper six cervical vertebrae (Fig. 159) and enter the skull through the foramen magnum. It joins the vertebral artery of the other side to form the **basilar artery** (Fig. 219) at the level of the pons. The vertebral artery occasionally arises from the cervical trunk. The basilar artery has been discussed as a component of the Circle of Willis.

The **Cervical Trunk** (Figs. 224, 233–235) springs from the anterior surface of the subclavian just before this vessel crosses the first rib. This trunk corresponds to the thyrocervical trunk (thyroid axis) in man, but since the thyroid element is taken over by a. cervicalis profunda in the rat (Fig. 227), it is necessary to reduce the name to cervical trunk. After giving off two or three small branches to the deep neck muscles, and the ascending cervical, it runs upward and somewhat laterally through the shoulder and neck region in the same general direction as the external jugular vein. It leaves the thorax through the space between the first rib, the manubrium of the sternum, and the clavicle (Fig. 234), and passes laterally parallel with, and covered by the clavicle for a short distance. Then passing upwards in company with the anterior jugular vein, which lies superficial to it, it at once gives off several branches before continuing across the upper margin of subscapularis and dividing into its final branches as it crosses omohyoideus (Fig. 236).

**Branches.—**

| | |
|---|---|
| Ascending cervical | Superficial cervical |
| Acromiodeltoid | Transverse scapular |
| Transverse cervical | |

The **ascending cervical artery** (Figs. 224, 235, 237) arises close to the root of the cervical trunk and divides into three branches. Two of these run medially and anteriorly, ventral to the plexus, and, accompanied by branches of the internal jugular vein (Fig. 235), one enters the prevertebral muscles at once, the other, somewhat larger, passes beneath the vagus and the carotid artery and extends upward through the neck giving branches to the trachea and œsophagus, and muscular branches to the prevertebral muscles of the neck. The third or **deep branch** (Figs. 235, 237) of the ascending cervical makes a sharp bend and passes dorsally through the plexus, between the eighth cervical and first thoracic nerves. It sends a twig to scalenus medius, then runs dorsally and posteriorly. It sends one branch anteriorly beneath scalenus to levator scapulae, the rhomboideus muscles, and to the scalenes and deep muscles of the neck. Another branch supplies serratus anterior, while the terminal portion passes beneath levator scapulae to the scaleni and to serratus anterior and posterior (Fig. 235).

The **acromiodeltoid artery** (Figs. 224, 232–235, 237) springs from the cervical trunk as the latter passes under the clavicle (Fig. 233). It extends laterally, covered by the external jugular vein, for a short distance, then becomes superficial in the angle between the cephalic and external jugular veins (Fig. 232). It then runs laterally with the cephalic vein, in the groove between pectoralis major and acromiodeltoideus. As it leaves the cervical trunk a minute branch turns back to the sternoclavicular joint (Fig. 233). It next gives a twig to the subclavius as it crosses that muscle (Fig. 233). Another branch runs under the clavicle to reach the shoulder joint (Figs. 237, 238), while the terminal portion gives branches to the pectoralis and deltoid muscles as it passes between them (Fig. 232). From the fact that this vessel runs along the boundary between the pectoralis and deltoid muscle in company with the cephalic vein, and since it supplies acromial, deltoid and pectoral branches, it would seem to correspond to the thoraco-acromial artery in man, but inasmuch as the thoraco-acromial artery is usually identified as the one accompanying the lateral anterior thoracic nerve, it has seemed necessary to reserve this term for the first branch of the axillary, which does bear the necessary relationship to the nerve (Fig. 237). The author has therefore ventured to use the term acromiodeltoid for the branch in question.

The **superficial cervical artery** (Figs. 224, 232–235, 237) leaves the cervical trunk opposite the acromiodeltoid artery, beneath the clavicle, and in the angle between the external jugular and anterior jugular veins (Fig. 233). It runs anteriorly with the latter under the clavicle, emerging in the angle between the cleidomastoid and sternomastoid muscles, and becomes quite superficial in its course through the neck. It supplies the cleidomastoid, sternomastoid, sternohyoid, sternothyroid and omohyoid muscles, the superficial cervical lymph glands and the parotid (Fig. 329) and sends several branches to the skin and platysma of the ventral and lateral cervical region (Fig. 232).

As the cervical trunk emerges anterior to the clavicle and becomes more superficial in the angle formed by the attachment of clavotrapezius to the clavicle, it divides into its terminal branches, the *transverse scapular* and *transverse cervical* arteries (Figs. 224, 233–237). The point at which these arteries separate varies. The transverse scapular may arise first, and the trunk then divides into superficial cervical and transverse cervical branches, but the superficial cervical usually arises first from the trunk leaving the transverse scapular and transverse cervical as the terminal divisions. Goeppert (1909), working on the mouse regards the ascending cervical as the direct continuation of the cervical trunk, and gives the name superficial cervical to the large remaining vessel with its several branches. In the albino rat it seems obvious that the ascending cervical is merely a small branch, and that the trunk continues well into the cervical region before giving off its relatively large terminal branches.

The **transverse scapular artery** (suprascapular) (Figs. 224, 236) curves laterally and posteriorly beneath the clavicle sending branches to the clavicle and shoulder joint, and to the supraspinatus, infraspinatus, subscapularis and omohyoideus muscles.

The **transverse cervical artery** (Figs. 224, 236) runs a little more anteriorly than the transverse scapular artery. It distributes a branch to clavotrapezius and cleidomastoideus, a branch to pectoralis and the skin of the chest, then curves dorsally around the edge of omohyoideus and continues along the upper border of supraspinatus, distributing branches to acromiotrapezius, spinotrapezius, occipitoscapularis, splenius, and supraspinatus muscles.

The **Axillary Artery** (Figs. 224, 235, 237, 238) is a continuation of the subclavian and is that portion included within the axilla, extending between the outer border of the first rib and the point where it leaves the axilla between the latissimus dorsi and the cutaneous maximus. Beyond this point it becomes the brachial artery.

**Branches.**—The branches of the axillary are:

> Thoraco-acromial
> Lateral thoracic
> Circumflexo-subscapular trunk

The **Thoraco-acromial Artery** (Figs. 165, 224, 237, 238) is a small branch arising from the axillary just outside the thoracic wall (Fig. 165). It sends branches ventrally to the anterior portion of rectus abdominis, to the superficial portion of pectoralis major, to the first or upper part of pectoralis minor, to the acromial region, and to the wall of the thorax. Its pectoral branches are by far the largest. As has been mentioned, in the description of the acromiodeltoid artery, if it were not for the fact that this artery in question is accompanied by the lateral anterior thoracic nerve, it would scarcely merit the name thoraco-acromial as compared with the acromiodeltoid. But since in comparative and embryological studies this relation to the nerve is regarded as the criterion for identifying the thoraco-acromial artery, one must use this term here. Inasmuch as the a. thoracalis suprema in man often arises as a branch of the thoraco-acromial, it would seem that in this case a portion of the highest thoracic may be represented by the large pectoral branches of this vessel, the remaining branches of the highest thoracic being represented by the acromiodeltoid branch of the cervical trunk.

The **Lateral Thoracic Artery** (long thoracic, or external mammary) (Figs. 165, 167, 224, 237, 238), springs from the middle of the axillary artery and runs downward along the thoracic wall with the medial anterior thoracic nerve (Figs. 165, 167) covered by latissimus dorsi and cutaneous maximus. Its branches supply a portion of rectus abdominis, the deep division of pectoralis major, the second and third portions of pectoralis minor, lymph nodes of the axilla, cutaneous maximus, and, by branches which perforate cutaneous maximus, the skin of the lateral thoracic region. Certain branches, namely those to subscapularis and serratus anterior, which in man arise from the lateral thoracic, are here replaced by branches from the subscapular artery (Fig. 238).

The **Circumflexo-subscapular Trunk** (Figs. 224, 237, 238, 240) incorporates all the remaining branches which in man generally arise independently from the axillary, namely, the *anterior humeral circumflex*, the *posterior humeral circumflex*, and the *subscapular* arteries. This relationship is occasionally seen in man but since it is the usual condition in the rat it seems best to give the name circumflexo-subscapular to this relatively constant trunk.

The circumflexo-subscapular trunk, which is short and thick, leaves the axillary artery as the latter crosses the border of teres major just lateral to the origin of the lateral thoracic artery (Fig. 238). It is directed anteriorly for a short distance and gives off a small branch to cutaneous maximus before dividing into its three main branches.

The **anterior humeral circumflex** *artery* (anterior circumflex artery) (Figs. 237, 238), leaves the circumflexo-subscapular trunk with the posterior humeral circumflex artery but branches laterally almost at once to supply the shoulder joint and the long head of biceps brachii.

The **posterior humeral circumflex artery** (posterior circumflex) (Fig. 238) arises from the circumflexo-subscapular trunk with the anterior humeral circumflex. Their common stem may be very short. Beyond the anterior humeral circumflex branch, the posterior humeral circumflex artery passes out of the axilla between the borders of subscapularis and teres major and beneath the head of biceps, winding around the humerus to the lateral surface of the shoulder, where it emerges from beneath the teres minor, between the lateral and long heads of triceps brachii, covered only by spinodeltoideus. Under cover of this muscle it sends branches to the shoulder joint, to spinodeltoideus, acromiodeltoideus, teres minor, the long head of biceps brachii, and to the three heads of triceps brachii (Fig. 239).

The **subscapular artery** is the largest branch of the circumflexo-subscapular trunk. It divides close to its origin into two main branches, the *scapular circumflex* and the *thoracodorsal arteries*.

The **scapular circumflex artery** (a. dorsalis scapulae) (Fig. 238) gives a short branch to teres major then curves around the axillary border of the scapula between subscapularis and teres major to the dorsal surface of the scapula where, under cover of spinodeltoideus (Fig. 239), it sends branches to that muscle, to infraspinatus, subscapularis, teres major and minor, and to the long head of triceps brachii.

The **thoracodorsal artery** (Figs. 224, 238, 239), with the thoracodorsal nerve (Fig. 237), follows the axillary border of the scapula giving numerous branches to the muscles of the scapular and lateral thoracic regions. Close to its origin it gives a branch to subscapularis and teres major, then as it courses dorsally along the axillary border (Fig. 239), it supplies branches to the axillary lymph nodes, to subscapularis, teres major, latissimus dorsi, serratus anterior, spinotrapezius, cutaneous maximus; branches which pierce spinotrapezius to reach the multilocular adipose tissue; and branches which perforate cutaneous maximus to supply the skin of the back.

The **Brachial Artery** (Figs. 167, 224, 232, 237, 240-242), is the continuation of the axillary artery through the brachium, from the point where it leaves the axilla (Fig. 169), between latissimus dorsi and cutaneous maximus, to the level of the elbow joint, where it curves around the insertion of biceps brachii and passes under pronator teres and flexor carpi radialis on the surface of the ulna, and divides into *ulnar* and *median arteries* (Fig. 242).

It is accompanied in this course by the basilic vein (Fig. 240) and placed at first caudal to the entire brachial plexus (Fig. 167), the medial anterior thoracic nerve cross-ing it superficially. It then runs deep to the ulnar and median nerves as far as the middle of the brachium, beyond which it lies superficial to the median (Fig. 240). In the anti-cubital fossa it has reached the lateral side of the latter nerve and passes beneath pronator teres and flexor carpi radialis to reach the volar surface of the ulna upon which it divides into the ulnar and median arteries (Figs. 232, 240, 242). There are three ways in which the artery may become superficially placed with regard to the median nerve, shown by Goeppert in 1909 to obtain in the mouse; it may pass around the ulnar side of the ulnar (Fig. 241), crossing both nerves, (dorsal to ulnar and median as described by Goeppert); it may pass around the ulnar side of the median, between the median and ulnar nerves (Fig. 240); or around the radial side of the median (Fig. 170), (ventral to ulnar and median nerves, according to Goeppert). For purposes of comparison the au-thor has designated these as cases I, II, and III respectively, and finds the following results based on ten specimens, or twenty examples, as compared with Goeppert's sixteen.

RELATION OF BRACHIAL ARTERY TO MEDIAN AND ULNAR NERVES

|  | CASE I | CASE II | CASE III |
|---|---|---|---|
|  | Around the ulnar side of the ulnar nerve | Around the ulnar side of the median and between the median and ulnar | Around the radial side of the median |
| White mouse (Goeppert).... | 25% | 25% | 50% |
| Albino rat................ | 10% | 50% | 40% |

Goeppert states that in only one specimen did he find two sides alike. The author found five specimens, or 50%, alike on the two sides.

**Branches.**—The branches of the brachial artery are:

|  |  |
|---|---|
| Profunda brachii | Ulnar collateral |
| Radial | Transversa cubiti |
| Ulnar recurrent | |

The **Arteria Profunda Brachii** (Figs. 232, 237, 239, 240) leaves the brachial artery at the lower margin of the pectoralis muscle and sends one *descending branch* to supply epitrochleo-anconeus, and the long and medial heads of triceps (Figs. 232, 240); one *ascending branch* to brachialis and all three heads of triceps (Fig. 240); then continues between the medial head of triceps and brachialis (Figs. 239, 240), supplying the latter and the lateral head of triceps.

The terminal portion of profunda brachii divides into a *posterior terminal branch* to the rete olecrani (Fig. 244) and an *anterior terminal branch* (Fig. 242) which accom-panies the superficial radial nerve into the antibrachium as the **a. collateralis radialis** (Fig. 244). Halfway down the antibachium it contributes a branch to the *ramus anastomoticus radialis* (Figs. 242, 244), and terminates by dividing into the **second** and **third dorsal metacarpal arteries**, which are reinforced by **perforating branches** from the corresponding volar digitals before dividing into **dorsal digital arteries** to the contiguous sides of digits two, three and four (Fig. 244).

The **Radial Artery** (Figs. 232, 241-243) springs from the brachial artery just above the bend of the elbow. One **recurrent ramus** (Figs. 240, 241), after supplying twigs to coracobrachialis, turns upward across biceps brachii, in company with an anastomotic vein to the cephalic, gives a cutaneous branch to accompany the musculocutaneous nerve, then runs superficially across the pectoralis muscle, to anastomose with the acromiodeltoid artery. Other twigs are given to the long head of biceps brachii, to flexor digitorum profundus, pronator teres and extensor carpi radialis. A second large branch, **ramus anastomoticus radialis** (Figs. 240-244), runs obliquely down the antibrachium, crosses the musculocutaneous nerve (Fig. 240) and curves around extensor carpi radialis. It receives a branch from a. collateralis radialis as it reaches the extensor surface of the lower half of the antibrachium, where it is reinforced by a large branch of the mediano-radial artery (Figs. 243, 244), then anastomoses with the vessels of the extensor surface of the manus and sends a branch through the proximal end of the first intermetacarpal space to the volar side of the manus before ending as the first **dorsal metacarpal artery** which divides into **dorsal digital arteries** to contiguous sides of the first and second digits (Fig. 244). Occasionally the recurrent and anastomotic rami arise separately from the brachial.

The **Ulnar Collateral Artery** (superior ulnar collateral) (Figs. 232, 240–242), springs usually from the medial surface of the brachial opposite the radial branch or branches, and turns toward the elbow. In the angle between triceps brachii and pronator teres it gives a medial superficial branch which crosses the medial epicondyle of the humerus to reach palmaris longus, and flexor carpi ulnaris (Fig. 240). The ulnar collateral then takes a deeper course to the back of the humerus where it gives branches to triceps, anconeus and the elbow joint (Fig. 240).

The **A. Transversa Cubiti** (Figs. 232, 240–242) arises from the deep surface of the brachial opposite the first branching of the median nerve, and turns laterally beneath the nerve, to run between coracobrachialis and the short head of the biceps to the radial surface of the forearm where it supplies the two heads of extensor carpi radialis.

The **Ulnar Recurrent Artery** (Figs. 232, 242) arises from the brachial artery just before it divides into its terminal ulnar and median branches, or it may arise from the ulnar artery in common with the musculo-anastomotic ramus (ulnar-radial branch). Leaving the deep surface of the brachial it runs directly toward the elbow, curving around the ulna to reach flexor carpi ulnaris and the ulnar head of flexor digitorum profundus.

The **Median Artery** (Figs. 232, 240-243) following the median nerve, runs parallel with, and between flexor carpi radialis and the superficial head of flexor digitorum profundus. It becomes superficial in the distal half of the antibrachium and gives the *mediano-radial branch* (Figs. 232, 242, 243) before reaching the carpus where it passes under the transverse carpal ligament to the manus. Here it divides into two short branches. The lateral one makes an anastomosis with the ulnar artery, the medial one with the mediano-radial. There is thus established a very imperfect **volar "arch"** which lies superficial to the flexor tendons of the digits and from which arise the first, second, third and fourth **volar metacarpal arteries** and the **proper volar digital artery** to the ulnar surface of **the fifth digit** (Fig. 243). Each of these four volar meta-carpal arteries sends a *perforating branch* (Figs. 243, 244) to the corresponding vessels of the dorsal surface, and divides into **proper volar digital arteries** to the adjacent sides of digits one, two, three, four, and five (Fig. 244). From the volar arch, or from the metacarpal vessels, small branches run to the interdigital pads.

The **mediano-radial artery** (Figs. 242, 243), leaving the median halfway down the antibrachium, follows the tendon of flexor carpi radialis to the manus, where it runs beneath falciformis, gives a **proper volar digital artery** (Fig. 243) **to** the radial side of **the first digit**, then unites with the termination of the ulnar artery to form the *volar arch* (Fig. 243), the branches of which have been described with the median.

**Branches.**—The mediano-radial sends its *first branch* across the flexor and extensor carpi radialis tendons to reinforce the radial anastomotic branch of the brachial as it runs with the superficial radial nerve to the extensor surface of the manus (Figs. 243, 244).

*A second* anastomotic vessel leaves the mediano-radial at the level of the transverse carpal ligament, runs obliquely downward over the distal end of the radius and under the falciform bone to the base of the first metacarpal where it runs beneath the extensor pollicis tendons (Fig. 243) to anastomose with a terminal branch of the dorsal inter-osseous and with a. collateralis radialis. From this latter connection spring the first and second **dorsal metacarpal arteries** (Fig. 244). A *third* anastomotic branch, arising slightly below the second, turns proximally at once, and gives a small vessel which runs beneath flexor carpi radialis and between the radial head of flexor digitorum pro-fundus and the radius, to anastomose with the volar interosseous (Figs. 242, 243). This third branch then continues up the antibrachium on the surface of the radial head of flexor digitorum profundus to join the musculo-anastomotic ramus of the ulnar artery (Fig. 242).

The **Ulnar Artery** (Figs. 232, 240, 242–244) one of the terminal branches of the brachial begins slightly beyond the elbow at the point of insertion of biceps brachii (Fig. 242). It immediately gives two large branches, then crosses the ulna, running between the ulnar and superficial heads of flexor digitorum profundus, and under flexor digitorum sublimis, to reach the ulnar nerve which it accompanies to the manus (Fig. 243), be-coming superficial halfway down the ridge of the ulna along the boundary between flexor digitorum sublimis and flexor carpi ulnaris. As it emerges from between these muscles it gives a *dorsal branch* (Fig. 243) which accompanies the dorsal branch of the ulnar nerve beneath the tendon of flexor carpi ulnaris to the extensor surface of the manus, where it supplies the **fourth dorsal metacarpal artery** and the **dorsal digital artery to** the ulnar side of **the fifth digit**. The terminal portion of the ulnar artery continues beneath the pisiform bone with the volar branch of the ulnar nerve and unites, through the median, with the mediano-radial artery to form the *volar arch* (Fig. 243). From this arch arises the small **proper volar digital artery to** the ulnar side of **the fifth digit**, and the **volar** metacarpal arteries of the first, second, third, and fourth interdigital spaces (Fig. 243). These have been described with the median artery.

**Branches.**—Close to its origin the ulnar artery gives off a large *musculo-anastomotic branch* (ulnar-radial) and the *common interosseous artery* (Fig. 232).

The **musculo-anastomotic ramus** (ulnar-radial) (Figs. 232, 242, 243) arises close to the bifurcation of the brachial into median and ulnar arteries, at the same level, or sometimes in common with the ulnar recurrent (Fig. 242). It gives small muscular twigs to flexor carpi radialis, palmaris longus, the superficial head of flexor digitorum profundus, and to flexor digitorum sublimis, then takes a deep and oblique course through the antibrachium running toward the radial side across the radial head of flexor digitorum profundus, beneath the median vessels and nerve and under flexor carpi radialis, to anastomose with a branch from the mediano-radial just above the carpus (Fig. 242).

The **common interosseous artery** (Figs. 232, 242) leaves the ulnar at right angles just below the musculo-anastomotic branch. It runs distally for a short distance on the flexor surface of the ulna, gives small deep muscular branches to pronator teres and flexor digitorum profundus, and sends a **dorsal interosseous recurrent branch** (Figs. 239, 242, 244) which runs proximally between the radius and ulna to the interval between the olecranon and the lateral epicondyle of the humerus (Fig. 239). The common interosseous then reaches the interosseous membrane and divides into a **dorsal** and **volar interosseous artery** (Figs. 232, 242) running on the extensor and flexor surfaces of the membrane respectively. The dorsal interosseous is a small vessel usually lost on the surface of the membrane, but occasionally traceable to the carpus where it anastomoses with the mediano-radial (Fig. 244). The volar interosseous runs the whole length of the membrane, finally anastomosing just above the carpus with a small branch from the mediano-radial (Fig. 243).

### Arteries of the Trunk

As the **Descending Aorta** passes through the trunk it is divided for purposes of description into *thoracic* and *abdominal* portions.

The **Thoracic Aorta** (Fig. 230) extends from the aortic arch to the diaphragm lying slightly to the left of the vertebral column at first but approaching the mid-line as it reaches the diaphragm. Its relationships to surrounding structures are similar to those in man except for the fact that in the rat the persistent left vena cava superior crosses the aorta to reach the heart and the azygos vein lies to the left instead of to the right of the aorta. Its visceral branches to the pericardium, bronchi, œsophagus and mediastinal lymph glands are, as a rule, taken over by branches of the costocervical trunk on the right side and by the pericardiacomediastinal artery on the left.

**Branches.**—The *parietal branches* of the thoracic aorta are:

Intercostals
Subcostals
Superior phrenic

Intercostal Arteries (Fig. 230). On the left about one-third of the specimens show two arteries curving anteriorly and dorsally around the longus colli muscle to supply a branch to the spinal cord and one to the dorsal musculature. These arteries evidently represent the persisting proximal portion plus the posterior branch of the third and fourth intercostal arteries. The muscular branches of these arteries to the intercostal muscles are represented by branches of the costocervical trunk. Approximately one-third of the specimens show the persistence of only the lower of these two arteries while one-third of the specimens show the usual nine pairs of aortic intercostals distributed segmentally from the fourth to the twelfth intercostal spaces. The lower two pairs run dorsal to quadratus lumborum. Each artery divides into an *anterior* and a *posterior ramus* (Fig. 230).

The **anterior ramus** runs obliquely upward toward the angle of the rib above, where it is continued outward in the costal groove, parallel with the intercostal vein and the intercostal nerve. From the angle of the rib outward, the arteries lie between the external and internal intercostal muscles.

The **posterior ramus** branches dorsally from the artery close to its origin and, running close to the vertebra and the neck of the rib, sends a **spinal branch** through the intervertebral foramen to the vertebra, spinal cord, and its membranes, and continues to the muscles of the back, sending cutaneous branches to the skin of the back.

The **Subcostal Arteries** (Fig. 230) form the tenth pair of the intercostal series but since they lie below the thirteenth rib they are known as the subcostals. Like the lower intercostals they are covered by quadratus lumborum at their origin.

The **Superior Phrenic Arteries**, arising from the lower extent of the thoracic aorta, immediately enter and supply the posterior part of the diaphragm.

The **Abdominal Aorta** (Fig. 246) begins at the point where it pierces the diaphragm and ends with its division into right and left common iliac arteries. It lies along the mid-line of the vertebral column slightly dorsal to the vena cava inferior, and crossed by the left renal vein.

Branches.—The branches of the abdominal aorta are as follows:

| Visceral branches | Parietal branches |
|---|---|
| Coeliac | Inferior phrenic |
| Superior mesenteric | Lumbar |
| Inferior mesenteric | Iliolumbar |
| Superior and inferior suprarenal | Middle caudal |
| Renal | Common iliac |
| Internal spermatic | |
| Ovarian | |

These will be described according to the order in which they leave the aorta.

The **Inferior Phrenic Arteries** (Figs. 246, 247, 249-251) enter the diaphragm from the anterior abdominal region. The two are seldom symmetrical in their origin. On the right the phrenic frequently is a branch of the inferior suprarenal while on the left it springs directly from the aorta. In either case the artery gives a **superior suprarenal branch** (Figs. 250, 251) to the suprarenal gland before turning upward to the diaphragm.

The **Coeliac Artery** (coeliac axis) (Fig. 246) is unpaired, arising from the ventral surface of the aorta at the level of the crus of the diaphragm. It has a short trunk which divides into three branches, the *left gastric*, the *lienal* and the *hepatic* (Fig. 246).

The **left gastric artery** (coronary) (Fig. 246) turns immediately toward the cardiac opening of the stomach from which it branches to both surfaces of the stomach.

The **lienal artery** (splenic) (Fig. 246) passes to the left and behind the stomach to the spleen, giving branches to the stomach and pancreas as well.

The **hepatic artery** (Fig. 246) turns right sending a branch downward toward the duodenum, before continuing to the liver. This branch divides into the *right gastric* (pyloric) artery to the lesser curvature of the stomach, and the *gastro-duodenal* branch to the greater curvature of the stomach, and to the duodenum and pancreas.

The **Superior Mesenteric** (Fig. 246), the largest of the abdominal aortic branches is unpaired. Upon leaving the aorta it crosses ventral to the vena cava inferior. It gives an **inferior pancreaticoduodenal** branch to the pancreas and the first loop of the duodenum; a series of **intestinal** branches to the jejunum and ileum; an **ileocolic** branch to the ileum, caecum, and first part of the colon; a **right colic** branch sometimes arising with the ileocolic, to supply the ascending colon; and a **middle colic** branch to the transverse colon. The latter anastomoses with a branch of the inferior mesenteric artery.

The **Renal Arteries** (Fig. 247) arise very close to the superior mesenteric, the right one, always somewhat higher than the left, and frequently above the superior mesenteric, crosses behind the vena cava inferior and runs somewhat anterior and dorsal to the renal vein. Both renal arteries give rise to **inferior suprarenal branches** (Figs. 247, 249) to the suprarenal gland, then divide into two branches as they enter the hilus of each kidney. The inferior phrenic artery may arise with the inferior suprarenal from the renal (Fig. 249). This is much more common on the right side.

The paired **Internal Spermatic Arteries** of the male (Figs. 246–248) leave the aorta a short distance below the renals, the right one arising slightly higher than the left, or they may arise from the renals. Each vessel runs obliquely downward, crossing the ventral surface of the ureters, and passes through the inguinal canal to the scrotum with the other components of the spermatic cord. It divides into two branches, the **epididymal** (Fig. 248) which supplies the head of the epididymis and the mass of fat surrounding the scrotal contents, and the **testicular** (Fig. 248), which follows an extremely tortuous course through the pampiniform venous plexus, finally dividing into terminal branches which ramify over the testis.

The paired **Ovarian Arteries** of the female (Fig. 249) have the same origin as the spermatics in the male, namely, from the aorta slightly below the renals and the right one higher than the left. They cross the ureters ventrally and just before reaching the ovaries, divide into **tubal, ovarian,** and **uterine branches** (Fig. 249) to the oviduct, ovary, and uterus respectively.

The **Lumbar Arteries** (Fig. 252) arise as five segmental pairs from the dorsal surface of the aorta. The first pair arises just below the crura of the diaphragm, the second pair from the level of the renal arteries. The third pair arises just posterior to the spermatics. The fourth pair arises below the right iliolumbar artery, while the fifth pair arises from the middle sacral artery. These arteries take a deep course, dorsally and laterally behind the psoas and quadratus lumborum, which they supply. They also send a *spinal branch* into the vertebral canal. On the right side the lumbar arteries are crossed by the vena cava inferior.

The paired **Iliolumbar Arteries** (Figs. 246, 247, 249, 252–254, 273, 274) may arise from the abdominal aorta, or the right may branch from the aorta while the left arises from the common iliac. The right one appears to be more stable, while the level of the left iliolumbar varies considerably, ranging from a position at the level of the posterior end of the kidney, above the right vessel and just below the internal spermatic (or ovarian), to a point of origin from the common iliac (Fig. 273). This seems to be irrespective of the sex of the animal.

The typical iliolumbar artery runs laterally across the ventral surface of the psoas major muscle at the outer margin of which it divides into an *iliac* and *lumbar branch* (Fig. 252). The **lumbar branch** runs antero-laterally across the quadratus lumborum muscle which it supplies, together with psoas, gives branches to the muscles of the abdominal wall and a *spinal branch* through the intervertebral foramen between the last lumbar and the first sacral vertebra to the cauda equina. The **iliac branch** (Figs. 252–254) directed more posteriorly, supplies iliacus, then, in company with the lateral femoral cutaneous nerve, crosses the anterior attachment of the inguinal ligament and penetrates the abdominal wall at the superior anterior spine of the ilium. Emerging under cover of cutaneous maximus it ramifies extensively through this muscle and the integument of the extensor surface of the thigh.

The **Inferior Mesenteric Artery** (Figs. 246, 249) branches from the ventral surface of the aorta close to the origin of the two common iliacs and divides into two main branches, the **left colic** (Figs. 246, 249) which runs anteriorly and is distributed along the wall of the descending colon; and the **superior hæmorrhoidal** (Fig. 249), the caudal continuation of the inferior mesenteric which supplies the rectum.

The **Middle Caudal Artery** (Fig. 252) corresponding to the middle sacral of man, is unpaired and continues the course of the aorta into the tail. From it arise the paired **lateral caudal arteries** (Fig. 252) extending the length of the tail.

### *Arteries of the Lower Extremity and Pelvis*

In many cases it was impossible to make a complete dissection of the lower extremity of both sides, as much of the material was injected through one femoral artery eliminating that side completely for identification of vessels.

The **Common Iliac Arteries** (Fig. 247), terminal branches of the aorta, separate just below the inferior mesenteric, run laterally along the boundary between the psoas and the ventral caudal muscles, and divide into *external iliac* and *hypogastric* arteries, or hypogastric trunk (Figs. 247, 255).

This branching of the common iliac is very variable, ranging all the way from one specimen in which there was no splitting into hypogastric and external iliac (in which case all pelvic branches arise from the common iliac), to a complete splitting of the two, extending high enough so that all pelvic arteries spring from the hypogastric trunk. Twelve specimens are not enough upon which to base any authoritative statement as to the average in such wide variation. Certain points should, however, be emphasized and these will be brought out in connection with the individual vessels.

Any attempt to follow the order of description commonly used for these vessels in man would result in a most unsatisfactory arrangement when applied to the rat owing to the much greater relative development of certain vessels and to difference of relationships correlated with the quadrupedal position. The author has therefore chosen to follow the arrangement which seems most logical from the standpoint of the dissector, and the vessels are described in approximately the order in which they are encountered.

The **External Iliac Artery** (Figs. 255, 256) is the continuation of the common iliac beyond the hypogastric trunk. It passes under the inguinal ligament and becomes the femoral in the leg.

**Branches.**—The only branch arising from the external iliac above the inguinal ligament is the *pudic-epigastric* trunk (Fig. 256). Bluntschli (1907) uses this term in his work on arteries of the lower catarrhine apes. The vessel which he describes arises below the inguinal ligament, as do the external pudic vessels in man, and has, in addition to pudic and epigastric components, a superficial circumflex iliac branch which is not included with this trunk in the rat.

The **Pudic-Epigastric Trunk** (Fig. 256) appeared in six of the ten rats examined and showed four or five component vessels, as follows:

> Deep circumflex iliac
> Inferior epigastric
> External spermatic in the male
> Superior (superficial) external pudendal
> Inferior (deep) external pudendal

In the four remaining specimens these vessels did not spring from a common trunk. The superior external pudendal arose with the deep circumflex iliac (Fig. 255), and the inferior epigastric and the inferior external pudendal arose together from the hypogastric trunk.

The **deep circumflex iliac** (Figs. 255, 273-275), leaves the pudic-epigastric trunk close to its origin and after giving off a small branch which crosses the external iliac vessels to supply the iliopsoas group of muscles, runs anteriorly in company with the lumbo-inguinal branch of the genitofemoral nerve (Fig. 273), between obliquus internus and transversus, which it supplies, together with obliquus externus and the integument.

The trunk then continues downward along the inguinal ligament with the external spermatic branch of the genitofemoral nerve (Fig. 273) and leaves the abdominal cavity through the subcutaneous inguinal ring. As it crosses the ligament one small twig, the **pubic branch** (Fig. 275) turns back over the ligament then runs beneath it to reach adductor longus and pectineus. It sends numerous small offshoots to the mass of abdominal fat and usually anastomoses with the pubic branch from the obturator. This anastomotic branch is the **obturator branch** (Fig. 275) of the inferior epigastric in man.

The **inferior epigastric artery** (Figs. 255, 260, 261, 264–267, 269), usually a branch of the external iliac in man, is never so in the rat, but arises usually from the pudic epigastric trunk, most frequently in common with the inferior external pudendal, but it may arise independently, or with the superior external pudendal. In the absence of the trunk, the inferior epigastric arises from the hypogastric, as is occasionally the case in man, except that, in man, it is a branch of the obturator, whereas, in the rat, it is a branch of medial femoral circumflex. This is due to the fact that the medial femoral circumflex in the rat is taken over by the hypogastric trunk and arises well above the inguinal ligament instead of from the femoral as in man. The branches of the inferior epigastric ramify extensively to all the slips of rectus abdominis (Figs. 266, 267).

The **external spermatic artery** (Figs. 255, 256, 259–263, 269) is associated with the inferior epigastric, generally springing from the pudic-epigastric trunk slightly above or below the inferior epigastric, or arising in common with the latter from the hypogastric. It supplies the cremaster muscle.

In the case of the external pudendal vessels the author has departed from the B.N.A. terms, superficial and deep, in favor of the older comparative anatomy terms superior and inferior since these seem more appropriate for the rat.

The **superior** (superficial) **external pudendal** (Figs. 257–266) varies greatly in its origin. In the majority of cases it arises from the pudic-epigastric trunk independently or occasionally in common with inferior epigastric, or inferior external pudendal arteries. In the remaining cases it arises from the external iliac, either independently or with the deep circumflex iliac. It supplies branches in the female to the inguinal portion of the mammary gland (Fig. 268) and surrounding fat; one branch, the **artery of the bulb of the vestibule** to the vagina (Figs. 264, 265); and an **anterior labial branch** to the labia (Fig. 264). It supplies the prepuce, the preputial gland and surrounding integument in the male.

The **inferior** (deep) **external pudendal** (Figs. 256, 259–266, 275, 282) is also variable in its origin. When it arises from the pudic-epigastric trunk it is usually associated with inferior epigastric but may arise independently or with superior external pudendal. If the elements of the trunk are separated this vessel arises with the inferior epigastric branch from the medial femoral circumflex branch of the hypogastric artery. It supplies an **inferior hæmorrhoidal artery** (Figs. 249, 264, 265, 276) to the anus; a small muscular branch to semitendinosus and gracilis. In addition to these it supplies, in the female, a **posterior labial branch** (Figs. 265, 275) to the labia of the vagina, and twigs to the fat and mammary gland of the inguinal region. In the male it supplies an **anterior scrotal branch** (Figs. 259, 261), and small arteries to the bulbo- and ischiocavernosus muscles and to the base of the penis (Fig. 282).

The **Hypogastric Artery** or **Trunk** (internal iliac artery) (Figs. 255, 256, 273–275) is a large trunk, arising from the common iliac at the level of the sacro-iliac joint.

It can scarcely be said to divide into anterior and posterior trunks as in man, since the iliolumbar, lateral sacral, and superior gluteal arteries which constitute the posterior trunk in man, never arise together in the rat. In fully half the specimens observed the superior gluteal arises as an independent vessel from the common iliac above the hypogastric trunk (Fig. 256). In the remaining half of the specimens it is a branch of the hypogastric (Fig. 275), but appears so much deeper in dissection that it will be described last.

| Visceral branches | Parietal branches |
|---|---|
| Superior vesical | Obturator |
| Inferior vesical | Medial femoral circumflex |
| Deferential—in the male | Inferior epigastric |
| Uterine—in the female | External spermatic |
| Vaginal—in the female | Inferior external pudendal |
| Middle hæmorrhoidal | Lateral femoral circumflex |
| Internal pudendal | Inferior gluteal |
| | Superior gluteal |

A slightly different order will be followed in description, the internal pudendal being described with the inferior and superior gluteal arteries since this follows more closely the order encountered in dissection.

The level at which these several branches leave the hypogastric trunk varies greatly, making it practically impossible to describe their relationships in more than a general way.

**The Superior Vesical Artery** (Figs. 249, 255, 263, 269–275) arises generally from the ventral surface of the common iliac at the root of the hypogastric artery or trunk (Fig. 263) in such a way that it is often difficult to say whether it should be listed as a branch of the latter, but since it represents the original stem of the fetal hypogastric it is included here under branches of the hypogastric. It is separated from the rest of the hypogastric trunk by the common iliac vein (Figs. 263, 269, 273). In the one case where there was very slight splitting of common iliac into external iliac and hypogastric, the superior vesical arose quite independently from the common iliac, and in a single instance it arose from the external iliac (Fig. 278) in common with the deep circumflex iliac and superior external pudendal.

The superior vesical runs ventrally, giving off the following branches:

In the male, the **inferior vesical artery** (Figs. 269–271, 277) which takes a deep course beneath the proximal end of the ductus deferens, gives a branch to the ventral prostate, then reaches the sides of the bladder and runs well up over the fundus; the **deferential artery** (the artery of the ductus deferens) (Figs. 247, 248, 269–271), which accompanies the ductus deferens to the testis; an **ureteric branch** to the ureter (Figs. 270, 271); and branches to the seminal vesical, coagulating gland, the posterior prostate, and the gland of the ductus deferens before ending on the dorsal surface of the base of the bladder (Fig. 270).

In the female, the superior vesical furnishes one large **uterine artery** (Figs. 249,

273–276) which runs upward in the mesovarium, supplying the uterus and anastomosing with the ovarian artery; **cervical branches** to the cervix of the uterus (Fig. 274); and several **vaginal arteries** (Figs. 249, 268, 273–275) which give off small **urethral** and **ureteric** vessels (Figs. 274, 275). The vaginal arteries in the female correspond to the vesical of the male. Some of these vaginal and urethral branches may be taken over by the hypogastric trunk, in which case they arise with the obturator.

The **Middle Hæmorrhoidal Artery** (Figs. 249, 255, 256, 265, 273, 275, 276) is supplied by the hypogastric artery as a branch of the obturator. It runs downward along the sides of the rectum anastomosing with the corresponding vessel of the opposite side. In fact these sometimes unite to form a single median vessel on the wall of the rectum (Fig. 276).

The **Obturator Artery** (Figs. 255, 269, 273–277) is the most ventral component of the hypogastric trunk and is one of its two terminal branches. The medial femoral circumflex forms the other one, with a few exceptions where the internal pudendal arises lower down the trunk and thus becomes included with the obturator.

The obturator gives rise to the *middle hæmorrhoidal artery* described above. It gives branches to the flexors and abductors of the tail with which it lies in contact. Occasional specimens show small ureteric, urethral, prostatic, or vaginal twigs (Fig. 273).

The **pubic branch** (Figs. 274, 275), arises near the origin of the obturator artery and is occasionally taken over by the medial femoral circumflex. It crosses the brim of the pelvis and sends twigs to obliquus internus and adductor longus, and may anastomose with the pubic branch of the inferior epigastric. This anastomotic branch is termed the **obturator branch** (Fig. 275) of inferior epigastric. Immediately after giving rise to the middle hæmorrhoidal, the obturator passes between the abductor caudae externus and internus, or through the latter, and leaves the pelvic cavity to reach the obturator membrane, which it supplies, together with the caudal abductors and obturator internus (Fig. 283). It then divides into *anterior branches* (Figs. 282–285) and *posterior branches* (Figs. 282, 283, 285) which pierce the obturator membrane. The **anterior branch** after giving off the **acetabular branch** (Figs. 283, 285), which passes through the anterior end of the obturator foramen with the corresponding nerve, to the hip-joint, sends small twigs to obturator externus and quadratus femoris, then turns medially between the ventral margin of the pubis and obturator externus to reach gemellus inferior and superior and piriformis. The **posterior branch** runs along the surface of the obturator membrane and penetrates it to supply obturator externus.

The **Medial Femoral Circumflex** (Figs. 255, 269, 273–275, 286–289, 297) which has its origin below the inguinal ligament as a branch of the femoral artery in man, arises at a much higher level in the rat, and with few exceptions forms one of the terminal branches of the hypogastric artery, the obturator forming the other. Occasionally the lateral femoral circumflex arises in common with the medial (Fig. 269), but it usually is a direct branch of the hypogastric (Fig. 277).

From the hypogastric, the medial femoral circumflex extends laterally parallel to the psoas muscles, then leaves the pelvic cavity through the notch just anterior to the ilio-pectineal eminence (Fig. 297). Passing between psoas and pectineus it crosses the shaft of the femur slightly below the lesser trochanter (Figs. 286, 288). Under cover of pectineus and adductor longus it sends branches to both these muscles, and to adductor magnus and brevis (Fig. 286). One branch turns upward, crosses the insertion of quadratus femoris, which it supplies (Fig. 288), then running over the dorsal border of this muscle extends downward along the shaft of the femur to the insertion of gluteus maximus and medius, and adductor brevis to which it gives twigs (Figs. 289, 290).

After giving rise to this vessel, the medial femoral circumflex continues posteriorly along the border between quadratus femoris and adductor brevis (Fig. 289) crossing the medial surface of caudo-femoralis, biceps femoris, and semimembranosus, to all of which it sends branches, as well as to semitendinosus and obturator externus (Fig. 288).

In a considerable number of cases in which there is no pudic-epigastric trunk, the medial femoral circumflex gives rise to certain of its components, namely, inferior epigastric, external spermatic and inferior external pudendal.

The **Lateral Femoral Circumflex Artery** (Figs. 255, 291–295, 297) like the medial, springs from the hypogastric artery in the rat, instead of from the femoral as in man. Occasionally it arises in common with the medial femoral circumflex, but as a rule takes its origin from a higher point of the hypogastric. It springs from the dorsal surface of this vessel and leaves the pelvic cavity almost at once, passing under psoas major and dividing into an *ascending* and a *descending* branch (Fig. 294), as it crosses the ventral border of the ilium to the thigh.

The **ascending branch** (Figs. 292–294) divides into a dorsal and a ventral one. The *ventral* component crosses the lateral surface of the iliacus beneath the glutei (Fig. 292) and becomes superficial on the medial aspect of the thigh, appearing from between iliacus and rectus femoris (Fig. 291). It sends twigs to iliacus before distributing its terminal branches to gluteus maximus, tensor fasciae latae, obliquus externus, vastus lateralis, rectus femoris and vastus medialis (Fig. 291).

The *dorsal* division of the ascending branch gives twigs to iliacus, then runs under gluteus medius, which it supplies, together with gluteus minimus and gemellus superior (Fig. 294).

The **descending branch** (Figs. 294, 295) of the lateral femoral circumflex artery runs along the tendon of insertion of iliacus, crosses the medial surface of the head of rectus femoris, then passes laterally between the latter and vastus lateralis under cover of the glutei. It supplies branches to iliacus, rectus femoris, vastus lateralis, vastus intermedius, gluteus medius, gluteus minimus and the hip joint.

The **Internal Pudendal Artery** (Figs. 178, 255, 263–265, 269, 277, 278, 282, 283, 289, 296, 297, 299, 301, 302), though logically a visceral branch of the hypogastric, is included at this point among the parietal branches because of its order in dissection.

It springs from the dorso-medial surface of the hypogastric trunk in close proximity but slightly below the superior gluteal artery. Its level of origin varies greatly depending upon the extent of the embryological splitting between the hypogastric and external iliac vessels. Where this split runs well up between these two vessels the internal pudendal arises lower down on the hypogastric (Fig. 277). Where the split is slight (Fig. 256) the internal pudendal artery may be left as a branch of the common iliac. The same thing is true of the superior gluteal artery and indirectly of the inferior gluteal which is a branch of the internal pudendal in the rat.

Springing from the dorso-medial surface of the hypogastric, the internal pudendal runs parallel with the superior gluteal artery but separated from it by the internal pudendal vein (Figs. 269, 277). These two arteries and the vein leave the pelvic cavity by passing between flexor caudae brevis and abductor caudae internus. The internal pudendal continues posteriorly beneath the caudal abductors along the medial surface of the pelvic bone in company with the lumbo-sacral trunk or sciatic nerve (Fig. 296). At the level of the posterior border of piriformis the artery gives off the **inferior gluteal artery** (Figs. 296, 297) which leaves the pelvis through the lower part of the greater sciatic foramen to follow the sciatic nerve for a short distance to the thigh (Fig. 299).

Beyond the origin of inferior gluteal the internal pudendal runs medially accompanied by the corresponding nerve from the sacral plexus (Fig. 296), and upon emerging from the ischiorectal fossa becomes the *artery of the penis* in the male (Figs. 263, 278–280, 297), the *artery of the clitoris* in the female (Figs. 264, 265, 275).

The **artery of the penis** after giving off a **urethral branch** (Fig. 279), a twig to ischiocavernosus, and the **artery of the urethral bulb** (Fig. 279), which also supplies the bulbo-urethral gland and levator ani, divides (Fig. 279) into the **dorsal artery of the penis** (Fig. 280) and the **deep artery of the penis** (artery of the corpus cavernosum). The **artery of the clitoris** (Figs. 264, 265, 275) divides into the **dorsal artery of the clitoris** (Figs. 264, 275), which gives off a **urethral branch** (Figs. 264, 265), and the **deep artery of the clitoris** (artery of corpus cavernosum of the clitoris) (Figs. 264, 265) from which a few vaginal branches arise (Fig. 264).

The **Inferior Gluteal Artery** (Figs. 289, 296–302) springs from the internal pudendal at the posterior border of piriformis, and leaves the pelvic cavity through the lower part of the greater sciatic foramen, giving off the *posterior scrotal artery* before accompanying the sciatic nerve to the thigh (Fig. 296).

The **posterior scrotal artery** (Figs. 189, 190, 282, 289, 296–298, 301) continues the direction of the main vessel and crosses the medial surface of caudo-femoralis, biceps femoris, and semitendinosus to which it supplies twigs. With the posterior scrotal branch of the perineal nerve, it runs between the base of the tail and the tuberosity of the ischium to the posterior wall of the scrotum (Figs. 189, 190). Occasionally the posterior scrotal artery may be a branch of internal pudendal as is generally the case in man (Figs. 296, 301).

In its course to the thigh, the inferior gluteal artery runs between piriformis and gemellus superior (Fig. 302) then along the dorsal surface of the sciatic nerve, across the lateral surface of obturator internus and gemellus inferior (Figs. 289, 302). It supplies a small twig to obturator internus and the gemelli, the **arteria comitans nervus** ischiadici (Fig. 302) to the sciatic nerve, then divides into branches to gluteus maximus, caudo-femoralis, the various parts of biceps femoris, and to the integument of the flexor surface of the thigh.

The **Superior Gluteal Artery** (Figs. 178, 255, 256, 269, 274, 275, 277, 292, 296–302) arises from the dorso-medial surface of the hypogastric above the internal pudendal. Like the internal pudendal, its position depends upon the extent of embryological splitting between the external iliac and the hypogastric, and in approximately one half of the cases, where this separation is not extensive, the superior gluteal arises independently from the common iliac (Fig. 256). Its relation to the superior vesical is difficult to state with any accuracy since the latter is so variable in its origin and the superior gluteal may arise above (Fig. 269) or below it (Fig. 277).

From its origin, which is covered by the common iliac vein, the superior gluteal artery runs posteriorly for a short distance, parallel with the internal pudendal artery and separated from it by the internal pudendal vein (Fig. 277). Together these three vessels pass between flexor caudae longus and abductor caudae internus, passing medially and coming to lie dorsal to the sciatic nerve. The superior gluteal then divides into a superior and an inferior branch (Fig. 292).

The **superior branch** (Figs. 292, 297, 300) turns anteriorly, sends a twig to flexor caudae longus, then forming a loop around the sciatic nerve, crosses the dorsal border of the ilium through the extreme anterior end of the sciatic notch, and runs forward between gemellus superior and piriformis, from which position it sends branches to both of these muscles and the gluteus medius.

The **inferior branch** (Figs. 292, 294, 297, 299, 300) forms the continuation of the main vessel. It gives several twigs to the caudal muscles as it runs posteriorly along the tail in contact with the inner surface of the lateral muscles of the hip and thigh, and parallel with the posterior scrotal branch of the inferior gluteal (Fig. 297).

In addition to the small caudal vessels just mentioned, the inferior branch of the superior gluteal gives off two arteries which might easily be confused with the inferior gluteal. The latter term has, however, been applied to the artery which accompanies the sciatic nerve since this seems the more reliable criterion in the light of embryological and primitive relationships.

One of these branches leaves the pelvis just below the posterior margin of piriformis, which it supplies, together with gluteus medius (Figs. 301, 302). Slightly posterior to this vessel is one somewhat larger which turns laterally between gluteus medius and caudofemoralis to reach gluteus maximus and the integument of the lateral surface of the hip (Figs. 301, 302). The terminal portion of the inferior branch of the superior gluteal artery runs on down the lateral surface of the tail giving a cutaneous branch to the dorsal integument of this region (Fig. 298).

### Arteries of the Lower Extremity

The **Femoral Artery** (Figs. 303–308) is the continuation of the external iliac down the medial side of the thigh. It extends from the inguinal ligament to the point where the vessel passes between adductor brevis and caudo-femoralis to the popliteal fossa where it is known as the *popliteal artery.*

Branches.—

Superficial circumflex iliac
Muscular
Superficial epigastric
Highest genicular
Saphenous (great saphenous)
Profunda femoris and perforating

**Superficial Circumflex Iliac** (Figs. 303, 305). As the femoral artery crosses the ventral border of iliacus, a large muscular branch is given off from its lateral surface. Under cover of the femoral artery and vein, inguinal branches are distributed to iliacus, psoas major and psoas minor, and to pectineus, while deeper ones accompany branches of the femoral nerve to rectus femoris, vastus lateralis, vastus medialis (Fig. 291), and vastus intermedius (Fig. 305). These deeper branches may be supplied in whole (Fig. 291) or in part (Fig. 303) by the lateral femoral circumflex artery, but are more commonly derived from the superficial circumflex iliac. Occasionally the highest genicular is taken over by these deeper branches (Fig. 303).

**Muscular Branch** (Figs. 303–308). About one third of the distance down the thigh, near the level of insertion of pectineus and adductor longus, the muscular branch springs from the medial surface of the femoral, passes either over (Figs. 305, 306), or under (Figs. 303, 304, 308) the femoral vein, and crosses the adductor muscles to the skin of the flexor region of the thigh. Upon leaving the femoral, this artery crosses the insertion of pectineus and adductor longus, passes beneath gracilis anticus, thus crossing adductor magnus, and semimembranosus, emerges between the two gracilis muscles to pass superficial to gracilis posticus and semitendinosus to the skin. In this course it gives small branches to adductor longus, adductor magnus and the two gracilis muscles (Fig. 304).

The **Superficial Epigastric** (Figs. 257, 303, 304) springs from the superficial aspect of the femoral artery halfway down the thigh, slightly below the muscular branch mentioned above. It is distributed to the skin of the medial and extensor surface of the thigh and inguinal region and ends in a terminal **anastomotic branch** which perforates the abdominal wall, at the boundary between obliquus externus and rectus abdominus, to anastomose with the inferior epigastric (Fig. 257). In the inguinal region it is distributed to the subcutaneous fat and to the mammary gland in the female.

The **A. Genu Suprema** (highest genicular) (Figs. 304, 305) arises from the deep surface of the femoral. In the majority of cases its origin is above the muscular branch to the adductors and the superficial epigastric (Fig. 304), but it may leave the femoral artery just above the knee (Fig. 305). It appears at a deep level in the dissection running downward along the shaft of the femur, between vastus medialis and the adductors to the knee. This artery in the rat represents the combined *muscular* and *articular branches* (Figs. 303–305) of the highest genicular in man, the saphenous seldom combining with them in the rat. Occasionally the muscular branch springs directly from the femoral, in which case the articular branch may arise with the superficial circumflex iliac branch of the femoral or with the superficial epigastric.

One **muscular branch** of the highest genicular runs medial to the shaft of the femur to the back of the knee, supplying caudofemoralis, adductor magnus and brevis (Fig. 305). Other muscular twigs run to vastus medialis and vastus intermedius, while the **articular branch** (Fig. 305) continues on to enter into the articular rete of the knee, representing or taking the place of the **superior medial genicular** (Figs. 304, 305) in all specimens so far examined.

The **Saphenous Artery** (Figs. 303–305, 307, 320, 322–327) is the most superficial and the most extensive branch of the femoral. It leaves that vessel just below the superficial epigastric, as the femoral takes a deeper course to the popliteal space. The saphenous, remaining superficial, runs with the saphenous nerve down the medial surface of the lower leg to the ankle, crossing adductor magnus, gracilis anticus and posticus, semimembranosus, and semitendinosus (Fig. 304). It extends along the crest and medial surface of the tibia, and communicates with the superficial plexus of the knee formed by the genicular arteries around the patella.

As the saphenous crosses the posterior border of semitendinosus it divides into two, and ultimately into five terminal branches (Figs. 303, 322), two of which run with the saphenous nerve, one to the extensor surface of the foot as the **medial tarsal** (Figs. 324, 326, 327), the other between tibialis posticus and the tibia as a **communicating branch** (Fig. 324). The other three pass behind the medial malleolus, with the lateral and medial plantar branches of the tibial nerve, to the flexor surface of the foot as *lateral plantar* and **superficial** and **deep branches** of the **medial plantar arteries** (Figs. 322, 325, 326). The distribution of these vessels will be discussed later with the arteries of the foot.

The **"Profunda Femoris" Artery** is absent in the rat. Its medial and lateral femoral circumflex components are given off from the hypogastric well above the inguinal ligament. Any true *perforating branches* are likewise absent but there is one **muscular branch** (Figs. 304–307) which represents the same elements though it does not perforate the adductors, and the term perforating has been used in one or two drawings to indicate this homology. It arises from the flexor surface of the femoral artery just before the latter runs under caudo-femoralis, and passes between adductor brevis and caudofemoralis on the lateral side, and adductor magnus on the medial side to reach semimembranosus. It supplies each of these muscles.

The **Popliteal Artery** (Figs. 306–311, 316–320, 323, 326) is very short. It is the continuation of the femoral from the point where it enters the popliteal fossa between adductor brevis and caudofemoralis, just above the medial epicondyle of the femur, to its division into anterior and posterior tibial arteries at the anterior margin of the popliteus muscle. In a single instance the popliteal artery ran medial to caudofemoralis (Fig. 307).

**Branches.—**

> Superior muscular
> External and internal sural
> Lateral superior genicular
> Medial superior genicular
> Middle genicular
> Lateral inferior genicular
> Medial inferior genicular

The **Superior Muscular Artery** (Figs. 306–313, 315–320) is the largest branch of the popliteal. It runs along the surface of the gastrocnemius muscle between the biceps femoris on the lateral and the adductors on the medial side, emerging between the hamstring muscles to the back of the thigh (Fig. 309).

It gives, first, the **superficial sural artery** (cutaneous or posterior saphenous) (Figs. 313–320, 325–327), which accompanies the small (posterior) saphenous vein and the sural branch of the peroneal nerve down the lateral surface of the lower leg, then gives off a small cutaneous branch (Figs. 316, 318, 326) before passing behind the lateral malleolus and beneath the peroneal tendons to the extensor surface of the foot where it gives off a **lateral tarsal branch** (Fig. 327) to the tarsus, then divides into **dorsal metatarsal arteries** to the third and fourth interdigital spaces (Fig. 327). These will be further described with the arteries of the foot.

The superior muscular artery also gives a small **branch to the tibial nerve** (Figs. 306, 311–313, 315–318, 326) and three larger **muscular branches** (Figs. 309–313, 316–318). The *first* proximal one of these vessels (Figs. 309, 310) often arises in common with the branch to the tibial nerve and may be found either above or below the origin of superficial sural (posterior saphenous). Its distribution is to the anterior and posterior portions of biceps femoris and the skin of the lateral surface of the thigh.

The *second muscular branch* (Fig. 309), which supplies the posterior and accessory portions of biceps femoris, arises opposite the boundary between these two muscles, and the *third*, or distal, *muscular branch* (Fig. 309) supplies the accessory head of biceps femoris and semitendinosus. The superior muscular artery, after sending a twig to the lymph node of the knee (Fig. 313), terminates as the **femoropopliteal artery** (Figs. 313–317) which accompanies the corresponding vein, turning upward between the posterior border of biceps femoris and semitendinosus and supplying branches to the integument at the back of the thigh.

The **External and Internal Sural Arteries** (Figs. 306, 313, 316–318, 320, 326) are muscular branches of the popliteal somewhat variable in their origin. They may arise by a common trunk (Figs. 308, 309, 311, 316, 320, 326), just behind the knee joint, at approximately the same point as the superior muscular branch or directly from it (Fig. 317), or they may arise as separate branches from the latter vessel (Fig. 313), or the internal sural may spring independently from the popliteal (Fig. 318), or in common with the tibial artery (Figs. 306, 319). The external sural supplies plantaris, soleus, and the lateral head of gastrocnemius (Fig. 320), while the internal sural supplies the medial head (Fig. 318).

The **Lateral Superior Genicular Artery** (Figs. 308–311, 317–319) branches from the superior muscular artery close to its origin. The artery crosses the distal end of adductor brevis to the lateral surface of the shaft of the femur just above the epicondyle under cover of biceps femoris. It sends superficial branches to adductor brevis, biceps femoris, vastus lateralis and vastus intermedius and a deep branch downward along the distal end of the femur to the knee, where it anastomoses with the other genicular vessels (Fig. 310).

The **Medial Superior Genicular Artery** was lacking in all specimens examined so far, its place being taken by the articular branch of the highest genicular which runs downward beneath vastus medialis along the medial surface of the distal end of the femur to the circumpatellar anastomosis (Fig. 304).

The **Middle Genicular Artery** (azygos articular artery) (Figs. 308, 309, 317, 319, 320), springs from the common trunk of the tibial and peroneal arteries as it runs between the heads of gastrocnemius. It usually arises in common with the lateral inferior genicular (Fig. 319), runs laterally, then makes a loop under the popliteal artery and vein, and enters the joint capsule. It gives small branches to the lateral head of gastrocnemius and adductor brevis.

The **Lateral Inferior Genicular Artery** (Figs. 317, 319) arising usually with the middle genicular from the deep surface of the popliteal, sends a branch beneath the lateral head of gastrocnemius and around the head of the fibula to the knee, and a branch which supplies popliteus, biceps femoris and the lateral head of gastrocnemius, before reaching the plexus around the patella.

The **Inferior Medial Genicular Artery** (Figs. 307, 311, 319, 320) springs from the popliteal artery just above the anterior border of the popliteus muscle, or from the anterior or posterior tibial arteries, terminal branches of the popliteal. In any case the inferior medial genicular takes a deep course beneath the medial head of gastrocnemius following the anterior margin of popliteus, and runs below the condyle of the tibia to the medial side of the knee where it supplies the joint and anastomoses with the other genicular arteries.

The distal portion of the popliteal artery lies between the condyles of the femur in contact with the joint capsule. Just above the anterior border of the popliteus it divides into *anterior* and *posterior tibial arteries*.

The **Posterior Tibial Artery** (Figs. 316–320, 323) runs above the anterior tibial deep to the popliteus muscle (Figs. 317, 323). Passing distally between flexor digitorum longus and flexor hallucis longus, the posterior tibial is covered by the medial head of gastrocnemius. It is joined by a muscular branch of the tibial nerve, then distributed to popliteus, flexor hallucis longus, flexor digitorum longus, and tibialis posticus.

Several arteries which are branches of the posterior tibial in man are here taken over by other vessels, leaving the posterior tibial much reduced. The peroneal is the principal one of these and is a branch of the anterior tibial in the rat.

The medial inferior genicular sometimes springs from the posterior tibial.

The **Anterior Tibial Artery** (Figs. 316–321, 323, 324, 327) runs deep to the popliteus muscle (Figs. 317, 323), and between the tibia and fibula close to their articulation (Fig. 324), coming to lie on the lateral surface of the interosseous membrane (Fig. 321).

Branches.—

> Anterior tibial recurrent
> Peroneal
> Muscular

The **anterior tibial recurrent branch** (Figs. 321, 324) arises just below the head of the fibula, sends twigs to peroneus longus and brevis and runs anteriorly, perforating tibialis anterior which it supplies, to reach the knee where it anastomoses with the genicular arteries.

The **peroneal artery** (Figs. 317, 319–321, 324) leaves the anterior tibial slightly distal to the anterior tibial recurrent. Turning toward the fibula it gives off the **fibular branch** (Fig. 321) which runs between peroneous longus and brevis, to the peroneus digiti quarti and quinti and soleus muscles. The peroneal then continues downward along peroneus longus and brevis to which it gives branches, is joined by the superficial peroneal branch of the common peroneal nerve, then runs with peroneus brevis along the fused portion of tibia and fibula, and between the tendons of peroneus digiti quarti and quinti a short distance above the lateral malleolus. Here it becomes quite superficial and passes behind the malleolus to be distributed to the lateral surface of the heel and foot (Fig. 324).

Beginning with that part of the artery which is accompanied by the superficial peroneal nerve and runs along the fused tibia and fibula, the distal portion of the peroneal probably corresponds to the *perforating branch* (anterior peroneal) in man, though it is in no sense a perforating artery in the rat. This is undoubtedly due to the fact that the peroneal artery in the rat is a branch of the anterior tibial which crosses the anterior border of the interosseous membrane before giving rise to the peroneal. The latter lies throughout its extent on the lateral side of the membrane, and it would seem much more appropriate to call its distal portion **anterior peroneal** rather than perforating. This vessel ends as the **superficial cutaneous artery** of the dorsum of the foot. (Fig. 327.)

Beyond the origin of the peroneal, the anterior tibial runs along the lateral surface of the shaft of the tibia on the interosseous membrane accompanied by the deep peroneal nerve (Fig. 321).

Its **muscular branches** (Figs. 321, 324) are given off at intervals to tibialis anterior, extensor digitorum longus, and extensor hallucis longus, and one or two perforate the interosseous membrane to anastomose with branches of the posterior tibial.

The artery passes under extensor proprius hallucis and runs with it beneath the annular ligament of the tarsus to the extensor surface of the foot where it ends as the **dorsal artery of the foot** (Figs. 324, 327).

The **Dorsal Artery of the Foot** (Figs. 324, 327) is a continuation of the anterior tibial artery and begins where the latter passes under the annular ligament to the tarsal region. A short distance below this level it is joined by an **arcuate branch** from the **medial tarsal** (Fig. 327), one of the terminal branches of the saphenous artery.

From the short arch thus formed the first and second **dorsal metatarsal arteries** arise (Fig. 327). The first runs distally along the interosseous muscle, to which it sends twigs. It supplies **dorsal digital arteries** (Fig. 327) to the adjacent sides of the first and second digits, and a **perforating branch** to the plantar metatarsal artery (Figs. 326, 327).

The second dorsal metatarsal artery accompanies the deep peroneal nerve beneath the extensor tendons of the digit, and runs down over the interossei. In the cleft between metatarsals two and three, it gives a **perforating branch** (Figs. 326, 327) to the corresponding plantar metatarsal vessel, then continues to the medial side of the third digit. It anastomoses with the superficial cutaneous branch of the anterior peroneal and with the second plantar metatarsal by an **anterior perforating branch** (Figs. 326, 327).

The **third and fourth dorsal metatarsal arteries** (Fig. 327) are the terminal branches of the superficial sural artery. The third dorsal metatarsal vessel passes beneath the extensor tendons of digits one and two, and reaches the interossei between the third and fourth metatarsals where it gives a **perforating branch** to the third plantar metatarsal artery before continuing to the medial side of the fourth digit. It sends an anterior perforating branch to the plantar digital artery of the third digit.

The fourth dorsal metatarsal artery runs under the extensor tendon of the fifth digit and under the tendon of peroneus digiti quarti, then sends a **perforating branch** to the fourth plantar metatarsal artery before dividing into **dorsal digital arteries** to the adjacent sides of digits four and five (Fig. 327).

The *superficial* and *deep branches* of the **medial plantar artery** (Figs. 322, 325, 326), are continuations of the saphenous artery. They pass behind the medial malleolus, with the medial plantar nerve, to reach the plantar surface of the foot.

The **superficial branch** (Fig. 326), accompanied by a branch of the nerve, runs along the medial border of the foot and supplies a **plantar digital artery** (Figs. 326, 327) to the medial side of the first digit, several superficial branches to the skin, and anastomotic vessels to the deeper plantar arteries.

The **deep branch** of the medial plantar (Figs. 322, 325, 326), extends distally parallel with, and medial to, flexor digitorum brevis. Where this muscle breaks up into tendons the artery turns laterally and unites with the lateral plantar artery to form the **plantar arch** (Fig. 326), from which the **plantar metatarsal arteries** arise. The plantar arch lies superficial to the flexor digitorum brevis in the rat instead of deep to it as in man.

The *lateral plantar artery* (Figs. 322, 325, 326) one of the five terminal branches of the saphenous artery, joins the plantar nerves in the interval between the calcaneus and medial malleolus. It crosses the medial plantar nerve, gives a calcaneal communicating branch, then runs with the lateral plantar nerve beneath the origin of flexor digitorum brevis to the lateral side of the foot. It sends a **superficial branch** to the skin. The **deep branch** of the lateral plantar (Fig. 326), joins the deep branch of the medial plantar to form the **plantar arch** (Fig. 326), which lies superficial to the flexor muscles and tendons of the foot, and therefore, should not be called the deep plantar arch as in man. In addition to the plantar metatarsal arteries, several small twigs are given off from the arch to anastomose with the superficial cutaneous branches of the lateral and medial plantar vessels.

Five **plantar metatarsal arteries** (Fig. 326) spring from the plantar arch. The first four divide in their four. respective interdigital spaces, send an **anterior perforating branch** to the corresponding dorsal metatarsal artery, then divide into **plantar digital arteries** (Fig. 326) to supply adjacent sides of the digits.

The fifth dorsal metatarsal supplies only the lateral side of the fifth digit. The medial side of the first digit is supplied by the superficial branch of the medial plantar.

Distally the two plantar digital arteries of each toe are united by three communicating branches, one which crosses the middle phalanx under the tendon of flexor digitorum longus, one which crosses the terminal phalanx beneath the insertion of this tendon, and supplies the nail, and a terminal loop imbedded in the apical pad of each digit (Fig. 326).

## Veins

Since the number of drawings showing veins only, is so small, it has been thought best to include them with the arteries of the same area.

### PULMONARY

**Pulmonary Veins** (Fig. 112) number one from each lobe of the lung, one on the left, four on the right. Each of these is formed by the confluence of vessels from the pulmonary lobules. Upon leaving the lung the pulmonary veins enter the left auricle, those from the right side running dorsal to the right vena cava superior, those of the left side ventral to the descending aorta and dorsal to the left vena cava superior. On the left the pulmonary veins are crossed by the curve of the azygos vein just as it enters the left vena cava superior.

In their course from the lobes of the lungs the pulmonary veins lie anterior and slightly dorsal to the bronchi, whereas the pulmonary arteries lie ventral and somewhat posterior to the bronchi.

### SYSTEMIC

The **Systemic Veins**, or those which convey the blood from the whole body except the lungs and digestive tract, back to the heart, will be described in the following order:

Veins of the heart
Veins of the head and neck
Veins of the upper extremity
Veins of the thorax
Veins of the abdomen and pelvis
Veins of the lower extremity

The **Veins of the Heart** drain the muscle itself, ascending over its walls to enter the right auricle.

The **Veins of the Head and Neck** will be described in the following order:

> The veins of the exterior of the head and face
> The veins of the neck
> The superficial veins of the brain

*Veins of the Exterior of the Head and Face*

| | |
|---|---|
| Supraorbital | Superficial temporal |
| Nasal | Internal maxillary |
| Frontal | Posterior facial |
| Angular | The pharyngeal plexus |
| Anterior facial | The pterygoid plexus |

The **Supraorbital Vein** (Fig. 205) begins above the eye where it communicates with the superficial temporal vein. Passing forward and downward with the angular artery past the anterior end of the zygomatic arch, it unites with the nasal and frontal veins to form the *angular vein* (Figs. 205, 207).

The **Nasal Vein** (Fig. 205), beginning among the terminal branches of the lateral nasal artery on the tip and wing of the nose, sends a *lateral nasal* tributary downward over the side of the nose to the superior labial vein, before continuing posteriorly along the margin of the nasal bone to join the supraorbital and frontal veins in front of the anterior end of the zygomatic arch, to form the *angular vein* (Fig. 205).

The **Frontal Vein** (Fig. 205) is a small vessel arising on the dorsal surface of the frontal bone and joining the supraorbital and nasal veins to form the angular vein.

The **Angular Vein** (Fig. 205), formed on the side of the face, a short distance in front of the corner of the eye, by the confluence of supraorbital, frontal and nasal tributaries, follows the course of the angular artery ventrally as far as the corner of the mouth where it is joined by the superior labial to form the *anterior facial vein*. Before joining the superior labial, the vein receives one or more muscular tributaries from the masseter muscle, and a vessel from the levator labii superioris and vibrissae pad.

The **Anterior Facial Vein** (Figs. 148–151, 203–208, 232–235) is a continuation of the angular vein downward from the corner of the mouth. It follows the course of the external maxillary artery receiving tributaries corresponding to the branches of the artery (Figs. 205, 207, 208). It runs ventrally along the border of the masseter muscle, around the inferior border of the mandible to its medial surface, where it runs in the groove between the masseter and the anterior belly of the digastric muscle as far as the hyoid. It then enters the cervical region and turns slightly laterally to reach the posterior facial vein with which it unites to form the external jugular (Figs. 148, 150, 232–235). In man the anterior facial vein unites with a branch from the posterior facial and enters the internal jugular, the largest vein of the neck. In the rat the external jugular is the principal vein of the neck, and receives both the posterior and anterior facial veins, neither of which connect with the internal jugular vein.

Tributaries.—*On the lateral surface* the anterior facial vein receives the following vessels: the *superior labial vein* (Fig. 205); a *muscular branch* (Figs. 205, 207), which takes its origin along the roof of the mouth in communication with the posterior and superior alveolar vein and runs forward and downward through the masseter muscle receiving a tributary from the corner of the mouth just before joining the anterior facial; a second *muscular* tributary, from the buccinator and masseter muscle, which communicates with vessels from the inferior palpebral and posterior superior alveolar vein and the pterygoid plexus (Fig. 207); an *anastomotic branch* (Fig. 214), which begins in conjunction with the submental, runs along the medial surface of the mandible to the anterior end of the alveolar ridge, where it crosses to the lateral surface and receives twigs from the lining of the mouth before opening into the anterior facial vein, or into the inferior labial; the *inferior labial vein* (Figs. 205, 207, 214) which receives the *mental branch* (Fig. 207) as it emerges through the mental foramen from the mandibular canal where it connects with the inferior alveolar branch of the internal maxillary (Fig. 208), and frequently receives the anastomotic branch from the submental (Fig. 207); and finally a *cutaneous vein* from the integument and platysma of the mid-ventral region of the lower jaw (Figs. 205, 207).

*Below the mandible* the anterior facial vein receives; the *submental vein* (Figs. 197, 198, 204), the *masseteric vein* (Fig. 204), four or five small *glandular* branches (Figs. 234, 330), from the lymph nodes of the submaxillary region, from the parotid gland, the platysma and integument of the neck; and just as the anterior facial vein enters the cervical region, it receives a communicating vein from the pharyngeal plexus (Figs. 203, 204, 208, 233). Beginning in the pharyngeal plexus through which it also communicates with the pterygoid plexus, this vein follows the course of the tonsillar artery, collecting blood from the lining of the pharynx, the root of the tongue, the superior constrictor of the pharynx, pterygoideus internus, styloglossus, hyoglossus, and digastricus (Figs. 203, 204). Reaching the tendinous junction of the anterior and posterior bellies of the digastric, it unites with the hyoid vein (Fig. 203) which emerges from beneath the tendon of digastricus (Fig. 204). The *hyoid vein* drains the superficial hyoid region, receiving twigs from anterior digastricus, sternohyoideus, mylohyoideus, and omohyoideus (Fig. 203). Proceeding for a short distance posteriorly, the vein receives a glandular tributary from the submaxillary and major sublingual glands, before emptying into the anterior facial vein (Fig. 204). The anterior facial then unites with the posterior facial to form the external jugular.

The **Superficial Temporal Vein** (Figs. 148–151, 201–203, 205–207) begins on the frontal region of the skull where it communicates with the supraorbital branch of the angular vein. Passing posteriorly and ventrally behind the orbit, it crosses the posterior root of the zygomatic arch and the tempero-mandibular joint, and unites with the internal maxillary to form the posterior facial vein (Figs. 205–207).

Tributaries (Fig. 205).—The superficial temporal vein receives the *orbital vein* made up of palpebral and lacrimal veins from the upper lid and intraorbital lacrimal gland respectively; the *transverse facial vein*, which follows the corresponding artery and receives the inferior palpebral vein; the *middle temporal vein* from the temporalis muscle; and unites with the *internal maxillary vein* to form the posterior facial vein.

The **Internal Maxillary Vein** (Figs. 205–207, 214) receives tributaries corresponding very closely to branches of the internal maxillary artery. Arising in a deep position under cover of the temporalis and masseter muscles, it appears more superficially in the space beween the neck of the mandible and the tympanic bulla (Fig. 206) where it communicates with the inferior alveolar vein before continuing posteriorly to unite with the superficial temporal to form the posterior facial vein.

**Tributaries.**—The internal maxillary vein receives the *transverse sinus* which emerges through the postglenoid foramen (Figs. 206, 207, 214); an *articular vein* from the temporomandibular joint and the tympanum (Figs. 201, 223); the *posterior deep temporal* (Figs. 205–207); the *anterior deep temporal* (Figs. 206, 214), and through it the *masseteric* and *buccinator vein* (Fig. 206); and a large *communicating vein* from the inferior alveolar vein (Figs. 205–207, 214), which is, in turn, a component of the pterygoid plexus. Through the anterior deep temporal the internal maxillary makes a second communication with the pterygoid plexus (Fig. 214), through the posterior superior alveolar vein. In some instances the anterior deep temporal vein, instead of joining the internal maxillary (Fig. 207), joins the posterior superior alveolar vein. (Figs. 214, 215), to enter the large anastomotic channel between the internal maxillary and the inferior alveolar veins.

The **Posterior Facial Vein** (Figs. 197, 201–203, 205–208, 232–236) is formed just below the ear by the union of the superficial temporal and the internal maxillary veins already described.

The identification of the venous components and tributaries of the posterior facial vein may be debatable since they do not conform with those of man sufficiently to allow any close comparison. It has seemed best, therefore, to name the veins, as far as possible, according to the arteries with which they are associated, and to designate them as main channels or as tributaries, according to their relative size and method of anastomosis.

**Tributaries.**—The posterior facial vein receives the *anterior auricular vein* from the ear and exorbital lacrimal gland (Figs. 197, 201, 205, 206, 233–235); the large *vein from the pterygoid plexus* (Figs. 197, 201, 202, 205, 206, 215); and a small *masseteric vein* (Figs. 205, 206); then, crossing the sternomastoid muscle, it receives the *posterior auricular vein* (Figs. 197, 205, 234, 235, 239), and unites with the anterior facial vein to form the external jugular (Fig. 205). This is quite different from the condition found in man where the internal jugular is the principal vein of the neck and receives the anterior facial and a branch of the posterior facial vein.

The **Pterygoid Plexus** (Figs. 197, 201, 202, 205–208, 214, 215), located medial to the mandible, is made up of the following veins; *posterior superior alveolar, inferior alveolar, pterygoid,* and *posterior meningeal* (Figs. 215, 216), and sometimes *anterior deep temporal* (Fig. 214).

The **posterior superior alveolar vein** (Figs. 207, 213–215), anastomoses with the veins of the orbit (Figs. 207, 213), thus furnishing connection with the sphenopalatine and infraorbital veins which in man are described as part of the pterygoid plexus, but which are tributaries of the ophthalmic vein in the rat.

The **inferior alveolar vein** (Figs. 197, 201, 205–208, 214, 215), begins in the mandibular canal where it is continuous with the mental branch of the inferior labial vein (Fig. 214). In its course through the canal it receives a branch from the lower incisor tooth, then emerging on the medial surface of the mandible, through the mandibular foramen, it receives a large communicating vein from the internal maxillary, then joins the vein from the pterygoid plexus.

The **vein from the pterygoid plexus** emerges as a distinct vessel in the region of the external pterygoid plate (Fig. 215), receiving, at this point, a *posterior meningeal vein*, leaving the interior of the skull through the petrotympanic fissure, and a small twig from pterygoideus externus. Continuing across the lateral surface of the tympanic bulla, it is joined just ventral to the external acoustic meatus, by the *inferior alveolar vein*, and by the *vein from the pharyngeal plexus* before entering the posterior facial vein (Figs. 201, 202, 205, 206).

The **Pharyngeal Plexus** is made up of a network of veins beginning in the mucous membrane of the pharynx and receiving tributaries from the pterygoid canal, the soft palate, and the auditory tube (Figs. 215, 216). It gives a communicating loop to the pterygoid plexus and its principal drainage is into the posterior facial vein by way of the vein from the pterygoid plexus.

The **vein from the pharyngeal plexus** (Figs. 197, 201, 202, 205, 206, 208, 215), receives the vessels corresponding to the branches of the ascending palatine artery, namely from the external and internal pterygoid muscles, from the auditory tube, and from the pharynx, and sends a small vein along the course of the ascending palatine artery to join the hyoid vein and open into the anterior facial vein (Figs. 203, 204, 216). The principal drainage from the pharyngeal plexus, however, takes a deeper course and proceeds laterally across the ventral surface of the bulla, receiving a tributary from the larynx corresponding to the laryngeal branch of the superior thyroid artery (Fig. 202), then joins the vein from the pterygoid plexus just below the ear. The latter vessel runs as a short thick trunk from this point to its junction with the posterior facial vein.

*Veins of the Neck*

External jugular
Posterior external jugular
Anterior jugular
Internal jugular
Vertebral
Inferior thyroid

The **External Jugular Vein** (Figs. 197, 203, 205, 232–236) which in man is considerably smaller than the internal jugular, in the rat is the principal vein of the neck and is formed by the union of the posterior facial and anterior facial veins along the boundary between the sternomastoid, and clavotrapezius muscles. Following this boundary fairly closely the vein proceeds downward through the cervical region in a very superficial position, crosses the middle of the clavicle and turns somewhat medially before uniting with the axillary vein to form the subclavian (Fig. 233).

Tributaries.—The external jugular vein receives, close to its origin, two or three small vessels from the integument of the neck and upper chest regions, and a small vessel from the submaxillary lymph node. As it reaches the shoulder region it receives a larger tributary, the *posterior external jugular* (Figs. 197, 232–235), made up of veins from spinotrapezius, spinodeltoideus, and acromiodeltoideus muscles and from the integument of the shoulder and back of the neck. A vein from the sternomastoid muscle joins the external jugular just above the clavicle. As it crosses the clavicle it is joined by the *cephalic vein* (Figs. 232–235, 239–241, 245) and almost immediately thereafter by the *anterior jugular vein* (Figs. 197, 233–235, 238), which appears at a slightly deeper level between the cleidomastoid and sternomastoid muscles, and passes under the clavicle before uniting with the external jugular. Within a very short interval the external jugular joins the axillary vein and together they form the subclavian. The point of entrance of the anterior jugular varies. In some cases it enters just medial to the union of the external jugular and axillary veins and is then a tributary of the subclavian (Figs. 233, 234).

The **Posterior External Jugular Vein** (Figs. 197, 232, 233, 238, 239) is made up of two main vessels, one from the integument of the back of the neck and shoulder region, the other from the spinotrapezius, spinodeltoideus, and acromiodeltoideus muscles (Fig. 239). Curving over the shoulder it runs ventral to the clavotrapezius muscle to enter the external jugular vein about the middle of its course.

The **Anterior Jugular Vein** (Figs. 197, 233–236) begins with the junction of *transverse cervical* and *transverse scapular veins*, which unite in the angle formed by the anterior border of the clavicle and the lateral border of clavotrapezius. Passing obliquely downward through the neck in company with the cervical trunk, the anterior jugular receives a *superficial cervical vein* (Figs. 233–235), which, with the corresponding artery, appears in the interval between the sternomastoid and cleidomastoid muscles and enters the anterior jugular as the latter passes under the clavicle to unite with the external jugular vein. The anterior jugular generally unites with the subclavian vein in the space between the clavicle and first rib (Fig. 234).

The **transverse cervical vein** (Figs. 233–236) begins with the confluence of small vessels from supraspinatus, splenius, occipitoscapularis, spinotrapezius, and acromiotrapezius muscles. Following the upper border of supraspinatus, the vein curves around the lateral border of omohyoideus to unite with the transverse scapular vein to form the anterior jugular (Fig. 236).

The **transverse scapular vein** (Figs. 233–236) originates in a confluence of tributaries which follow and correspond to the terminal branches of the transverse scapular artery. It thus collects blood from omohyoideus, subscapularis, infraspinatus, and supraspinatus muscles, from the clavicle and shoulder joint, then unites with the transverse cervical vein on the ventral surface of the omohyoid muscle and together they form the anterior jugular vein (Fig. 236).

The **superficial cervical vein** (Figs. 233–235) drains only the region supplied by the muscle branches of the corresponding artery, while the parotid gland, the cervical lymph nodes, and cervical platysma and integument are drained by several small vessels which open separately into the external jugular vein (Fig. 330).

The **Internal Jugular Vein** (Figs. 197, 198, 209, 219, 227, 235) in the rat, is relatively small as compared with man. Made up of tributaries from the occipital and inferior petrosal sinuses (Fig. 219), it leaves the skull through the posterior lacerated foramen (Figs. 209, 219), runs dorsal to the stylohyoid and digastric muscles, and emerging from beneath the latter receives the occipital vein (Figs. 197, 198), then, crossing ventral to the levator claviculae muscles, continues its course down the neck, receiving the pharyngeal vein and the thyroid veins (Figs. 197, 227, 235). At the base of the neck it crosses ventral to the common carotid artery and unites with the subclavian vein to form the vena cava superior (Fig. 197).

**Tributaries.**—The internal jugular, in the rat, has no communication with the anterior or posterior facial veins as in man. Its tributaries in the neck region are the *occipital vein* (Figs. 197, 198), made up of branches corresponding to those of the occipital branch of the external carotid artery; the *pharyngeal vein* (Fig. 198) from the structures supplied by the ascending pharyngeal artery, namely, from stylopharyngeus, styloglossus, stylohyoideus, sternohyoideus, thyreohyoideus, digastricus posterior, omohyoideus, and sternothyroideus muscles, and from the pharynx; the *thyroid vein*, made up of superior and middle thyroid tributaries (Figs. 197, 227).

At the base of the neck the internal jugular receives a vessel made up of tributaries from scalenus, rectus capitis and longus colli muscles (Fig. 235); and finally, as it crosses the subclavian artery, a deep vein (Fig. 235) following the deep branch of the ascending cervical artery and collecting from the rhomboideus, levator scapulae, serratus anterior, serratus posterior, and scalenus muscles, and the deep muscles of the neck.

In man the *inferior thyroid vein* (Fig. 227) enters the innominate vein, but, in the rat, where two venae cavae superiores persist, this vein accompanies the inferior thyroid branch of a. profunda cervicalis, and enters the vena cava superior.

The **Vertebral Vein** (Figs. 197, 220, 221, 227, 235) arises in the venous plexus within the vertebral canal. Passing out of the canal through a groove or foramen in the dorsal arch of the atlas, just dorsal to the articular surface for the occipital condyle, it immediately turns to enter the vertebrarterial canal in the transverse process of the atlas. It continues downward through this canal in successive vertebrae as far as the sixth cervical. Passing through the sixth cervical it emerges from beneath the carotid tubercle of that vertebra (Figs. 197, 235) and crosses the subclavian artery to enter the vena cava superior just medial to the internal jugular vein.

The **Inferior Thyroid Vein** (Fig. 227) is a small vessel which anastomoses with the superior thyroid vein. It accompanies the inferior thyroid artery downward along the lateral surface of the trachea. Like the artery it has an œsophageal branch. Passing dorsal to the innominate artery and costocervical vessels it opens into the vena cava superior a short distance below the vertebral vein. The inferior thyroid may enter the internal jugular vein.

### Superficial Veins of the Brain

Only the superficial veins of the cerebral hemispheres and cerebellum will be described.

> Superior cerebral
> Inferior cerebral
> Anterior cerebral
> Superior cerebellar
> Inferior cerebellar

The **Superior Cerebral Veins** (Figs. 218, 220–222) averaging about nine or ten in number, arise on the lateral surface of the hemispheres, and running medially over the upper surface of the brain, enter the superior sagittal sinus. The anterior pair, extending around the frontal pole, communicates with the rather large inferior cerebral veins.

Blood from the lateral surface of the hemispheres, in the region supplied by the middle cerebral artery, is collected by tributaries of the superior and inferior cerebral veins (Fig. 220).

The **Inferior Cerebral Veins** (Figs. 218, 220) are made up of tributaries from the ventrolateral surface of the hemispheres. They communicate with the first pair of superior cerebral veins (Fig. 218), and unite with the superior petrosal sinus in the angle between cerebrum and cerebellum (Fig. 218). The superior petrosal sinus then communicates directly with the transverse sinus (Fig. 220).

The **Anterior Cerebral Vein** (Fig. 219) accompanies the azygos anterior cerebral artery in the longitudinal fissure. It receives blood from the corpus callosum. Running downward around the genu of the latter structure it forms a slight plexus located between the optic nerves and the brain, then passe on to open, with the ophthalmic vein, into the cavernous sinus.

The **Superior Cerebellar Veins** (Figs. 220, 221, 231) arise from small tributaries between the gyri of the cerebellum. These unite to form two veins which run anteriorly, one along either border of the median lobe, or vermis, to reach the transverse sinus.

The **Inferior Cerebellar Veins** (Figs. 220, 221, 231) arise as small branches on the inferior surface of the cerebellum. These tributaries unite to form two veins which run anteriorly between the lateral and parafloccular lobes of the cerebellum to enter the transverse sinus. They also communicate with the vertebral veins.

The **Sinuses of the Dura Mater** are divided into a postero-superior and an antero-inferior group. The *postero-superior sinuses* are:

> Superior sagittal     Straight        Occipital
> Inferior sagittal     Transverse

The **Superior Sagittal Sinus** (superior longitudinal sinus) (Figs. 220, 221) follows the line of the sagittal suture of the skull along the attachment of the falx cerebri from the foramen caecum to the parieto-occipital suture where it meets and becomes continuous with the transverse sinus.

The **Inferior Sagittal Sinus** (inferior longitudinal sinus) runs in the free border of the falx cerebri, receiving small vessels from the medial surface of the hemispheres. It opens into the straight sinus.

The **Straight Sinus** (sinus rectus, tentorial sinus) (Figs. 222, 231) occupies the line of junction of the falx cerebri with the tentorium, and opens into the transverse sinus. It receives the *great cerebral vein* (vein of Galen) (Fig. 231).

The **Transverse Sinuses** (lateral sinuses) (Figs. 201, 202, 206, 207, 214, 220–223), begin at the junction of the superior sagittal and straight sinuses, and extend laterally from the midline, along the attachment of the tentorium, to reach the postglenoid foramen through which they pass to unite with the internal maxillary vein just before it joins the superficial temporal vein, or they may possibly be regarded as tributaries of the anastomotic vein between the internal maxillary and inferior alveolar veins. The superior petrosal and superior cerebellar veins are the chief tributaries of the transverse sinuses. Further study may show that a portion of this transverse sinus in the rat is homologous with the petrosquamous sinus of comparative anatomy. This is sometimes present in man. "In rare cases it pierces the skull behind the condyle of the mandible, and terminates in the posterior facial vein." (Cunningham '31). This is essentially the condition in the rat.

The **Confluence of the Sinuses** (torcular Herophili) (Fig. 222), is formed by the junction of the superior sagittal, straight, and transverse sinuses.

The *antero-inferior sinuses* are:

> Cavernous
> Superior petrosal
> Inferior petrosal

The **Cavernous Sinus** (Fig. 219) is formed anteriorly by the *anterior cerebral* and *ophthalmic veins*, and forms the origin, posteriorly, of the *superior* and *inferior petrosal sinuses*. No attempt has been made to work out the intricacies of the *intercavernous sinus* or the *basilar plexus* in such a small animal.

The **Ophthalmic Vein** formed by the union of *superior* and *inferior ophthalmic veins* (Fig. 213), enters the cranium through the anterior lacerated foramen and, joining the anterior cerebral vein, enters the cavernous sinus (Fig. 219).

The **superior ophthalmic vein** (Fig. 213), beginning in the medial angle of the orbit, where it anastomoses with tributaries of the angular vein, runs through the orbit with the ophthalmic artery, gives an anastomotic branch to the posterior superior alveolar vein, then unites with the inferior ophthalmic vein. The *anterior ethmoidal vein* enters the anastomotic branch mentioned above.

The **inferior ophthalmic vein** (Fig. 213), formed by the union of *infraorbital* and *sphenopalatine* tributaries, which follow the course of the corresponding arteries, runs along the infraorbital groove to the posterior region of the orbit where it receives the *descending palatine vein* (Fig. 213), before uniting with the superior ophthalmic to form the ophthalmic vein. In man the infraorbital and sphenopalatine veins are usually components of the pterygoid plexus. In the rat they communicate with the plexus through an anastomosis with the superior alveolar vein (Fig. 213) but are distinctly components of the inferior ophthalmic vein.

The **Superior Petrosal Sinus** (Figs. 219–221), beginning in the cavernous sinus, runs posteriorly along the tentorial fold (Fig. 219), to the angle between the cerebral hemisphere and the cerebellum, where it joins the transverse sinus (Fig. 221).

The **Inferior Petrosal Sinus** (Fig. 219), having its origin in the cavernous sinus, runs posteriorly along the floor of the cranium to the posterior lacerated foramen where it ends in the internal jugular vein.

*Veins of the Upper Extremity*

(Figs. 232, 240, 241, 245)

Since in only about 20% of the specimens examined were the veins of the manus and antibrachium sufficiently injected to be demonstrable, no authoritative statement can be made. Where they have been traced it is seen that **dorsal metacarpal veins** unite to form the **dorsal venous rete** (Fig. 245), from which two principal veins, the *basilic* and *cephalic* take their origin, the basilic from the ulnar side, the cephalic from the radial side of the network, while, on the volar surface, **proper volar digital veins** form a **volar venous plexus** from which the *median vein* takes origin.

The **Cephalic Vein** (Figs. 232, 239–241, 245), beginning in the radial end of the dorsal venous rete (Figs. 232, 245), curves around the radial surface of the antibrachium to reach the cubital region, where it is connected with the basilic vein by the *v. mediano-cubiti* (Fig. 240). From the elbow it passes upward along the biceps muscle as far as the boundary between the pectoral and acromiodeltoid muscles. Following this boundary it reaches and unites with the external jugular vein at the level of the clavicle (Fig. 232).

The **Median Vein** (median basilic) (Figs. 232, 240), begins in the volar venous plexus of the manus and running upward through the antibrachium, at first with the mediano-radial branch of the median artery, then with the median artery and nerve, receives the **interosseous vein**, then passes beneath flexor carpi radialis and pronator teres to the cubital region where it forms the main tributary of the basilic vein.

The **Basilic Vein** (brachial) (Figs. 232, 240, 245), arising from the ulnar end of the dorsal venous rete (Fig. 245), and corresponding to the ulnar vein of older terminology, curves around the ulnar surface of the antibrachium, then takes a deeper course with the ulnar artery and nerve, between the extensor digitorum communis and extensor carpi ulnaris muscles to reach the bend of the elbow where it communicates, as a rule, with the cephalic vein by way of the *v. mediano-cubiti* (Fig. 240), and receives the *median vein* as its main tributary.

In its course through the brachium from elbow to axilla the basilic vein follows the brachial artery and receives tributaries corresponding to its branches (Fig. 232), namely, ulnar recurrent, transversa cubiti, ulnar collateral, and profunda brachii veins. Each of these follows fairly closely the artery of the same name, lying slightly superficial to it. The brachial vessels run under the lower border of pectoralis at the boundary between the biceps and medial head of triceps, and pass between latissimus dorsi and cutaneous maximus to enter the axilla as the axillary vein and artery.

The **Axillary Vein** (Fig. 235) is the proximal continuation of the basilic through the axilla, from the point where it enters the axilla between latissimus dorsi and cutaneous maximus to the first rib where it joins the external jugular and enters the thorax as the subclavian.

**Tributaries.**—The axillary vein receives the following veins; circumflexo-subscapular, lateral thoracic, and thoraco-acromial.

The **Circumflexo-subscapular Vein** (Fig. 238) enters the axillary along the boundary between latissimus dorsi and teres major. Its tributaries correspond to the arteries of the circumflexo-subscapular arterial trunk, namely, subscapular, posterior humeral circumflex, and anterior humeral circumflex veins, all of which parallel their respective arteries very closely.

The **Lateral Thoracic Vein** (Fig. 238) ascends over the lateral thoracic wall under cover of latissimus dorsi and cutaneous maximus, in company with the lateral thoracic artery and joins the axillary slightly proximal to the circumflexo-subscapular vein.

The **Thoraco-acromial Vein** (Fig. 238) enters the axillary just before the latter crosses the first rib. It accompanies the thoraco-acromial artery and the lateral anterior thoracic nerve.

The **Subclavian Vein** (Figs. 233–235), formed by the confluence of the axillary and external jugular veins, is extremely short. It enters the thorax between the first rib and the clavicle and immediately joins the internal jugular to form the vena cava superior.

## Veins of the Thorax

The right and left **Venae Cavae Superiores** (Figs. 111, 112, 225, 226, 228, 332), formed by the confluence of internal jugular and subclavian veins at the level of the first rib, pass ventral to the root of the subclavian arteries and run downward to enter the right atrium. The right vein is short, opening directly into the anterior portion of the atrium. The left vein extends farther posteriorly, crossing the arch of the aorta, the root of the lung, the pulmonary vessels and the bronchus, to enter the atrium from below with the vena cava inferior (Fig. 112).

Since the persistence of two *venae cavae superiores* (Fig. 112) is the normal condition in the rat, their tributaries present a fairly symmetrical picture and with a few exceptions any description will therefore apply to both sides. Where there are differences the variations will be given.

**Tributaries.**—The tributaries of the vena cava superior are the vertebral, internal mammary, pericardiacomediastinal, bronchial, superior intercostal, superior phrenic, and on the left side only, the azygos.

The **Vertebral Vein** has been described with veins of the neck.

The **Internal Mammary Vein** (Figs. 226, 228, 230) ascends along the course of the corresponding artery, collecting from tributaries which parallel the branches of the artery, namely from the superior epigastric, musculophrenic, perforating, anterior intercostal, sternal, and pericardiacomediastinal veins.

The **Pericardiacomediastinal Veins** (Fig. 225) drain the regions supplied by branches of the pericardiacomediastinal artery. They may follow the pattern of the corresponding arteries, but more often appear as separate small veins opening independently into the internal mammary or vena cava superior. The thymic vein (Fig. 225) is usually distinct and opens into the internal mammary just below the first rib. The small veins from the thoracic lymph nodes enter close to the thymic. The vein from the pericardial and anterior mediastinal region accompanies the artery only as far as the vena cava superior into which it opens.

The **Bronchial Veins** (Fig. 225) from the trachea, œsophagus, lungs, bronchi, and associated lymph glands enter the vena cava superior.

The **Superior Intercostal Vein** (Figs. 226, 230). On the right side the tributaries of the superior intercostal vein correspond very nearly to the branches of the costocervical arterial trunk. Beginning as a small vessel in the fourth intercostal space, parallel with the fifth rib, the vein runs anteriorly medial to the artery collecting from the third, second, and first intercostal spaces, and turning abruptly to gain the vena cava superior.

On the left, this vein, like the artery is more variable. Furthermore, it is extremely difficult to trace as it is seldom filled with the injection mass. Where it can be traced it is made up of tributaries from the first and second intercostal spaces and sometimes from the third. The vein from the third intercostal space usually joins an intercostal vein opening into the azygos.

The **Superior Phrenic Vein** (Figs. 225, 228) which parallels the pericardiacophrenic artery through most of its course, enters the vena cava superior a short distance from its termination.

The **Azygos Vein** (Fig. 230) lies on the left of the aorta and vertebral column, quite the reverse of the condition in man. This is probably associated with the persistence of the left vena cava superior.

In the lower part the azygos is extremely difficult to trace owing to incomplete injection. In those specimens in which it is at least partially injected it may be found entering the thoracic cavity concealed by the quadratus lumborum muscle. Proceeding anteriorly it receives tributaries from each of the intercostal spaces up to and including the third on the left side. On the right side it first receives the *hemiazygos* (if this vein is present) then successive intercostals as far as the fifth intercostal space. If the hemiazygos is lacking, the two sides are practically symmetrical in the lower portion, the right intercostals crossing the vertebral column dorsal to the aorta and each opening separately into the azygos. Any number of variations may be seen in the relation of the azygos to the intercostal arteries of the aorta. It may lie ventral to them all or may pass dorsal to one or more of them.

The **hemiazygos**, concealed at first by the quadratus lumborum, runs anteriorly on the right side receiving tributaries from the lower three or four intercostal spaces then crossing under quadratus lumborum and the aorta to enter the azygos, generally in the ninth or tenth intercostal space.

Apparently an accessory hemiazygos is lacking, since it did not appear in any cases so far studied.

The **Veins of the Vertebral Column** are not described since microscopic dissections have seemed beyond the scope of the present work and all descriptions based on such dissections are therefore omitted.

*Veins of the Abdomen and Pelvis*

The **Vena Cava Inferior** (Figs. 112, 247), collecting tributaries from the parts of the body below the diaphragm, enters the posterodorsal region of the right atrium with the left vena cava superior (Figs. 112, 332).

It begins with the junction of the two common iliac veins slightly posterior to the bifurcation of the aorta (Fig. 247), at the boundary between the psoas major muscle and the caudal flexors. Lying dorsal to the aorta at first, it proceeds anteriorly to a position at the right of the aorta, until, at the level of the kidney it lies ventral to it, then crosses the crus of the diaphragm dorsal to the liver and penetrates the diaphragm.

Anterior to the diaphragm the vena cava inferior, separated from the aorta by the œsophagus, runs through a notch in the post-caval lobe of the right lung, joins the left vena cava superior and enters the right atrium (Fig. 112).

**Tributaries.—**

| | |
|---|---|
| Lumbar | Renal |
| Iliolumbar | Inferior phrenic |
| Spermatic or ovarian | Hepatic |

The **lumbar veins** (not shown), corresponding to the lumbar arteries, join the vena cava inferior as a series of four segmental pairs of vessels draining the iliopsoas and quadratus lumborum muscles.

The **iliolumbar veins** (Figs. 247, 249, 253, 254, 273), paralleling the arteries, arise by the junction of an iliac and a lumbar branch. The *iliac branch* (Figs. 249, 253, 254, 273) arises in the integument of the extensor surface of the thigh, receives branches from the cutaneous maximus of the hip and thigh, then, in company with the lateral femoral cutaneous nerve and iliolumbar artery penetrates the abdominal wall in the region of the superior anterior spine of the ilium (Figs. 253, 254), crosses the anterior attachment of the inguinal ligament and joins the lumbar portion of the vein.

The *lumbar branch* (Figs. 249, 273), collecting from the muscles of the abdominal wall and from psoas and quadratus lumborum muscles, joins the iliac branch and forms the iliolumbar vein.

The iliolumbar vein of the right side is stable like the artery, and so far as observed, invariably opens into the vena cava inferior about one third of the distance between the entrance of the common iliac and renal veins (Figs. 247, 249, 273). The left iliolumbar also corresponds to the artery and may enter the common iliac vein (Fig. 273) or the vena cava inferior (Figs. 176, 247, 249) at any point up to the level of the kidney and just below the spermatics.

The **testicular vein** (internal spermatic) (Fig. 248), arising by a confluence of vessels about the anterior end of the testis enters into the formation of the **pampiniform plexus** (Fig. 248) which wraps about the testicular artery as a core. Emerging from the inguinal canal the vein of the left side contributes to the common iliac by anastomosing with the superior vesical (Fig. 270), then continues in the form of two concomitant vessels along the testicular artery, finally opening into the left renal vein (Figs. 246, 247). On the right the vein opens into the vena cava inferior with no evidence of anastomosis with the common iliac so far as seen.

The left **ovarian vein** enters the left renal, and the right ovarian the vena cava inferior (Fig. 249).

The **renal veins** (Figs. 247, 249) like the arteries are more widely separated than in man, and the left renal opens into the vena cava inferior well below the right which is quite the reverse of the condition in man. As a rule the left renal, which is longer than the right, crosses ventral to the aorta but may in rare cases pass dorsal to it (Fig. 249). The left renal vein usually receives the combined *inferior suprarenal* and *inferior phrenic veins* anteriorly (Fig. 250), as well as the ovarian or testicular vein posteriorly. The right renal receives only the inferior suprarenal vein (Fig. 249).

The **inferior phrenic vein** (Figs. 249, 250) of the right side, collecting from the inferior surface of the diaphragm, opens directly into the vena cava inferior. It may be joined by a small branch from the suprarenal gland. The inferior phrenic of the left side unites with the left superior suprarenal and enters the left renal vein (Fig. 250).

The **hepatic vein** (Figs. 246, 247), collecting branches from the liver, enters the vena cava inferior just as this vessel pierces the diaphragm.

The **Portal System** (Fig. 246) comprises those veins which receive blood from the digestive tract below the diaphragm, and from its associated organs the spleen and pancreas, and carry it to the liver where it passes through capillaries before entering the systemic system via the hepatic veins, and the vena cava inferior.

**Tributaries.—**

> Lienal (splenic)
> Superior mesenteric
> Pyloric

The **Lienal Vein** (splenic vein) (Fig. 246) has its origin in several veins from the spleen. It runs parallel with the lienal artery receiving the following tributaries; the pancreatic vein from the portion of the pancreas lying along the dorsal layer of the greater omentum; the **left gastroepiploic**, running with the artery of the same name from right to left along the greater curvature of the stomach, collecting tributaries from the stomach and omentum; and the **short gastric** from the fundus of the stomach. After passing dorsal to the stomach, the lienal vein is joined by the **coronary vein** which in man is a distinct branch of the portal.

In the rat the coronary, beginning in the region supplied by the left gastric artery, follows the course of this artery along the lesser curvature of the stomach toward the cardiac end where it receives a small *œsophageal branch* before turning right and leaving the stomach to unite with the lienal vein. Having received the coronary, the lienal vein runs ventral to the aorta and dorsal to the gastroduodenal vessels to reach the portal vein.

The **Superior Mesenteric Vein** (Fig. 246) collects blood from the small intestine, the caecum, and from the whole length of the colon.

**Tributaries.**—The superior mesenteric vein receives not only tributaries corresponding to the branches of the superior mesenteric artery but also one corresponding to the inferior mesenteric artery. Its tributaries are **middle colic branch** from the transverse colon, plus the **inferior mesenteric** from the descending colon and rectum; a **right colic** branch from the ascending colon; an **ileocolic** branch from the first part of the colon, from the caecum, and ileum; a series of about sixteen **intestinal** branches from the ileum and jejunum; and the **inferior pancreaticoduodenal vein** (accompanying the artery of the same name) from the head of the pancreas and the second portion of the duodenum. Inasmuch as the superior and inferior pancreaticoduodenal veins do not unite as in man, the two distinct names must be retained. In the rat the superior pancreaticoduodenal veins are tributaries of the pyloric veins.

The inferior mesenteric vein, beginning as the superior hæmorrhoidal vein along the rectum, ascends parallel with the left colic branch of the inferior mesenteric artery along the descending colon as far as the transverse colon. Here it unites with the middle colic vein, and with it, enters the superior mesenteric vein between the inferior, pancreatic, duodenal and lienal tributaries of the latter.

The **Pyloric Vein** (Fig. 246) collects from the region of the gastroduodenal branch of the hepatic artery. Beginning along the pyloric region of the stomach it runs from left to right, and upon leaving the stomach receives the **superior pancreaticoduodenal vein** made up of pancreatic and duodenal tributaries from those organs. In man this latter vein unites with the inferior pancreaticoduodenal. The pyloric enters the portal vein a short distance above the lienal vein and just below the liver.

### Veins of the Pelvis

In only one specimen observed so far was there a distinct **hypogastric vein** with tributaries corresponding to the branches of the hypogastric artery. In all other specimens examined the various elements of the hypogastric entered the common iliac vein independently or combined with one other vessel. If the common iliac vein is formed by the external iliac vein uniting with the hypogastric elements from the pelvis, then the *common iliac vein* begins with the junction of the external iliac and the medial femoral circumflex vein, which is the most laterally situated of the pelvic veins.

The **Common Iliac Vein** (Figs. 246, 247, 255, 273), formed by the junction of external iliac and medial femoral circumflex veins, begins a short distance above the inguinal ligament and runs medially with the common iliac artery along the boundary between the ventral caudal and the psoas muscles, to unite in the midline with the common iliac of the opposite side and form the *vena cava inferior* (Figs. 112, 246, 247, 255). The common iliac vein may be dorsal or ventral to the hypogastric artery and its branches (Fig. 273), but is always dorsal to the superior vesical artery and the pudic-epigastric trunk.

**Tributaries.**—The common iliac vein receives tributaries which correspond in general to the branches of the hypogastric artery, namely, the *superior vesical, obturator, medial femoral circumflex, lateral femoral circumflex, internal pudendal,* and *superior gluteal veins.* Certain ones of these are occasionally combined and these variations will be described under the individual vessels. The order of description is based upon the order of dissection.

The **Superior Vesical Vein** in the male (Figs. 255, 263, 269, 270), has approximately the same distribution as the corresponding artery. Beginning on the base of the bladder (Fig. 270) it runs with the artery along the ureter, from which it receives a twig. It collects a branch from the gland of the ductus deferens (Fig. 269), then receives the **deferential vein** (Fig. 270), which begins in the region of the epididymis, where it anastomoses with veins from the pampiniform plexus (Fig. 248), and follows the ductus deferens to the point where it crosses the ureter. A short distance above this branch, the superior vesical is joined by a large vein made up of the following tributaries; the *inferior vesical* (Fig. 269) from the prostate and the ventral surface of the bladder, a large branch from the seminal vesical, a vein from the gland of the ductus deferens, and a branch from the coagulating gland and posterior prostate. Crossing the ureter the superior vesical anastomoses with the proximal end of the pampiniform plexus and with the internal spermatic vein before entering the common iliac as its highest tributary.

In the female (Figs. 273–275) the superior vesical vein begins with one or two tributaries from the bladder and ureter. These are joined by veins from the **vaginal plexus** and by a large vein from the **pudendal plexus** situated beneath the symphysis pubis (Fig. 275). Occasionally the latter may join the obturator vein. An exceptionally large **uterine vein** (Fig. 273), beginning in the region of the ovary, where it anastomoses with the ovarian vein from the vena cava inferior or renal vein, runs downward in the mesovarium the entire length of the uterus, forming the main channel of the **uterine plexus** (Figs. 273–276). In the region of the cervix the uterine vein joins the superior vesical which crosses the ureter and enters the ventral surface of the common iliac vein.

The **Obturator Vein** (Figs. 255, 269, 273–277) begins in the deeper muscles of the pelvis as small vessels corresponding to the anterior and posterior branches of the artery (Figs. 284, 285). Piercing the obturator membrane the vein runs upward with the artery between the caudal abductors to enter the pelvic cavity where it receives the **middle hæmorrhoidal vein** (Figs. 249, 255, 265, 269, 275–277) from the rectum and caudal muscles before joining the common iliac vein. In one specimen the large vein from the pudendal plexus, which as a rule joins the superior vesical, contributed to the obturator, and in two instances the medial femoral circumflex vein joined the obturator.

The **Medial Femoral Circumflex Vein** (Figs. 255, 273–277, 287, 306) receives vessels which follow almost exactly the branches of the corresponding artery (Fig. 306), uniting as one vein on the medial surface of the femur just below the lesser trochanter under pectineus and adductor longus muscles (Fig. 287). From here it runs with the iliacus and psoas muscles and emerges from beneath the anterior border of pectineus to the pelvic cavity where it joins the common iliac vein between the pudic epigastric and the obturator. Occasionally it joins the obturator.

The **Lateral Femoral Circumflex Vein** (Figs. 255, 277, 303) is made up of two vessels which collect from the regions supplied by the divisions of the corresponding artery. One of these veins accompanies the ascending branch of the artery and is formed by a dorsal tributary from iliacus, gluteus medius and minimus, and gemellus superior, and a ventral tributary from quadriceps femoris, gluteus maximus and iliacus. This ventral tributary in one specimen crossed the medial surface of the thigh to join the femoral vein with the superficial circumflex iliac vein from the inguinal region (Fig. 291). Following the descending branch of the artery is a vein which arises by twigs from extensor quadriceps femoris, gluteus medius and minimus, and the hip-joint. This vein joins the ascending vessel just anterior to the origin of rectus femoris to form the lateral femoral circumflex vein, which crosses the ventral border of the ilium to reach the pelvic cavity where it emerges from beneath iliacus to reach the dorsal surface of the common iliac between the internal pudendal and medial femoral circumflex veins.

The **Internal Pudendal Vein** (Figs. 249, 255, 263, 276, 277) begins in the **pudendal plexus** (Figs. 263, 275, 281) which is in the form of a complete venous circle situated between the ventral surface of the urethra and the inner surface of the symphysis pubis. Opening into this circle of veins posteriorly is the median **dorsal (deep) vein of the penis** (Figs. 263, 280) in the male, **dorsal (deep) vein of the clitoris in the female** (Figs. 264, 275, 281); a pair of anastomotic veins which run from the inferior external pudendal veins beneath the pubic arch (Figs. 263–266, 275) to enter the plexus laterally; the **internal pudendal veins** (Figs. 263, 281) opening at the same point; and two veins which descend on either side of the urethra to enter the anterior half of the circle (Figs. 263, 275, 281). These two veins anastomose through the vesical and vaginal plexuses (Fig. 275) with the superior vesical veins or occasionally with the obturator (Fig. 274, shown but not labelled).

As the internal pudendal vein leaves the plexus it receives branches from ischiocavernosus and the urethral bulb in the male (Fig. 263), and from the urethra and vaginal plexuses in the female (Figs. 275, 276). It then runs under the posterior border of abductor caudae externus to enter the ischiorectal fossa with the internal pudendal artery (Fig. 277). As the vein enters the fossa it receives a tributary from the hæmorrhoidal plexus and one from the integument of the perineal region. The latter is the **posterior scrotal vein** (Fig. 296) in the male, the **posterior labial** in the female.

As it reaches the obturator internus tendon the internal pudendal is joined by the *inferior gluteal vein* (Fig. 296), then continues through the ischiorectal fossa, lying between the superior gluteal and internal pudendal arteries, medial and ventral to the lumbosacral trunk. In this position it receives tributaries corresponding to the branches of the superior gluteal artery. In only one specimen seen so far was there a distinct superior gluteal vein.

The **Inferior Gluteal Vein** (Figs. 296, 301) drains the region supplied by the branches of the inferior gluteal artery, namely the integument at the back of the thigh and the hamstring muscles, and runs with the artery along the sciatic nerve to the lower extent of the great sciatic foramen where it joins the internal pudendal vein. The posterior scrotal vein in the male, or the vein from the perineum in the female may join either the inferior gluteal or the internal pudendal vein.

The **Superior Gluteal Vein** (Figs 276, 301) as a distinct vessel, was only found in one side of a single specimen. In this instance it entered the dorsal surface of the common iliac vein just proximal to the internal pudendal vein. In all other cases the venous tributaries accompanying the branches of the superior gluteal artery entered the internal pudendal vein at intervals along its course through the ischiorectal fossa. One such vein collects blood from the integument of the hip, from gluteus maximus, gluteus medius and piriformis. Small veins from the caudal muscles follow branches of the superior gluteal artery and enter the internal pudendal vein, while the highest tributary of this vein is a vessel which receives branches from gluteus medius, piriformis and gemellus superior, and with the superior branch of the superior gluteal artery crosses the dorsal border of the ilium through the anterior end of the sciatic notch to enter the deep surface of the internal pudendal vein close to its junction with the common iliac.

The **External Iliac Vein** (Figs. 247, 255, 274, 275, 303) is the continuation of the femoral vein above the inguinal ligament. It receives the *pudic-epigastric vein*, which is the companion vessel of the pudic-epigastric arterial trunk, and may also receive the *deep circumflex iliac vein* as an independent branch. Immediately above this point the external iliac joins the medial femoral circumflex to form the *commom iliac vein*.

The **Pudic-Epigastric Vein** (Figs. 257, 260, 264, 266, 268, 273, 275) is the largest tributary of the external iliac vein. It is made up of branches corresponding to the arteries of the pudic-epigastric trunk. In man these branches, with the exception of the deep circumflex iliac and inferior epigastric veins, enter the femoral vein in the femoral triangle. In the rat they are combined in one main vessel, entering the abdominal cavity through the subcutaneous inguinal ring (Fig. 262), and joining the external iliac vein above the inguinal ligament (Fig. 275).

**Tributaries.**—The veins which contribute to the pudic-epigastric vein are the *deep circumflex iliac* (Figs. 269, 303 shown, not labelled), *inferior epigastric* (Figs. 260,

264-267), *external spermatic* in the male (Fig. 261), *superior external pudendal* (Figs. 265, 266), and *inferior external pudendal* (Figs. 264-266).

The **deep circumflex iliac vein** (Figs. 274, 275, 277, 303) begins in the abdominal muscles and follows the artery of the same name into the pelvis, receiving a tributary from the iliopsoas muscles before entering either the external iliac vein (Fig. 275) or epigastric vessel (Fig. 303).

The **inferior epigastric vein** (Figs. 255, 260, 261, 266, 267) takes its origin in the slips of rectus abdominis where it forms connections with the other epigastric veins. It runs downward to join the pudic-epigastric as the latter enters the abdominal cavity through the subcutaneous inguinal ring.

The **external spermatic vein** (Figs. 255, 261, 262) in the male, arises in the cremaster muscle and joins the pudic-epigastric vein a short distance above the inferior epigastric.

The **superior external pudendal vein** (Figs. 257-266) in the male, collects blood from the prepuce, preputial gland and surrounding integument (Figs. 257-262), runs upward for a short distance along the penis to the suspensory ligament, then turns downward, receives small twigs from the scrotum and enters the pudic-epigastric vein just above the anterior margin of pyramidalis. In the female the superior external pudendal is formed by a vessel from the bulb of the vestibule and a branch from the labia (Figs. 264-266). As it runs anteriorly it receives veins from the fat and mammary gland of the inguinal region, then joins the inferior external pudental to form the pudic-epigastric vein. A large communicating vessel connects the two external pudendal veins (Figs. 264-266).

The **inferior external pudendal vein** (Figs. 259-266, 275) is the most extensive tributary of the pudic-epigastric. The **inferior hæmorrhoidal vein** (Figs. 262, 264, 265, 276), starting in the **hæmorrhoidal plexus** of the anal region (Figs. 262, 276), runs anteriorly and is joined by a large vessel from the **vaginal plexus** in the female (Figs. 275, 276), or by the **anterior scrotal veins** in the male (Figs. 259, 261), to form the inferior external pudendal. A large **anastomotic vein** (Figs. 262-266) emerges from the pelvis through the pubic arch, connecting the inferior external pudendal vein, through the **pudendal plexus** (Figs. 263, 275, 281), with the internal pudendal vein. The inferior and superior external pudendal then unite to form the pudic-epigastric vein which receives the various tributaries mentioned above and proceeds anteriorly through the subcutaneous inguinal ring and across the brim of the pelvis to join the external iliac vein a short distance above the inguinal ligament.

### Veins of the Lower Extremity

In the rat the veins of the lower extremity can not be as sharply divided into a superficial and deep series as in man, since some deep veins drain into the superficial veins, the great and small saphenous, while some superficial branches are tributaries of the deep veins, the peroneal, anterior tibial and posterior tibial. The veins which have their origin on the plantar surface of the foot will be described first, followed by those from the dorsum.

*Veins of the Foot and Lower Leg.*—In no specimens examined so far were the *superficial veins* of the foot sufficiently injected to be distinguishable as a continuous system.

The *deep veins* (Fig. 326) of the foot are first distinguishable as the four **plantar metatarsal veins** which follow the metatarsal arteries and unite to form the **plantar venous arch.** This arch runs parallel with the arterial arch obliquely across flexor digitorum brevis, and is continued as the **medial plantar vein**, which accompanies the medial plantar nerve and the deep branch of the medial plantar artery up the side of the foot. The vein receives several tributaries from the superficial venous plexus, remnants of which are usually demonstrable. The two blood vessels leave the nerve shortly above the medial malleolus and the vein unites with the **lateral plantar vein** before opening into the **great** (large) **saphenous vein** at the distal border of semitendinosus (Fig. 326).

The **lateral plantar vein** (Fig. 326), begins in the superficial *palmar venous plexus*, through which it is connected with the medial plantar vein. It appears as a well defined single vein just medial to the calcaneus, in company with the lateral plantar artery which it follows up the medial surface of the lower leg, where it unites with the medial plantar vein as a tributary of the great saphenous.

## Dorsal Veins

From the dorsum of the foot the principal venous channels are the *superficial cutaneous*, the *medial tarsal*, the *anterior tibial*, and the *small saphenous vein*.

The **superficial cutaneous vein** (Fig. 327) accompanies the corresponding artery and the superficial peroneal nerve and becomes the *anterior peroneal vein* in the lower leg. This corresponds to the perforating branch of the peroneal in man. Running proximally, deep to extensor digitorum longus and anterior tibialis muscles, along the fusion of tibia and fibula, this vein becomes the **peroneal vein** (Fig. 321), and after receiving tributaries from the peroneal muscles enters the anterior tibial vein along the interosseous membrane a short distance below the head of the fibula. The anterior tibial enters the popliteal vein.

The first and second *dorsal metatarsal veins* follow the corresponding arteries and unite to form a short *dorsal venous arch* from which the *medial tarsal* and *anterior tibial veins* arise.

The **medial tarsal vein** (Fig. 327), runs with the artery under extensor proprius hallucis and anterior tibialis tendons over the medial malleolus and up the lower leg, receiving a communicating vein from the flexor surface of the tibia before uniting with the veins from the plantar surface of the foot to form the *great saphenous vein*.

The **Anterior Tibial Vein** (Fig. 321), starting in the *dorsal venous arch*, runs with the dorsal artery of the foot and the deep peroneal nerve under the transverse crural ligament and beneath the tendon of anterior tibialis in company with the tendon of extensor hallucis proprius. Where tibia and fibula separate the vein runs under the extensor hallucis proprius muscle to follow the lateral surface of the shaft of the tibia along the border of the interosseous membrane. In its course up the lower leg it receives tributaries which perforate the membrane, and small veins from the surrounding muscles. A short distance below the head of the fibula it receives the *peroneal vein* (Fig. 321), several *muscular tributaries* from anterior tibialis, and finally the *anterior tibial recurrent vein* (Fig. 321) before crossing the anterior border of the interosseous membrane to join the *posterior tibial vein* with which it forms the popliteal (Figs. 320, 323).

The **Posterior Tibial Vein** (Figs. 317–320, 323), like the artery, is much reduced owing to the fact that the peroneal is a branch of the anterior rather than the posterior tibial as in man. The posterior tibial begins in the flexor hallucis longus and flexor digitorum longus, beneath the medial head of gastrocnemius (Fig. 323), follows the artery proximally, and joins the *anterior tibial vein* just above the anterior border of the popliteus muscle, to form the *popliteal vein* (Figs. 320, 323).

The **Popliteal Vein** (Figs. 303, 306, 309, 311, 313, 316–320, 323, 326), beginning with the junction of the anterior and posterior tibial veins at the upper margin of the popliteus muscle (Figs. 317, 323), runs for a short distance through the popliteal fossa, then passes between adductor brevis and caudofemoralis just above the medial epicondyle of the femur, to become the *femoral vein* of the thigh (Fig. 303).

**Tributaries.**—Within the fossa the popliteal vein receives the following: *small saphenous vein* (Fig. 318), *medial inferior genicular*, *lateral inferior genicular*, and *middle genicular veins* (Figs. 317, 319), all of which follow their respective arteries.

The **Small Saphenous Vein** (Figs. 314–316, 318, 320, 325–327), begins with the junction of the third and fourth *dorsal metatarsal veins* which accompany their respective arteries. These venous tributaries unite just distal to the cruciate ligament and the vein then crosses the lateral end of the ligament, passes under the peroneal tendons and behind the malleolus to the lateral surface of the lower leg (Fig. 327). Extending proximally along the lateral head of gastrocnemius it runs between the outer and inner hamstring muscles to the popliteal space, where it forms a tributary of the popliteal vein (Fig. 318).

**Tributaries.**—The small saphenous vein receives the following tributaries; *superior muscular* (Fig. 316), *internal* and *external sural* (Fig. 316), and *lateral superior genicular veins* (Fig. 317).

The **superior muscular vein** (Figs. 316–318), from the hamstring muscles receives the *femoropopliteal vein* from the back of the thigh and the *lateral marginal vein* from the foot. Veins accompanying the first and second muscular branches of the superior muscular artery open independently into the small saphenous vein.

The *lateral marginal vein* (Figs. 314, 316, 318) is a cutaneous vessel having its origin over the dorso-lateral margin of the foot. It takes an oblique course upward, winding around the leg above the lateral malleolus and crossing the superficial sural artery to the border of the hamstring muscles where it unites with the femoropopliteal vein from the back of the thigh and enters the superior muscular tributary of the small saphenous vein.

*Veins of the Thigh.*—The **Femoral Vein** (Figs. 303–305) is the continuation of the popliteal. It enters the thigh between caudofemoralis and adductor brevis, then comes to a superficial position on the medial surface of the thigh along the boundary between adductor and extensor muscles. In this position it runs upward with the femoral artery and leaves the thigh by passing under the inguinal ligament, beyond which point it becomes the *external iliac vein* of the pelvis.

**Tributaries.**—The femoral receives veins corresponding to the branches of the femoral artery, namely, the *great* (large) *saphenous* (Figs. 304, 305, 320, 322, 325–327), *superficial epigastric* (Figs. 259, 262, 303, 304), *highest genicular* (Figs. 304, 305), *superficial circumflex iliac* (Figs. 303, 305), and one principal *muscular vein* (Figs. 303, 305). These accompany the arteries of the same name. In one specimen the vein from the region supplied by the ventral branch of the ascending portion of the lateral femoral cutaneous artery formed a strong anastomosis with the muscular veins from the inguinal region, and drained into the femoral instead of into the lateral femoral circumflex vein.

**LIST OF FIGURES**

**VII. CIRCULATORY SYSTEM**

## LYMPHATIC SYSTEM

(Figs. 328–339)

Inasmuch as the lymphatic system of the rat has been worked out and described by Thesle Job, 1915, it has not seemed desirable to duplicate his work, and his diagram of the system is shown in Fig. 328. The lymph nodes which he names as being constant have been shown topographically. With the exception of those associated with the digestive tract and the caudal node these are paired, only one of a pair being pictured.

The list is as follows:

| | |
|---|---|
| Elbow nodes (Fig. 334) | Intestinal nodes (Fig. 337) |
| Knee nodes (Fig. 336) | Axial nodes (Fig. 335) |
| Lumbar nodes (Fig. 338) | Thoracic group (Fig. 332) |
| Inguinal nodes (Fig. 334) | Posterior cervical nodes (Fig. 331) |
| Renal nodes (Fig. 339) | Anterior cervical nodes (Fig. 331) |
| Cisterna chyli (Fig. 246) | Submaxillary nodes (Figs. 329, 330) |
| Cisternal group of nodes (Fig. 338) | Caudal node (Fig. 338) |

Job states that the "thoracic duct (Fig. 333) runs dorso-laterally along the vena cava superior and left innominate to its junction with the venous system in the jugulo-subclavian. None showed it entering the right jugulo-subclavian as in other forms."

## LIST OF FIGURES

### LYMPHATICS

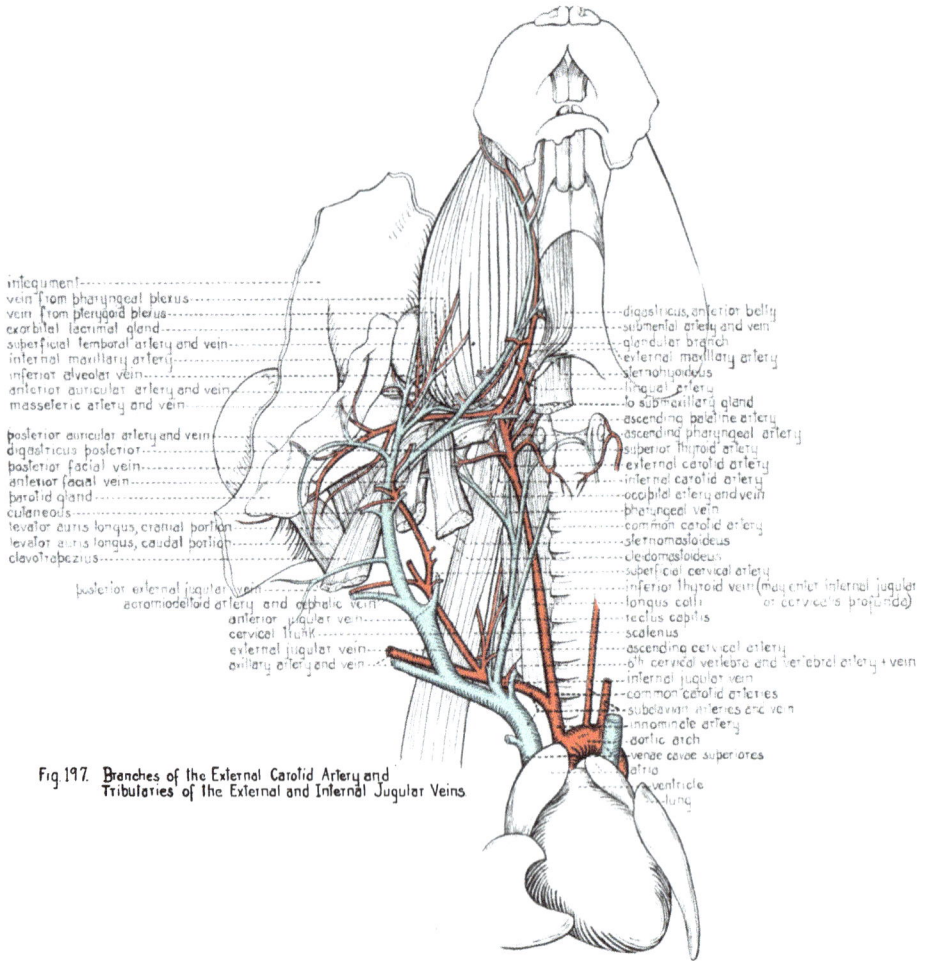

integument
vein from pharyngeal plexus
vein from pterygoid plexus
exorbital lacrimal gland
superficial temporal artery and vein
internal maxillary artery
inferior alveolar vein
anterior auricular artery and vein
masseteric artery and vein

posterior auricular artery and vein
digastricus posterior
posterior facial vein
anterior facial vein
parotid gland
cutaneous
levator auris longus, cranial portion
levator auris longus, caudal portion
clavotrapezius

posterior external jugular vein
acromiodeltoid artery and cephalic vein
anterior jugular vein
cervical trunk
external jugular vein
axillary artery and vein

digastricus, anterior belly
submental artery and vein
glandular branch
external maxillary artery
sternohyoideus
lingual artery
to submaxillary gland
ascending palatine artery
ascending pharyngeal artery
superior thyroid artery
external carotid artery
internal carotid artery
occipital artery and vein
pharyngeal vein
common carotid artery
sternomastoideus
cleidomastoideus
superficial cervical artery
inferior thyroid vein (magenior internal jugular
longus colli            or cervicalis profunda)
rectus capitis
scalenus
ascending cervical artery
6th cervical vertebra and vertebral artery + vein
internal jugular vein
common carotid arteries
subclavian arteries and vein
innominate artery
aortic arch
venae cavae superiores
atria
ventricle
lung

Fig. 197. Branches of the External Carotid Artery and
Tributaries of the External and Internal Jugular Veins.

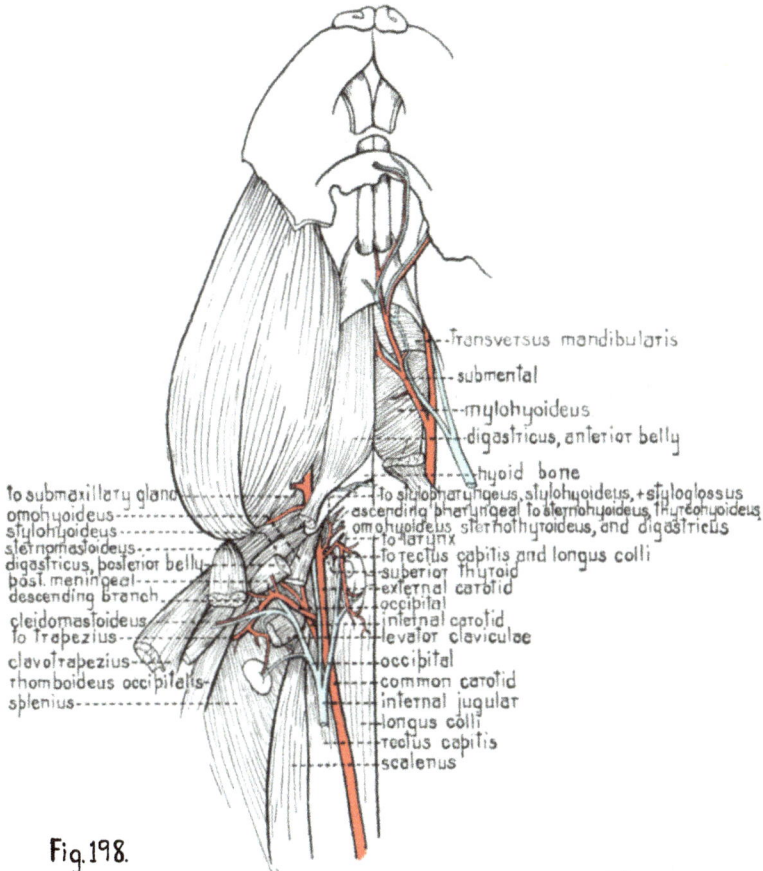

Fig. 198.

The Occipital, Ascending Pharyngeal, and Superior Thyroid Arteries
and Corresponding Tributaries of the Internal Jugular Vein.

digastricus, anterior belly
ascending palatine
sterno- and omohyoideus
lingual
thyreohyoideus
to the pharynx
ascending pharyngeal
superior thyroid
to rectus capitis and longus colli
external carotid
sternothyreoideus
sternohyoideus

masseter
hyoid bone
external maxillary
posterior auricular
to styloglossus
to stylohyoideus
styloglossus
stylohyoideus
occipital
digastricus, post. belly
internal carotid
omohyoideus
sternomastoideus

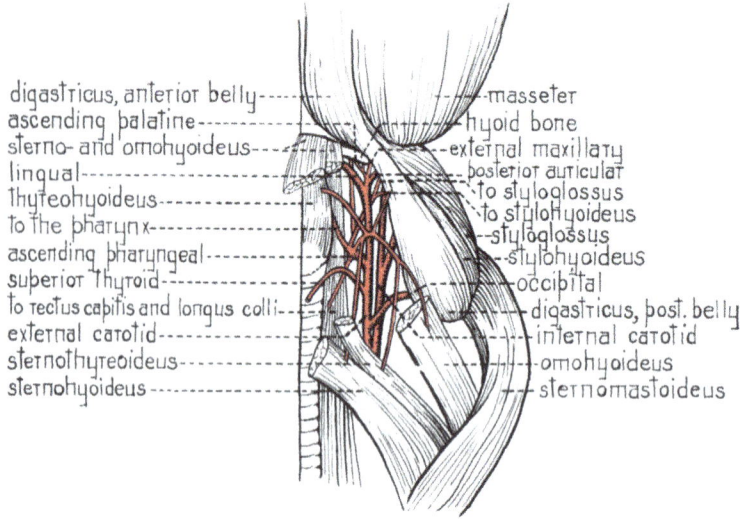

Fig. 199. Branches of the Ascending Pharyngeal Artery.

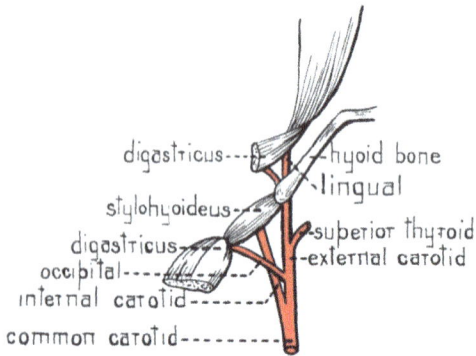

digastricus
stylohyoideus
digastricus
occipital
internal carotid
common carotid

hyoid bone
lingual
superior thyroid
external carotid

Fig. 200. Detail showing relation of external and internal
carotid arteries to hyoid bone and stylohyoideus.

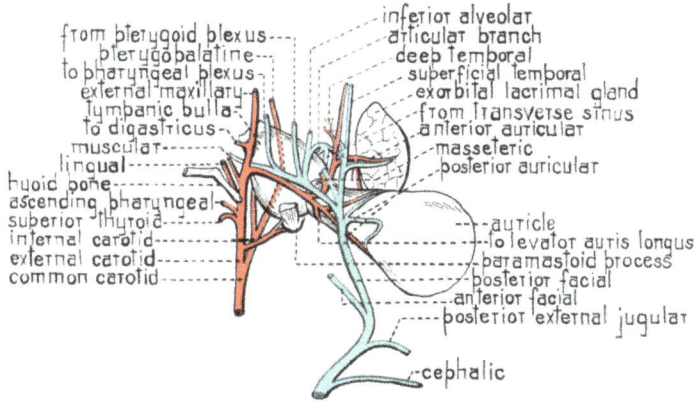

from pterygoid plexus
pterygopalatine
to pharyngeal plexus
external maxillary
tympanic bulla
to digastricus
muscular
lingual
hyoid bone
ascending pharyngeal
superior thyroid
internal carotid
external carotid
common carotid

inferior alveolar
articular branch
deep temporal
superficial temporal
exorbital lacrimal gland
from transverse sinus
anterior auricular
masseteric
posterior auricular

auricle
to levator auris longus
paramastoid process
posterior facial
anterior facial
posterior external jugular

cephalic

Fig. 201.  Detail, showing vessels in the region
of the tympanic bulla and external ear.

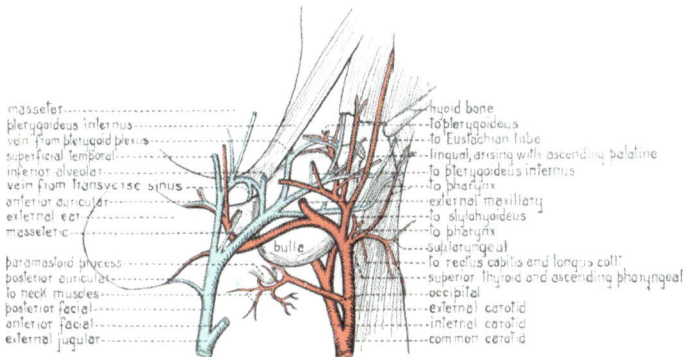

masseter
pterygoideus internus
vein from pterygoid plexus
superficial temporal
inferior alveolar
vein from transverse sinus
anterior auricular
external ear
masseteric

paramastoid process
posterior auricular
to neck muscles
posterior facial
anterior facial
external jugular

bulla

hyoid bone
to pterygoideus
to Eustachian tube
lingual, arising with ascending palatine
to pterygoideus internus
to pharynx
external maxillary
to stylohyoideus
to pharynx
sublaryngeal
to rectus capitis and longus colli
superior thyroid and ascending pharyngeal
occipital
external carotid
internal carotid
common carotid

Fig. 202.  Deep Branches of the External Carotid Artery and Corresponding Veins.   Right.

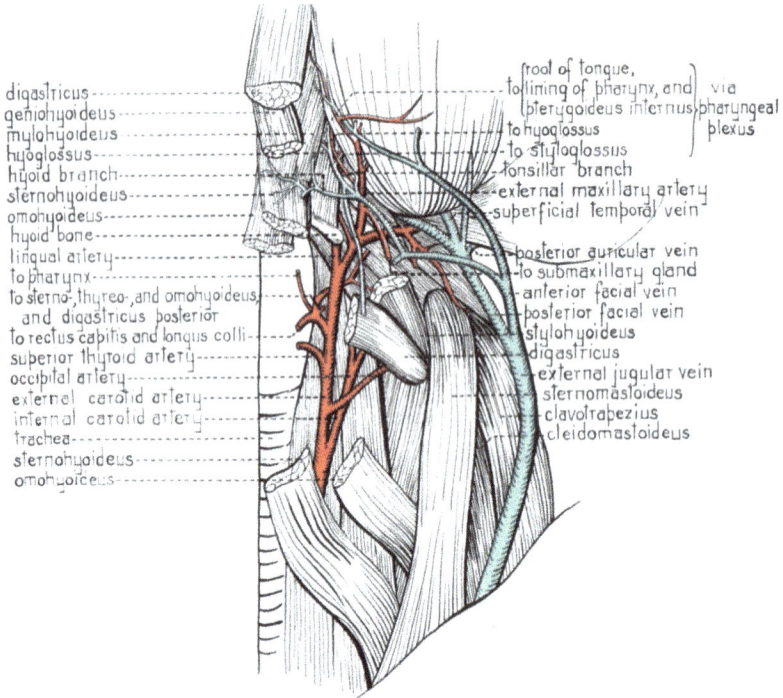

diagastricus
geniohyoideus
mylohyoideus
hyoglossus
hyoid branch
sternohyoideus
omohyoideus
hyoid bone
lingual artery
to pharynx
to sterno-, thyreo-, and omohyoideus,
  and digastricus posterior
to rectus capitis and longus colli
superior thyroid artery
occipital artery
external carotid artery
internal carotid artery
trachea
sternohyoideus
omohyoideus

root of tongue,
to lining of pharynx, and } via
(pterygoideus internus) pharyngeal plexus
to hyoglossus
to styloglossus
tonsillar branch
external maxillary artery
superficial temporal vein
posterior auricular vein
to submaxillary gland
anterior facial vein
posterior facial vein
stylohyoideus
digastricus
external jugular vein
sternomastoideus
clavotrapezius
cleidomastoideus

Fig.203. Branches of the External Carotid Artery and Tributaries
of the Anterior and Posterior Facial Veins.

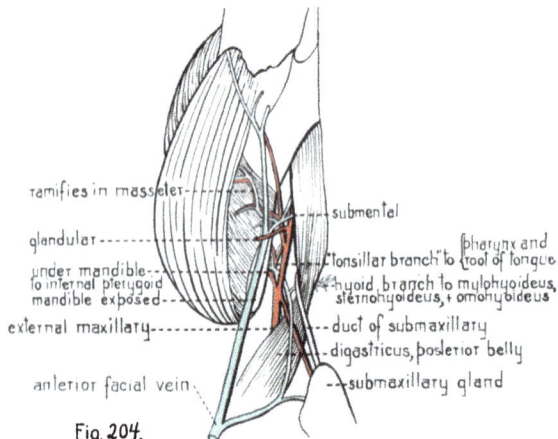

ramifies in masseter
glandular
under mandible
to internal pterygoid
mandible exposed
external maxillary
anterior facial vein

submental
tonsillar branch to {pharynx and root of tongue
hyoid branch to mylohyoideus, sternohyoideus, + omohyoideus
duct of submaxillary
digastricus, posterior belly
submaxillary gland

Fig. 204.
Cervical Branches of the External Maxillary Artery

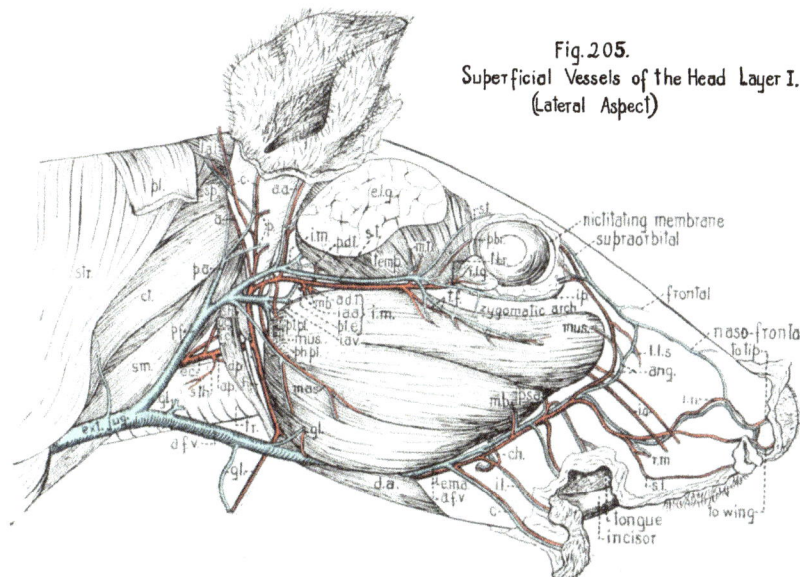

Fig. 205.
Superficial Vessels of the Head Layer I.
(Lateral Aspect)

| | | | |
|---|---|---|---|
| a. | auricular | l.l.s. | to levator labii superioris and vibrissae pad |
| a.a. | anterior auricular | l.n. | lateral nasal |
| a.d.t. | anterior deep temporal | mas | masseteric |
| a.f.v. | anterior facial vein | m b. | to masseter and buccinator |
| ang. | angular | m.t. | middle temporal |
| a.p. | ascending pharyngeal | mus. | muscular |
| c. | cutaneous | oc. | occipital |
| ch. | to lining of cheek | p. | parotid |
| cm. | cleidomastoideus | p.a. | posterior auricular |
| c.t. | clavotrapezius | p.br. | palpebral branch of zygomaticoorbital |
| d.a. | digastricus anterior | p.d.t | posterior deep temporal |
| d.p. | digastricus posterior | p.f. | posterior facial |
| e.c | external carotid | ph.pl. | to pharyngeal plexus |
| e.l.q | exorbital lacrimal gland | pl. | platysma |
| e.m.a | external maxillary artery | p.s.a. | to posterior superior alveolar |
| gl. | glandular | pt.e. | to pterygoideus externus and internus |
| h. | hyoid bone | pt. pl. | vein from pterygoid plexus |
| i.a.a. | inferior alveolar artery | r.m. | to roof of mouth and gum of incisor |
| i.a.v. | inferior alveolar vein | s.h. | stylohyoideus |
| i.c | internal carotid | s.l. | superior labial |
| i.l. | inferior labial | sm. | sternomastoideus |
| i.l.q. | intraorbital lacrimal gland | sp. | splenius |
| i.m. | internal maxillary | s.t. | superficial temporal |
| i.o. | infraorbital to vibrissae, from internal carotid | s.th. | superior thyroid |
| | | str. | spinotrapezius |
| i.p. | inferior palpebral | temp. | temporalis |
| l.a.l. | levator auris longus | tr. | trachea |
| l.br. | lacrimal branch of zygomaticoorbital | t.f. | transverse facial |

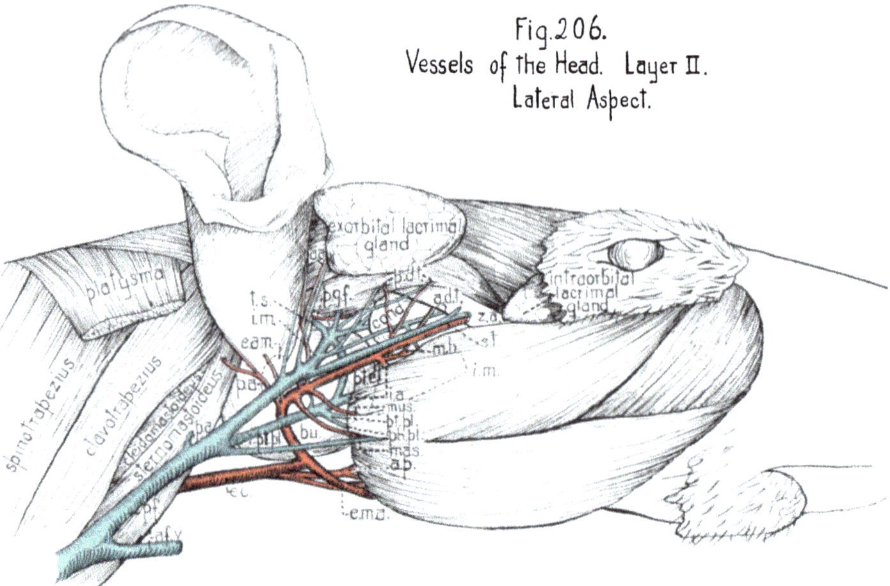

Fig.206.
Vessels of the Head. Layer II.
Lateral Aspect.

a.a. anterior auricular
a.d.t. anterior deep temporal, supplies
temperomandibular joint and runs
beneath zygomatic arch to orbit
a.f.v. anterior facial vein
a.p. ascending pharyngeal
bu. bulla
cond. condyle
e.a.m. external auditory meatus
e.m.a. external maxillary artery
i.a. inferior alveolar

i.m. internal maxillary
m.b. to masseter and buccinator
mus. muscular
p.a. posterior auricular
p.d.t. posterior deep temporal
p.f. posterior facial
p.g.f. postglenoid foramen
ph.pl. to pharyngeal plexus
pt.e. to pterygoid externus and internus
pt.pl. vein from pterygoid plexus
z.a. zygomatic arch
t.s. transverse sinus

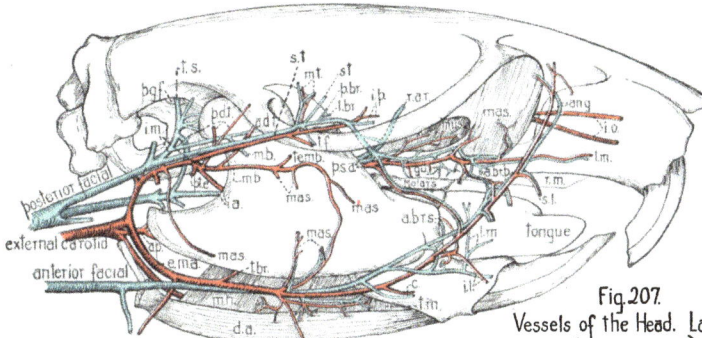

Fig. 207.
Vessels of the Head. Layer III.
(Lateral Aspect)

a.br.p. anastomotic branch to palatine
a.br.s. anastomotic branch to submental
a.d.t. anterior deep temporal
ang. angular
a.p. ascending pharyngeal
c. cutaneous
d.a. digastricus anterior
e.m.a. external maxillary art.
gu. gums
i.m. internal maxillary
i.o. infraorbital of int. carotid
i.p. inferior palpebral

l.br. lacrimal branch
l.m. to lining of mouth
mas. to masseter
mb. to masseter and buccinator
m.h. mylohyoid
m.t. middle temporal
p.br. palpebral branch
p.d.t. posterior deep temporal
p.g.f. postglenoid foramen
p.s.a. posterior superior alveolar

p.t.e. to pterygoideus externus + internus
r.a.r. runs over alveolar ridge
           to orbit
r.m. roof of mouth
s.l. superior labial
s.t. superficial temporal
temp. to temporalis
t.br. tonsillar branch
t.f. transverse facial
t.m. transversus mandibularis
t.s. transverse sinus

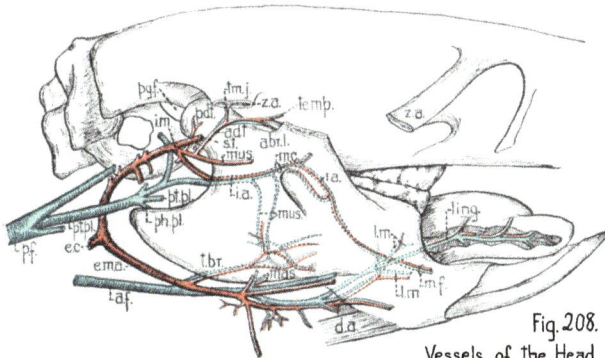

Fig. 208.
Vessels of the Head. Layer IV
(Lateral Aspect)

a.br.l. anastomotic branch to lacrimal
a.d.t. anterior deep temporal
a.f. anterior facial
d.a. digastricus anterior
e.c. external carotid
e.m.a. external maxillary artery
i.a. inferior alveolar
ling. lingual
l.m. to lining of mouth
mas. to masseter
m.c. mandibular canal
i.m. internal maxillary

m.f. mental foramen
mus. muscular
p.d.t. posterior deep temporal
p.f. posterior facial
p.g.f. postglenoid foramen
ph.pl. from pharyngeal plexus
pt.pl. from pterygoid plexus
t.br. tonsillar branch
temp. to temporalis
t.m.j. to temporomandibular joint
z.a. zygomatic arch, cut
s.t. superficial temporal

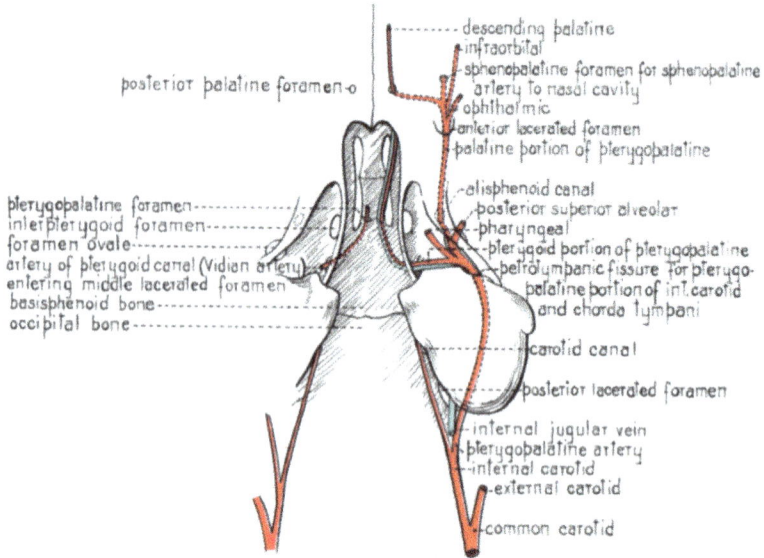

descending palatine
infraorbital
sphenopalatine foramen for sphenopalatine
artery to nasal cavity
ophthalmic
anterior lacerated foramen
palatine portion of pterygopalatine

posterior palatine foramen-o

alisphenoid canal
posterior superior alveolar
pharyngeal
pterygoid portion of pterygopalatine
retrotympanic fissure for pterygopalatine portion of int. carotid
and chorda tympani

pterygopalatine foramen
interpterygoid foramen
foramen ovale
artery of pterygoid canal (Vidian artery)
entering middle lacerated foramen
basisphenoid bone
occipital bone

carotid canal

posterior lacerated foramen

internal jugular vein
pterygopalatine artery
internal carotid
external carotid
common carotid

Fig. 209.  Branches of Internal Carotid Artery
in Relation to the Ventral Aspect of the Cranium.

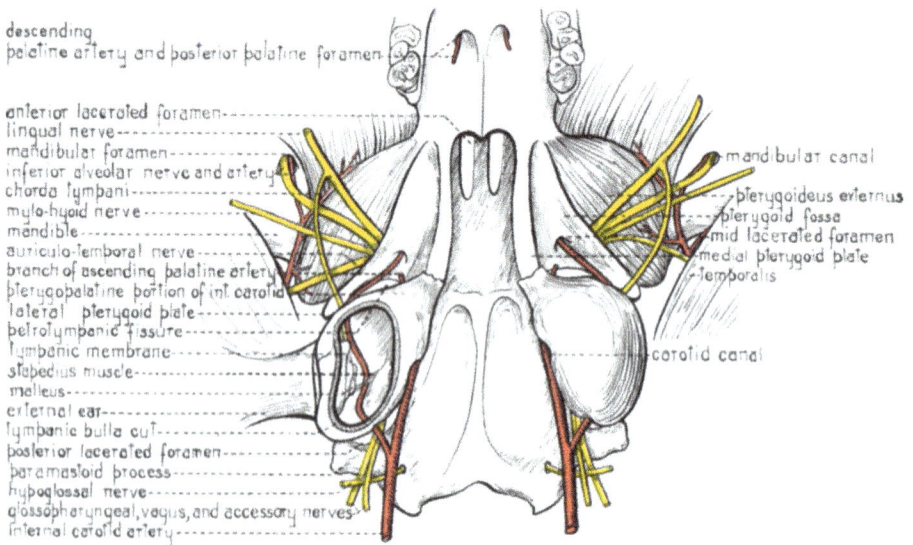

descending
palatine artery and posterior palatine foramen

anterior lacerated foramen
lingual nerve
mandibular foramen
inferior alveolar nerve and artery
chorda tympani
mylo-hyoid nerve
mandible
auriculo-temporal nerve
branch of ascending palatine artery
pterygopalatine portion of int. carotid
lateral pterygoid plate
retrotympanic fissure
tympanic membrane
stapedius muscle
malleus
external ear
tympanic bulla cut
posterior lacerated foramen
paramastoid process
hypoglossal nerve
glossopharyngeal, vagus, and accessory nerves
internal carotid artery

mandibular canal
pterygoideus externus
pterygoid fossa
mid lacerated foramen
medial pterygoid plate
temporalis

carotid canal

Fig. 210.   Internal Carotid Artery and Chorda Tympani

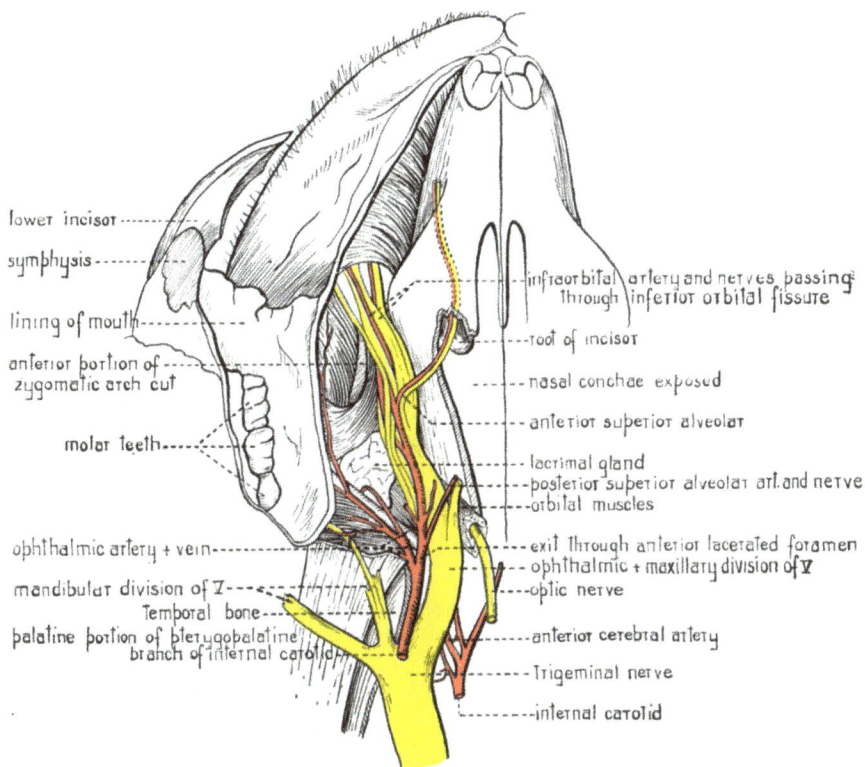

lower incisor

symphysis

lining of mouth

anterior portion of
zygomatic arch cut

molar teeth

ophthalmic artery + vein

mandibular division of Ⅴ
Temporal bone
palatine portion of pterygopalatine
branch of internal carotid

infraorbital artery and nerves passing
through inferior orbital fissure

root of incisor

nasal conchae exposed

anterior superior alveolar

lacrimal gland

posterior superior alveolar art. and nerve

orbital muscles

exit through anterior lacerated foramen

ophthalmic + maxillary division of Ⅴ

optic nerve

anterior cerebral artery

trigeminal nerve

internal carotid

Fig. 211  Base of skull dissected to show arteries and nerves entering orbit.

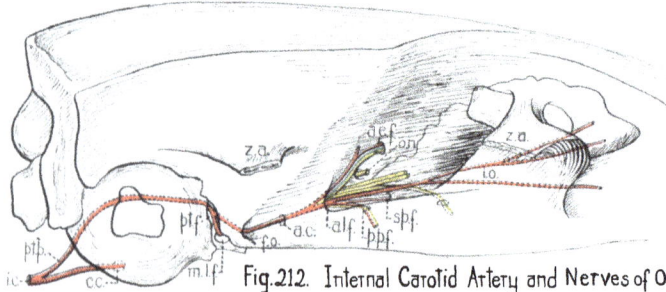

Fig. 212. Internal Carotid Artery and Nerves of Orbit.

| | | | |
|---|---|---|---|
| a.c. | alisphenoid canal | o.n. | optic nerve |
| a.e.f. | anterior ethmoidal foramen for nasociliary branch of trigeminal and anterior ethmoidal artery and vein | p.p.f. | posterior palatine foramen for descending palatine artery + nerve |
| a.l.f. | anterior lacerated foramen | p.t.f. | petrotympanic fissure |
| c.c. | carotid canal | p.t.p. | pterygopalatine portion of int. carotid |
| f.o. | foramen ovale | s.p.f. | sphenopalatine foramen for nasal nerves and sphenopalatine artery to nasal cavity |
| i.c. | internal carotid | | |
| i.o. | infraorbital | | |
| m.l.f. | middle lacerated foramen for artery of pterygoid canal | z.a. | zygomatic arch, cut. |

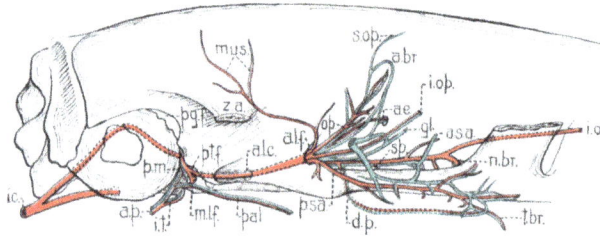

Fig. 213. Internal Carotid Artery and Veins of the Orbit

| | | | |
|---|---|---|---|
| a.br. | anastomotic branch | mus. | muscular |
| a.e. | anterior ethmoidal | n.br. | nasal branch of infraorbital |
| a.l.c. | alisphenoid canal | o.p. | ophthalmic |
| a.l.f. | anterior lacerated foramen | pal. | palatine branch of ascending palatine |
| a.p. | ascending palatine of ext. car. | | |
| a.s.a. | anterior superior alveolar branch of infraorbital | p.g.f. | postglenoid foramen |
| | | p.m. | posterior meningeal branch of ascending palatine |
| d.p. | descending palatine, running through palatine foramen | p.s.a. | posterior superior alveolar |
| gl. | to Harderian gland | p.t.f. | petrotympanic fissure |
| i.c. | internal carotid | s.o.b. | superior ophthalmic |
| i.o. | infraorbital | s.p. | sphenopalatine, runs along wall of nasal cavity |
| i.o.p. | inferior ophthalmic, thru Harderian gland to inner corner of eye | | |
| i.t. | inferior tympanic branch of ascending palatine | t.br. | terminal branch continues to the incisive canal which it enters to anastomose with the sphenopalatine |
| m.l.f. | middle lacerated foramen for the artery of the pterygoid canal and the pharyngeal artery to the soft palate | z.a. | zygomatic arch, cut |

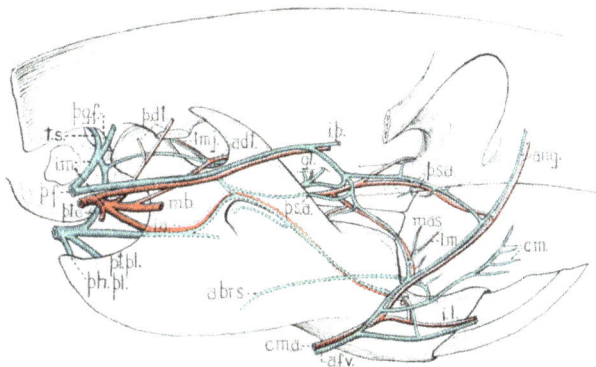

Fig. 214. Alveolar Vessels and their Anastomoses

| | | | |
|---|---|---|---|
| a.br.s. | anastomotic branch to submental | l.m. | to lining of mouth |
| a.d.t. | anterior deep temporal | mas. | to masseter |
| a.f.v. | anterior facial vein | m.b. | masseteric and buccinator branch |
| ang. | angular | p.d.t. | posterior deep temporal |
| c.m. | to corner of mouth | p.f. | posterior facial |
| e.ma | external maxillary artery | p.g.f. | postglenoid foramen |
| g.l. | from the orbit thru the Harderian gland | ph.pl. | to pharyngeal plexus |
| i.a. | inferior alveolar | p.s.a. | posterior superior alveolar |
| i.l. | inferior labial | pt.e. | to pterygoideus externus |
| i.m. | internal maxillary | pt.pl. | to pterygoid plexus |
| i.p. | inferior palpebral | tm.j | to temporomandibular joint |
| | | t.s. | transverse sinus |

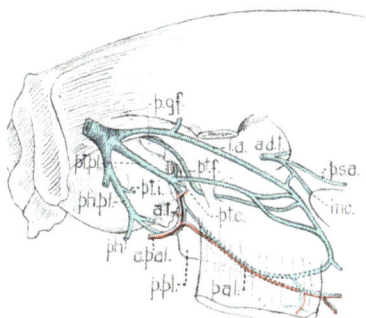

Fig. 215. Anastomosis of the Pterygoid and Pharyngeal Plexuses

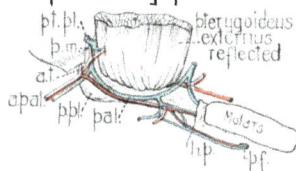

Fig. 216.
Detail showing palatine branch
of ascending palatine

| | | | |
|---|---|---|---|
| a.d.t. | anterior deep temporal | p.g.f. | postglenoid foramen |
| a.pal. | ascending palatine | ph. | pharyngeal branch |
| a.t. | to auditory tube, inf. tympanic | ph.pl. | from pharyngeal plexus |
| h.p. | hard palate | pp.l. | pterygoid plate |
| i.a. | inferior alveolar | pt.e. | to pterygoideus externus |
| inc. | to incisor | pt.i. | to pterygoideus internus |
| p.m. | postr. meningeal | pt.f. | petrotympanic fissure |
| pal. | palatine, covered by pterygoideus | pt.pl. | from pterygoid plexus |
| p.f. | palatine foramen | p.s.a. | posterior superior alveolar |

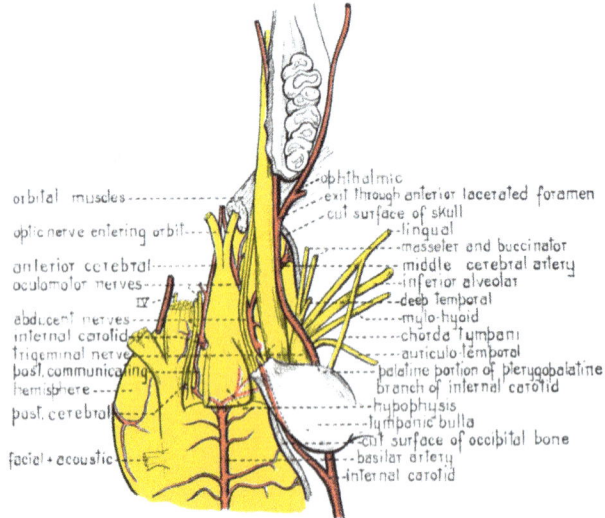

Fig. 217 Portions of skull removed to show relation of chorda tympani and trigeminal nerve to base of brain and branches of internal carotid.

orbital muscles
optic nerve entering orbit
anterior cerebral
oculomotor nerves
abducent nerves
internal carotid
trigeminal nerve
post. communicating
hemisphere
post. cerebral
facial + acoustic

ophthalmic
exit through anterior lacerated foramen
cut surface of skull
lingual
masseter and buccinator
middle cerebral artery
inferior alveolar
deep temporal
mylo-hyoid
chorda tympani
auriculo-temporal
palatine portion of pterygopalatine
branch of internal carotid
hypophysis
tympanic bulla
cut surface of occipital bone
basilar artery
internal carotid

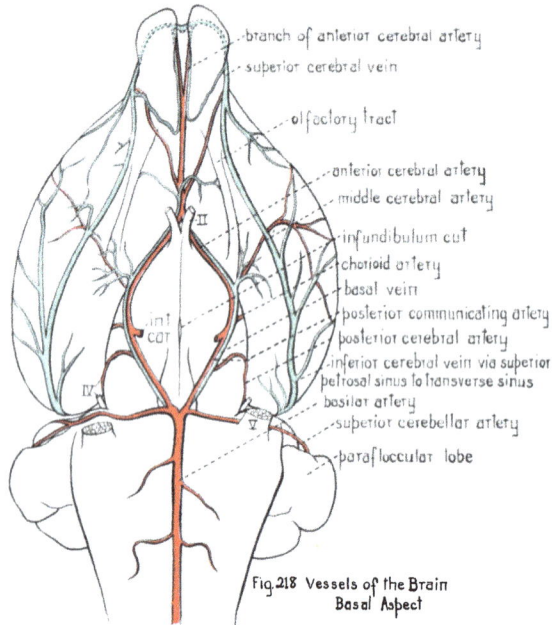

branch of anterior cerebral artery
superior cerebral vein
olfactory tract
anterior cerebral artery
middle cerebral artery
infundibulum cut
choroid artery
basal vein
posterior communicating artery
posterior cerebral artery
inferior cerebral vein via superior petrosal sinus to transverse sinus
basilar artery
superior cerebellar artery
paraflocculor lobe

Fig. 218 Vessels of the Brain
Basal Aspect

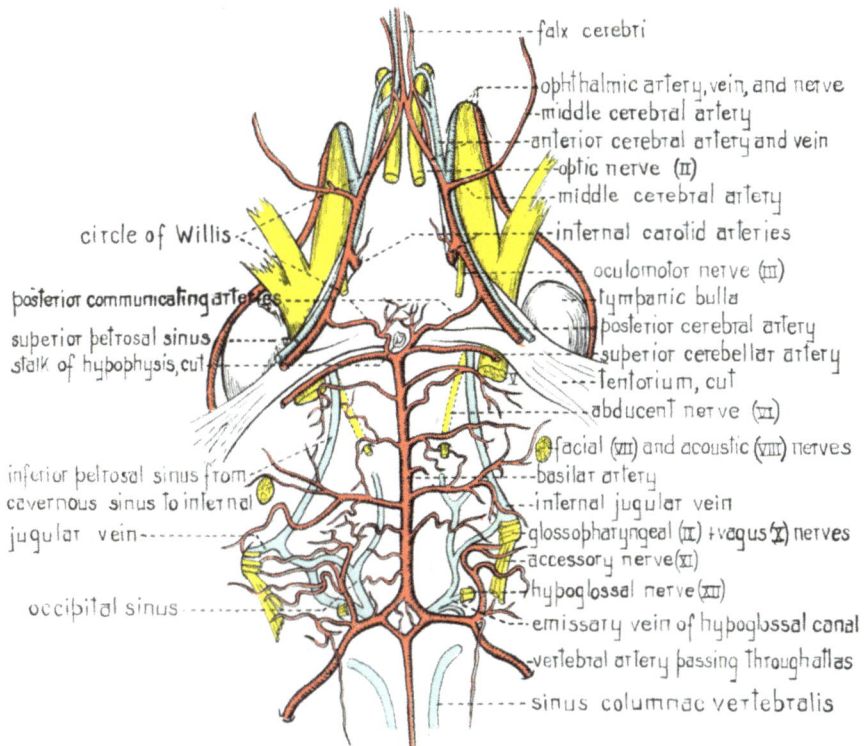

Fig. 219. Floor of cranium with brain removed, leaving membranes
to show arteries and veins of the base of the brain and
their relation to the exit of the cranial nerves.

inferior cerebral vein

middle cerebral artery
superior cerebral veins

superior sagittal sinus

super. petrosal sinus to Transverse sinus

(torcular Herophili), confluence of sinuses
transverse sinus

inferior cerebellar vein
superior cerebellar artery + vein
paraflocular lobe

posterior inferior cerebellar artery
inferior cerebellar vein
vertebral vein

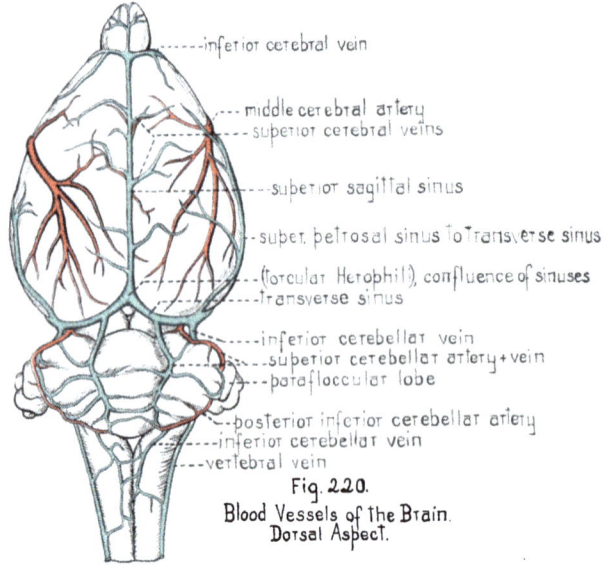

Fig. 220.
Blood Vessels of the Brain.
Dorsal Aspect.

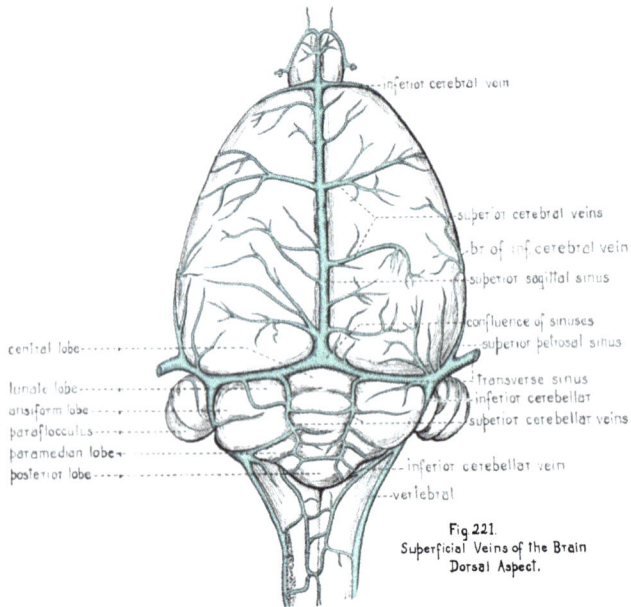

inferior cerebral vein

superior cerebral veins
br. of inf. cerebral vein
superior sagittal sinus

confluence of sinuses
superior petrosal sinus

central lobe

lunate lobe
ansiform lobe
paraflocculus
paramedian lobe
posterior lobe

transverse sinus
inferior cerebellar
superior cerebellar veins

inferior cerebellar vein
vertebral

Fig 221.
Superficial Veins of the Brain
Dorsal Aspect.

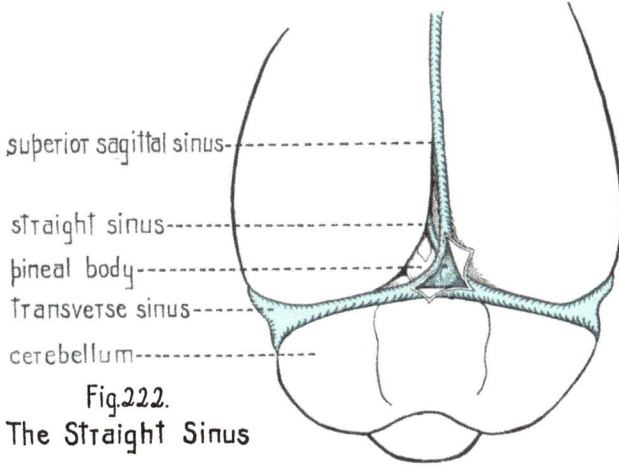

superior sagittal sinus -----------
straight sinus------------
pineal body------------
Transverse sinus-----
cerebellum------------

Fig.222.
The Straight Sinus

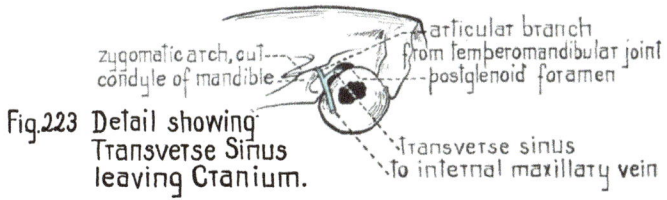

zygomatic arch,cut---
condyle of mandible----

articular branch
from temberomandibular joint
---bostglenoid foramen

Transverse sinus
to internal maxillary vein

Fig.223 Detail showing
Transverse Sinus
leaving Cranium.

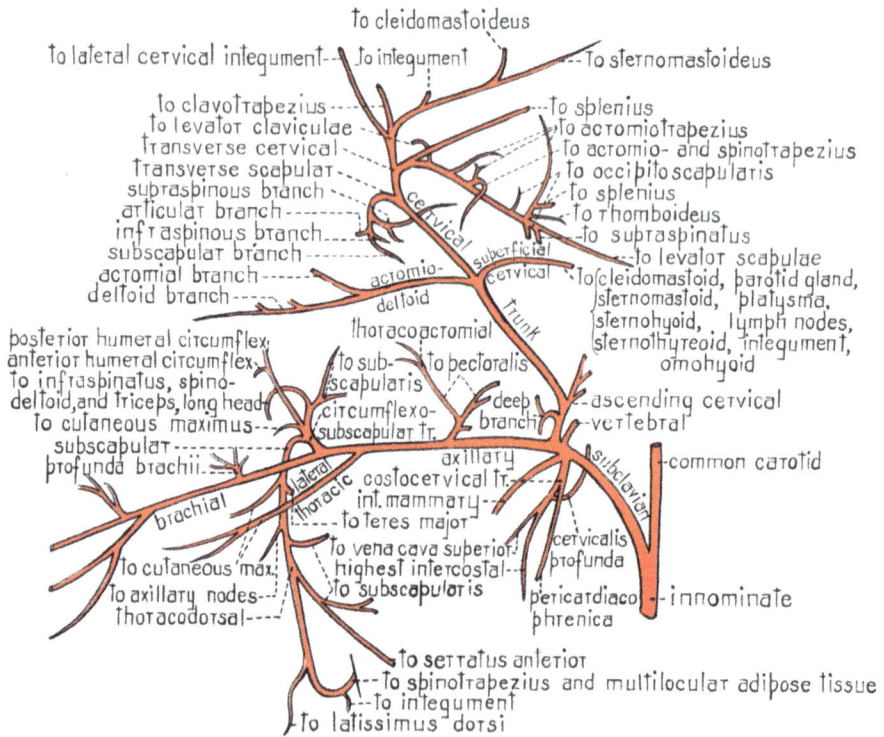

Fig. 224. Schema of branches of the subclavian artery.

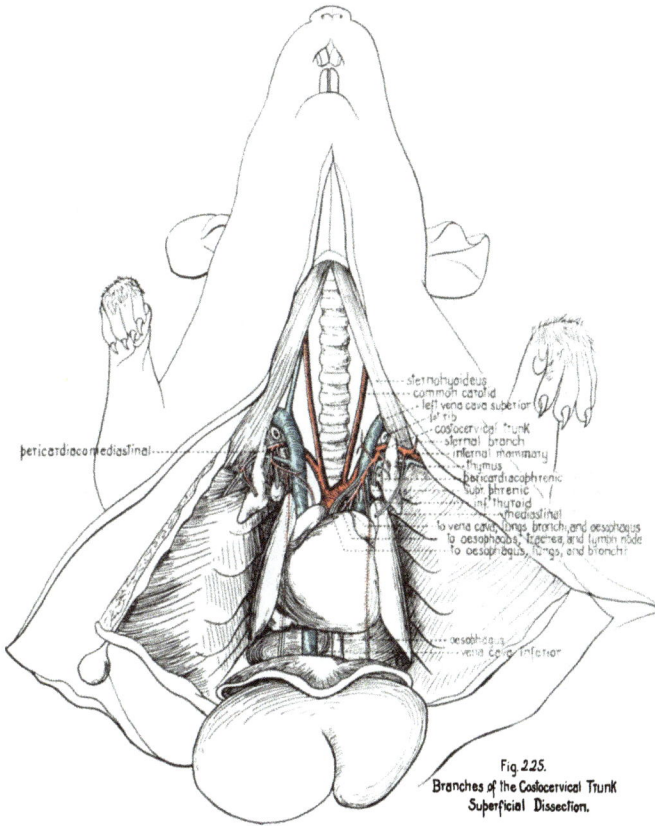

sternohyoideus
common carotid
left vena cava superior
1st rib
costocervical trunk
sternal branch
internal mammary
thymus
pericardiacophrenic
subr. phrenic
inf. thyroid
mediastinal
to vena cava, lungs, bronchi, and oesophagus
to oesophagus, trachea, and lymph node
to oesophagus, lungs, and bronchi

pericardiacomediastinal

oesophagus
vena cava inferior

Fig. 225.
Branches of the Costocervical Trunk
Superficial Dissection.

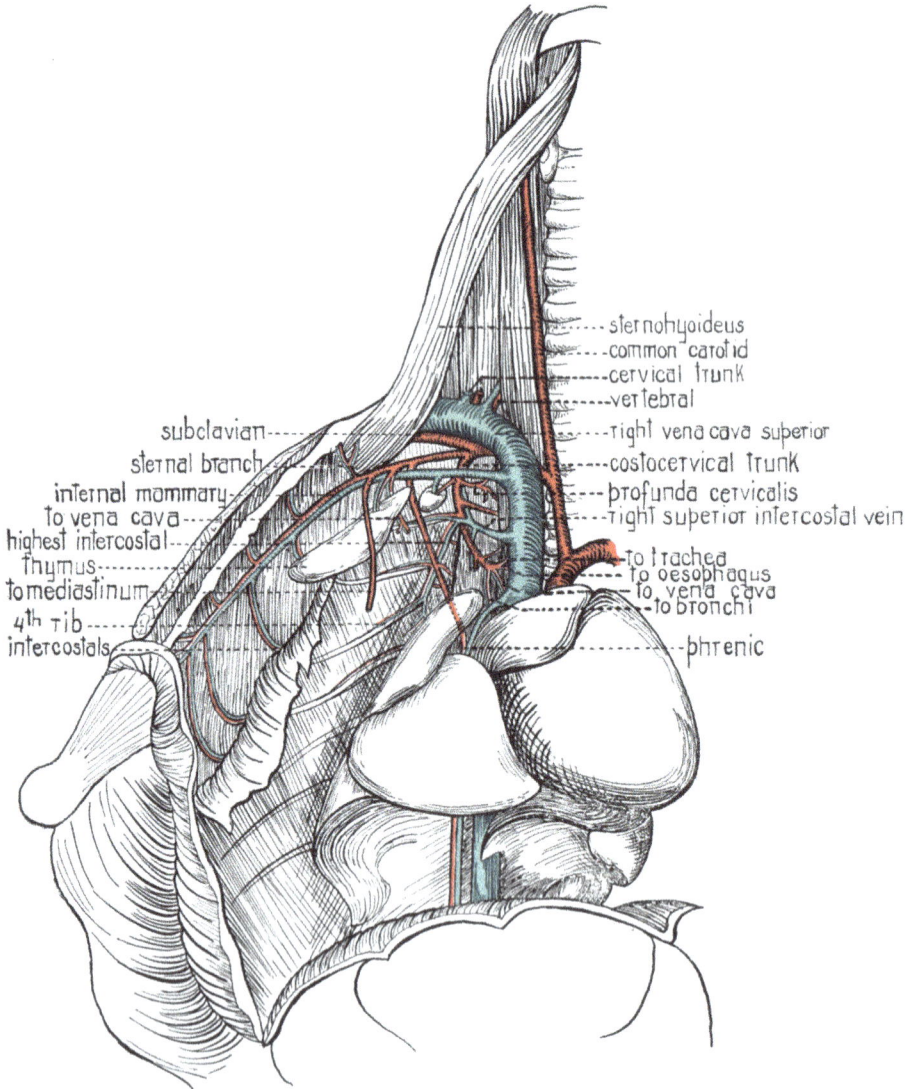

sternohyoideus
common carotid
cervical trunk
vertebral
right vena cava superior
costocervical trunk
profunda cervicalis
right superior intercostal vein
to trachea
to oesophagus
to vena cava
to bronchi
phrenic

subclavian
sternal branch
internal mammary
to vena cava
highest intercostal
thymus
to mediastinum
4th rib
intercostals

Fig.226  Branches of the Costocervical Trunk.    Deeper Dissection

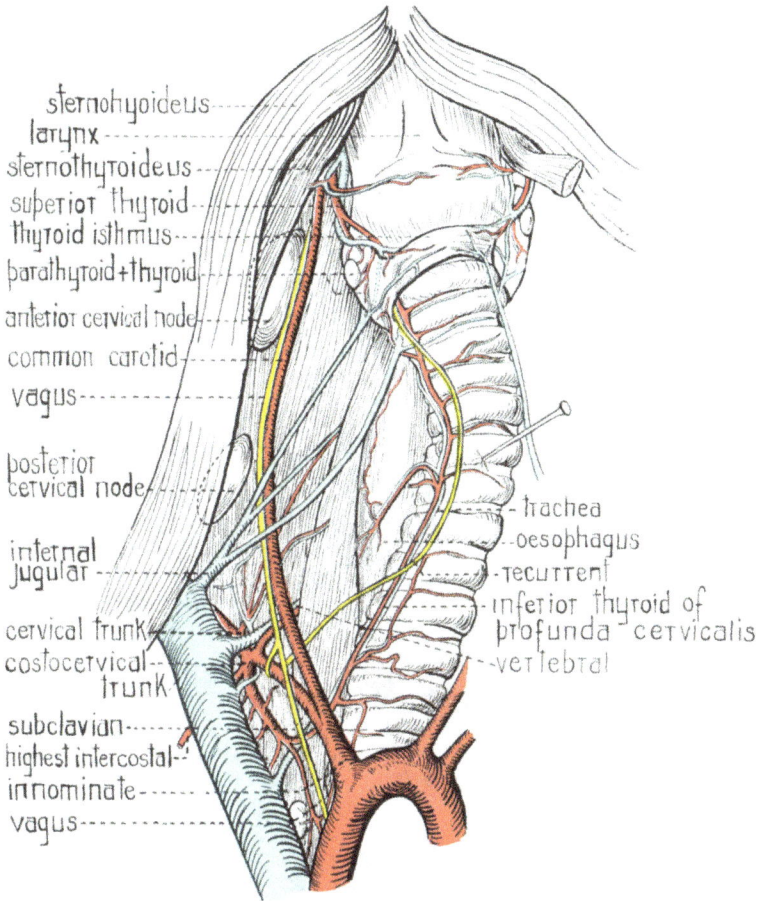

sternohyoideus
larynx
sternothyroideus
superior thyroid
thyroid isthmus
parathyroid + thyroid
anterior cervical node
common carotid
vagus

posterior
cervical node

internal
jugular

cervical trunk
costocervical
trunk

subclavian
highest intercostal
innominate
vagus

trachea
oesophagus
recurrent
inferior thyroid of
profunda cervicalis
vertebral

Fig. 227. Trachea drawn to left side to show Inferior Thyroid Vessels.

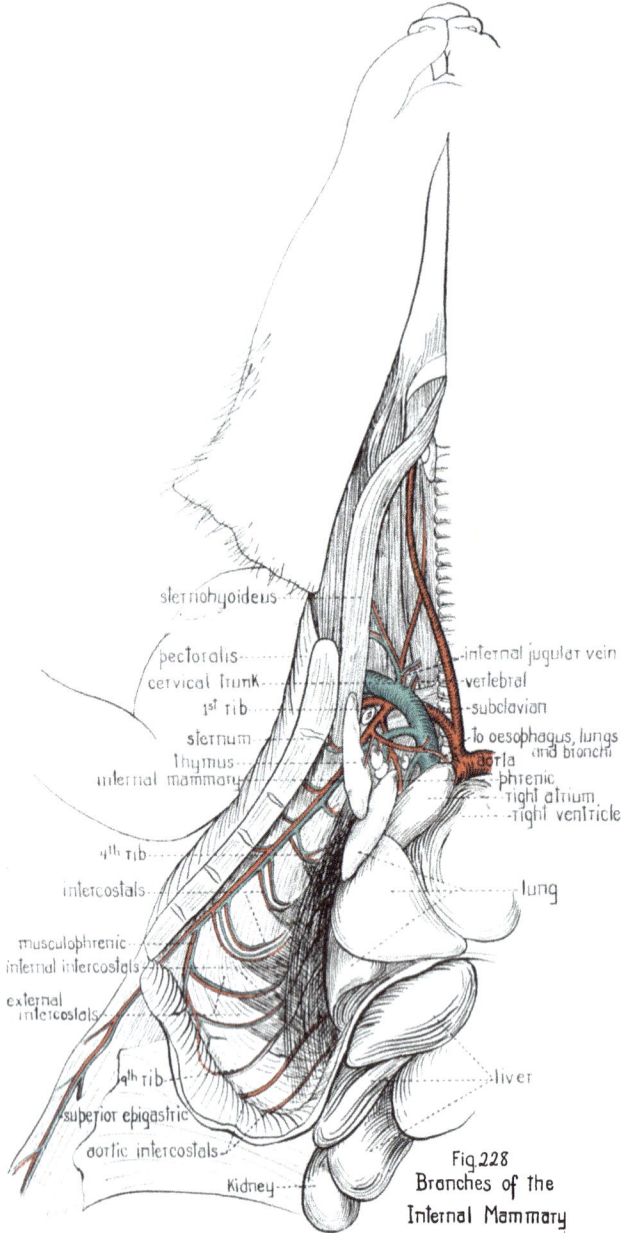

sternohyoideus

pectoralis
cervical trunk
1st rib
sternum
thymus
internal mammary

4th rib

intercostals

musculophrenic
internal intercostals

external
intercostals

8th rib
superior epigastric
aortic intercostals
kidney

internal jugular vein
vertebral
subclavian
to oesophagus, lungs
and bronchi
aorta
phrenic
right atrium
right ventricle

lung

liver

Fig. 228
Branches of the
Internal Mammary

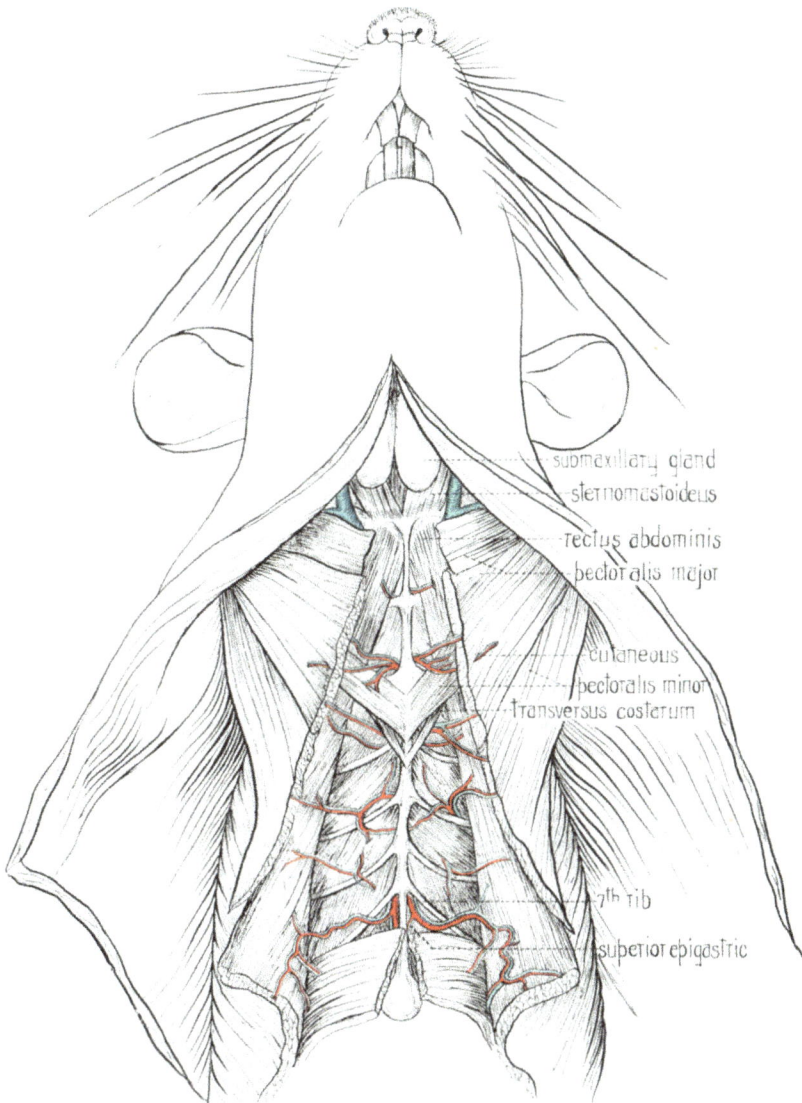

Fig. 229. Perforating Branches of Internal Mammary

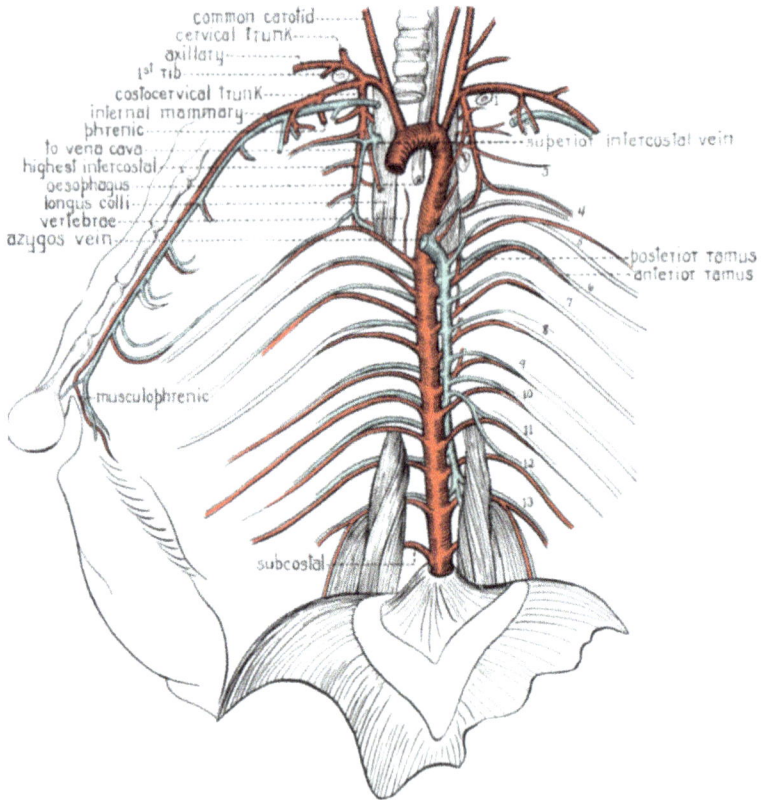

Fig.230. Intercostal Arteries, Intercostal and Azygos Veins

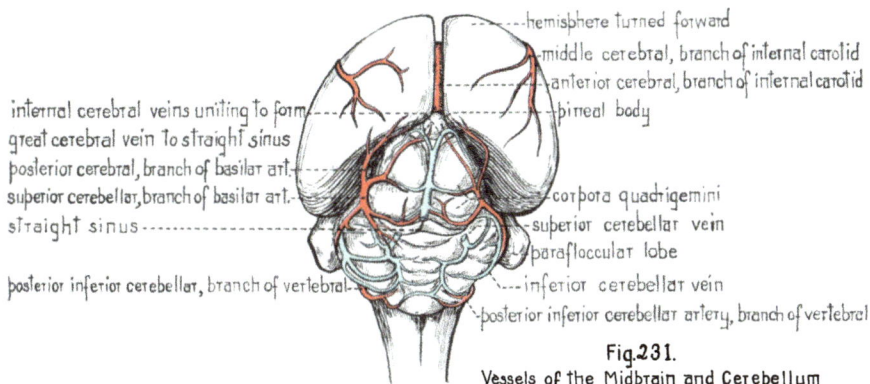

hemisphere turned forward
middle cerebral, branch of internal carotid
anterior cerebral, branch of internal carotid
pineal body

internal cerebral veins uniting to form
great cerebral vein to straight sinus
posterior cerebral, branch of basilar art.
superior cerebellar, branch of basilar art.
straight sinus

corpora quadrigemini
superior cerebellar vein
parafloccular lobe
inferior cerebellar vein
posterior inferior cerebellar artery, branch of vertebral

posterior inferior cerebellar, branch of vertebral

Fig. 231.
Vessels of the Midbrain and Cerebellum
Dorsal Aspect.
Branches of the Vertebral and Basilar Arteries.

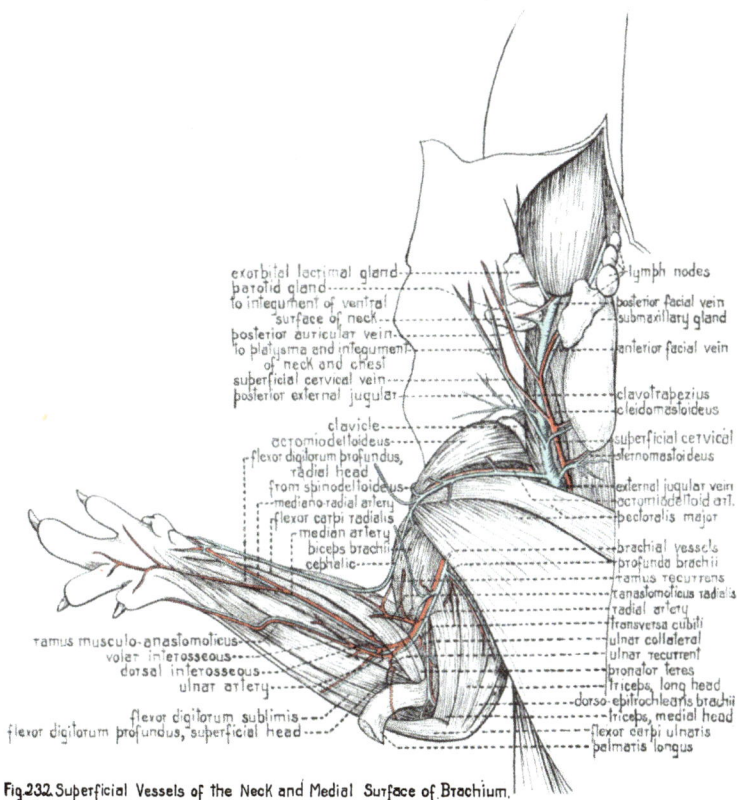

exorbital lacrimal gland
parotid gland
to integument of ventral
surface of neck
posterior auricular vein
to platysma and integument
of neck and chest
superficial cervical vein
posterior external jugular

clavicle
acromiodeltoideus
flexor digitorum profundus,
radial head
from spinodeltoideus
mediano-radial artery
flexor carpi radialis
median artery
biceps brachii
cephalic

lymph nodes
posterior facial vein
submaxillary gland
anterior facial vein

clavotrapezius
cleidomastoideus
superficial cervical
sternomastoideus

external jugular vein
acromiodeltoid art.
pectoralis major
brachial vessels
profunda brachii
ramus recurrens
anastomoticus radialis
radial artery
transversa cubiti
ulnar collateral
ulnar recurrent
pronator teres
triceps, long head
dorso-epitrochlearis brachii
triceps, medial head
flexor carpi ulnaris
palmaris longus

ramus musculo-anastomoticus
voler interosseous
dorsal interosseous
ulnar artery

flexor digitorum sublimis
flexor digitorum profundus, superficial head

Fig. 232. Superficial Vessels of the Neck and Medial Surface of Brachium.

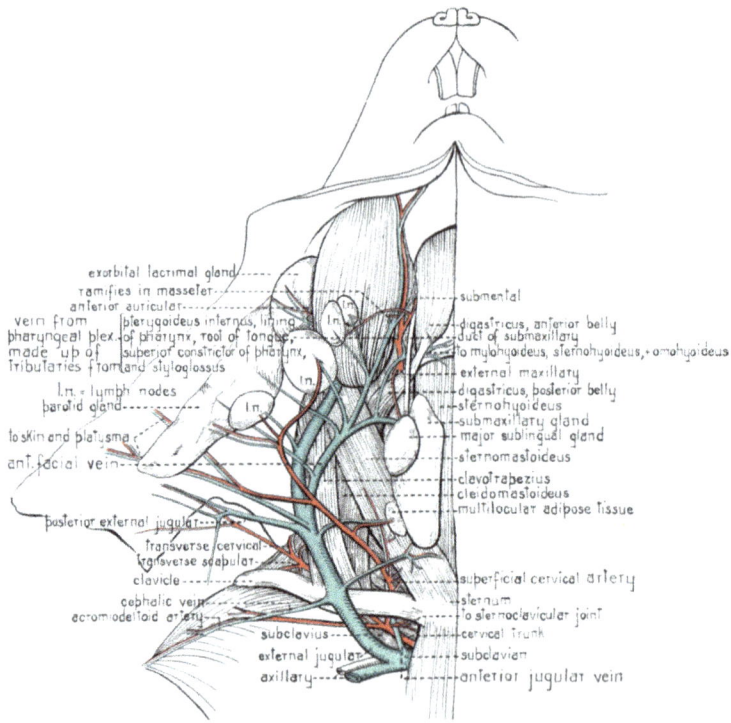

Fig.233. Superficial Vessels of the Neck.   Cervical Trunk.

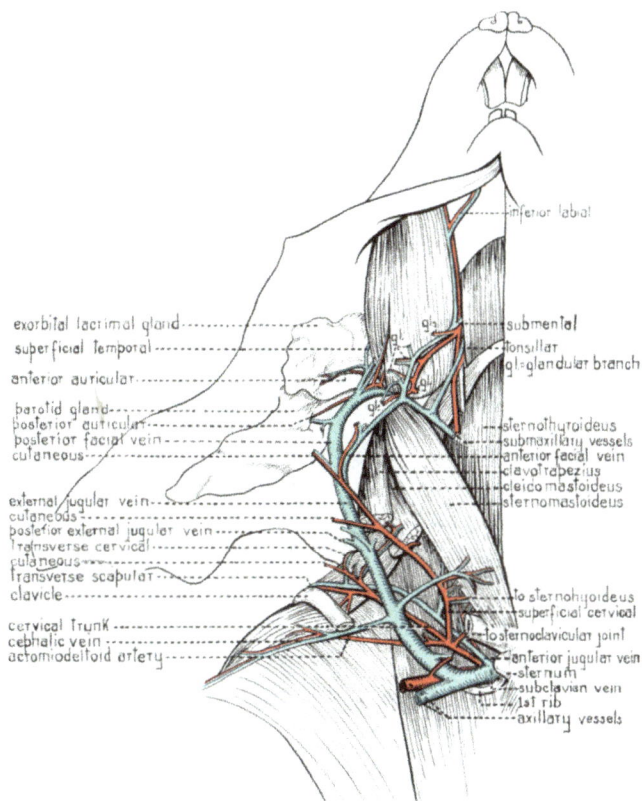

exorbital lacrimal gland
superficial temporal
anterior auricular

parotid gland
posterior auricular
posterior facial vein
cutaneous

external jugular vein
cutaneous
posterior external jugular vein
transverse cervical
cutaneous
transverse scapular
clavicle

cervical trunk
cephalic vein
actomiodeltoid artery

inferior labial

submental
tonsillar
glt-glandular branch

sternothyroideus
submaxillary vessels
anterior facial vein
clavotrapezius
cleicomastoideus
sternomastoideus

to sternohyoideus
superficial cervical
to sternoclavicular joint
anterior jugular vein
sternum
subclavian vein
1st rib
axillary vessels

Fig. 234. Clavicle cut to show branches of the cervical trunk.

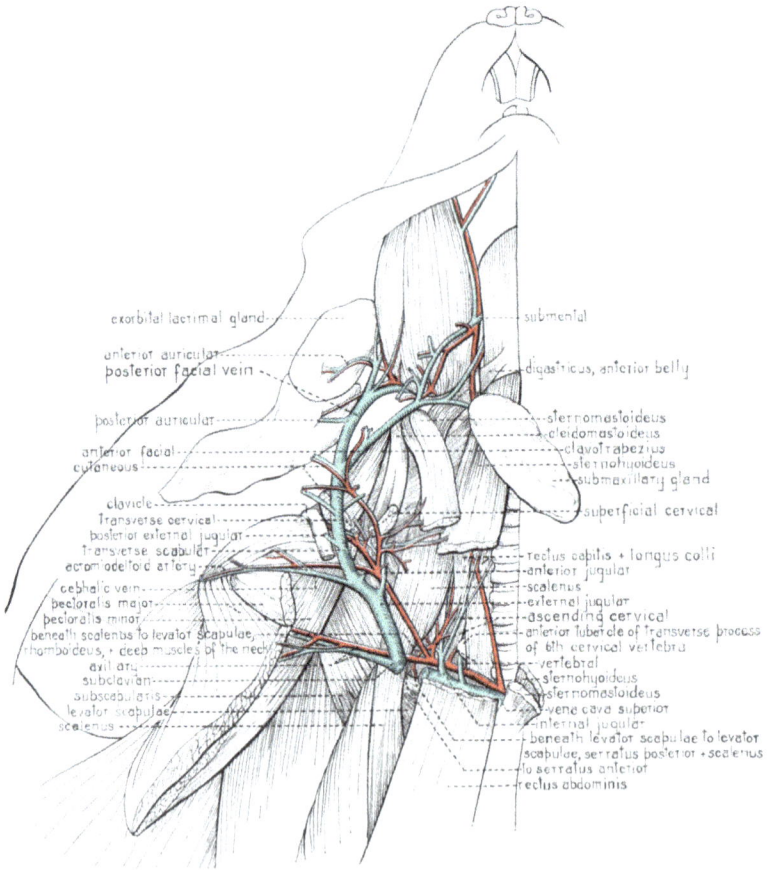

Fig. 235.  Branches of Subclavian Artery, Cervical Trunk, and Corresponding Veins.

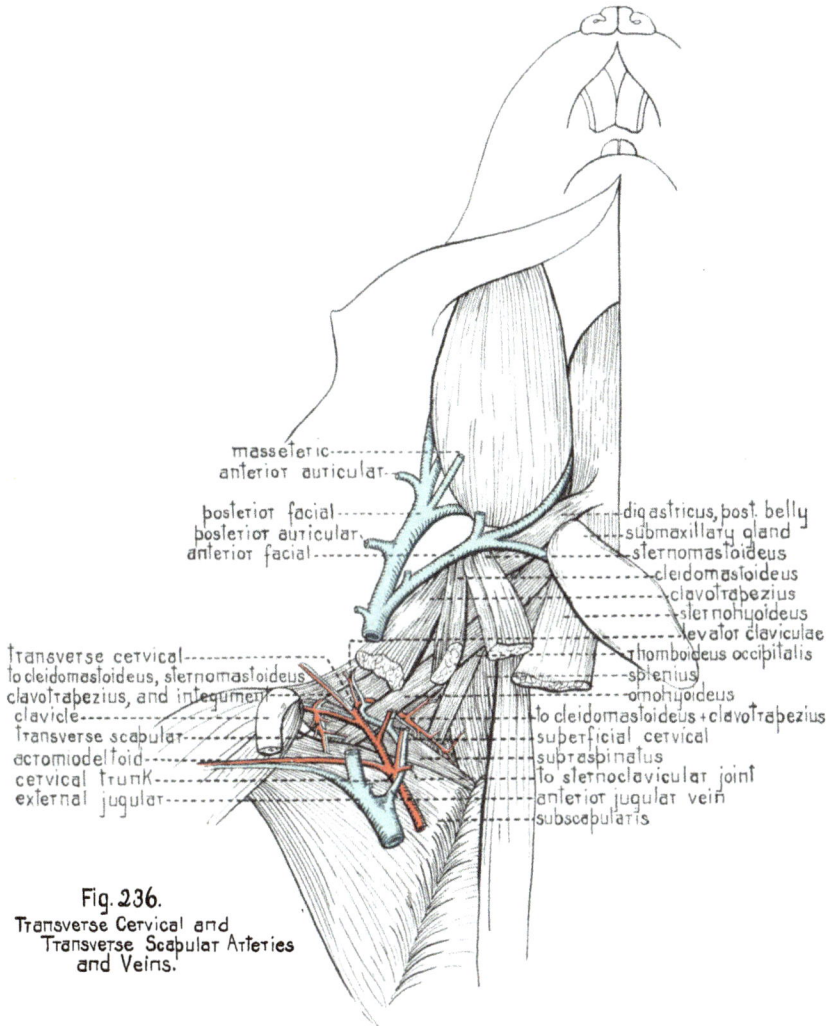

masseteric
anterior auricular

posterior facial
posterior auricular
anterior facial

digastricus, post. belly
submaxillary gland
sternomastoideus
cleidomastoideus
clavotrapezius
sternohyoideus
levator claviculae
rhomboideus occipitalis
splenius
omohyoideus
to cleidomastoideus + clavotrapezius
superficial cervical
supraspinatus
to sternoclavicular joint
anterior jugular vein
subscapularis

transverse cervical
to cleidomastoideus, sternomastoideus
clavotrapezius, and integument
clavicle
transverse scapular
acromiodeltoid
cervical trunk
external jugular

Fig. 236.
Transverse Cervical and
Transverse Scapular Arteries
and Veins.

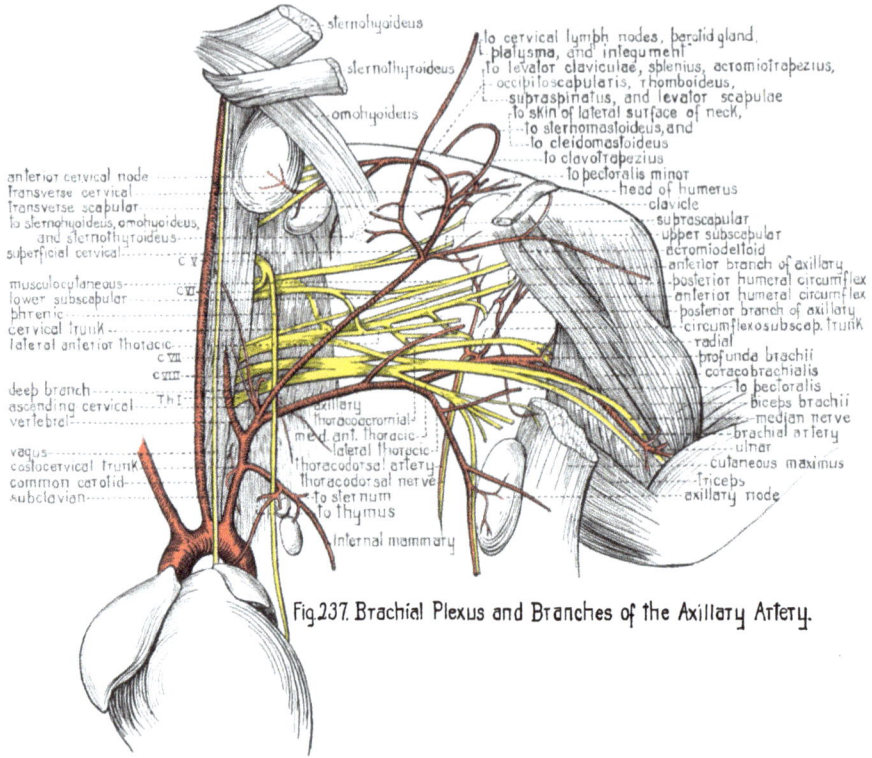

Fig. 237. Brachial Plexus and Branches of the Axillary Artery.

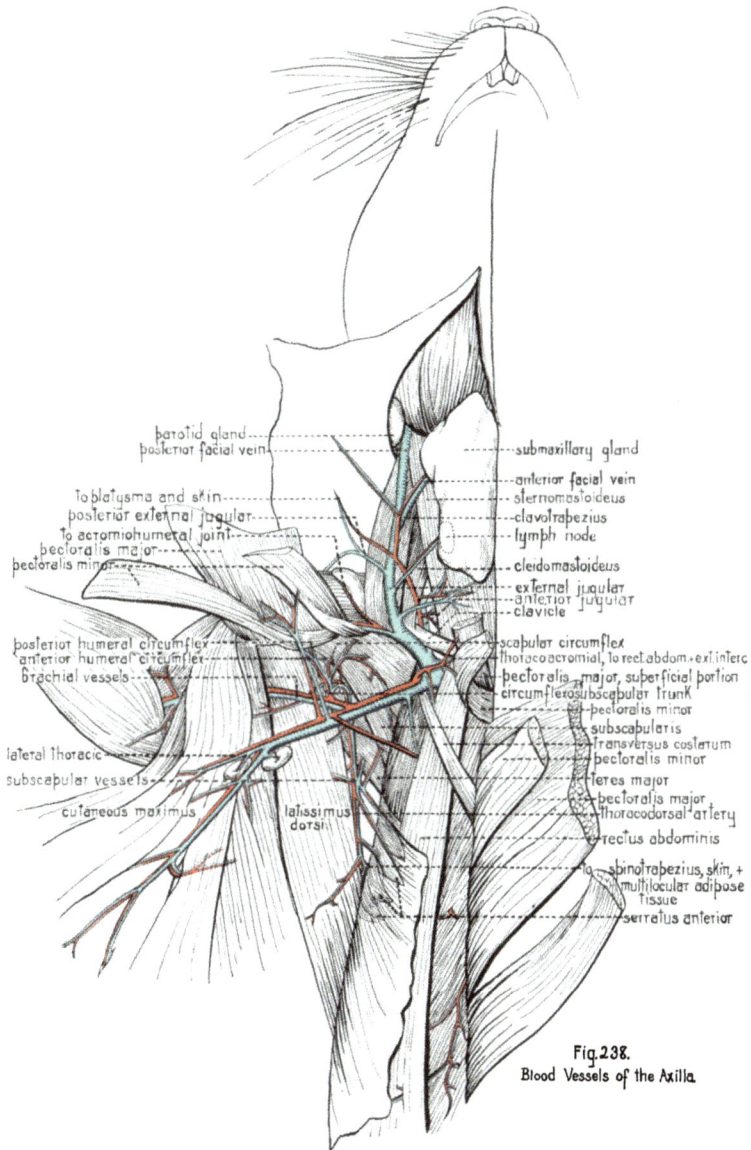

parotid gland
posterior facial vein

to platysma and skin
posterior external jugular
to acromiohumeral joint
pectoralis major
pectoralis minor

posterior humeral circumflex
anterior humeral circumflex
brachial vessels

lateral thoracic
subscapular vessels

cutaneous maximus

submaxillary gland
anterior facial vein
sternomastoideus
clavotrapezius
lymph node

cleidomastoideus

external jugular
anterior jugular
clavicle

scapular circumflex
thoracoacromial, to rect.abdom.+ext.interc.
pectoralis major, superficial portion
circumflexosubscapular trunk
pectoralis minor
subscapularis
transversus costarum
pectoralis minor
teres major
pectoralis major
thoracodorsal artery
rectus abdominis

to spinotrapezius, skin, +
multilocular adipose
tissue
serratus anterior

latissimus
dorsi

Fig. 238.
Blood Vessels of the Axilla

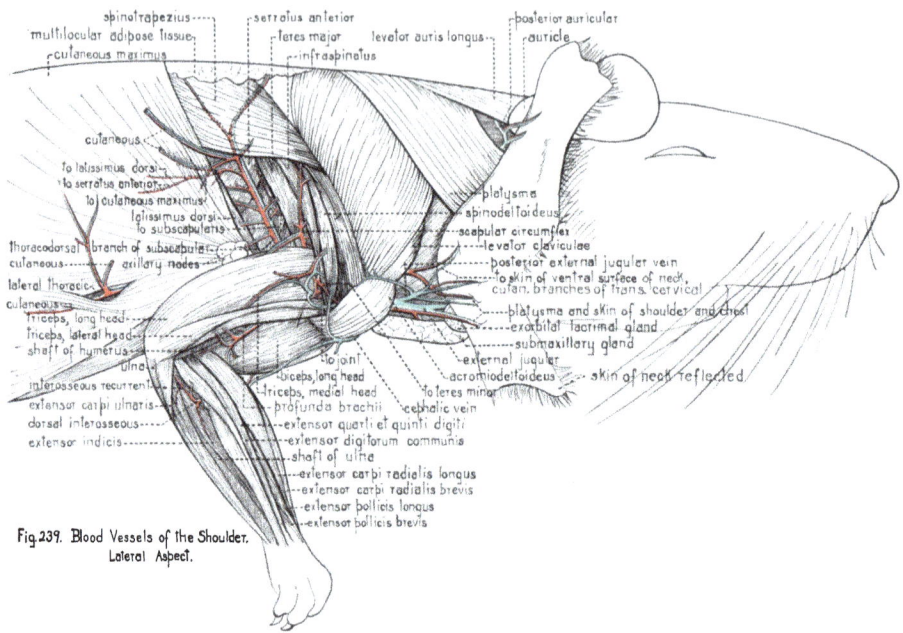

Fig. 239. Blood Vessels of the Shoulder.
Lateral Aspect.

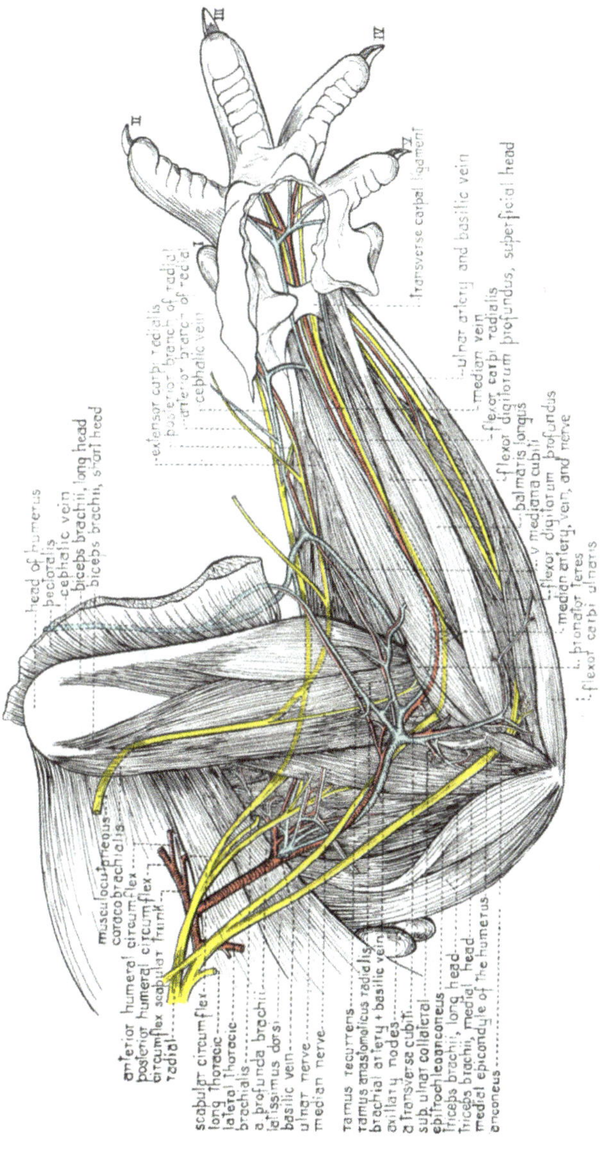

Fig. 240. Medial Aspect of the Brachium showing the Brachial Vessels
and Radial, Median, and Ulnar Nerves.

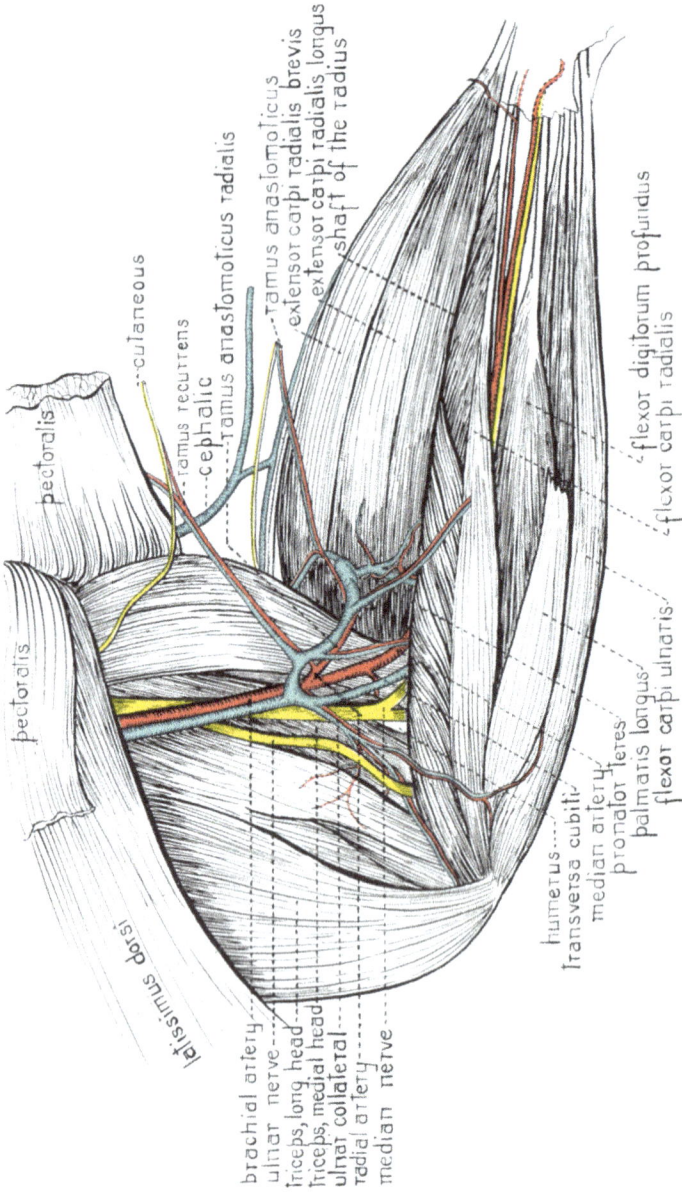

Fig. 241. Medial Aspect of the Brachium, showing the Brachial Artery crossing Ulnar and Median Nerves. (Case I)

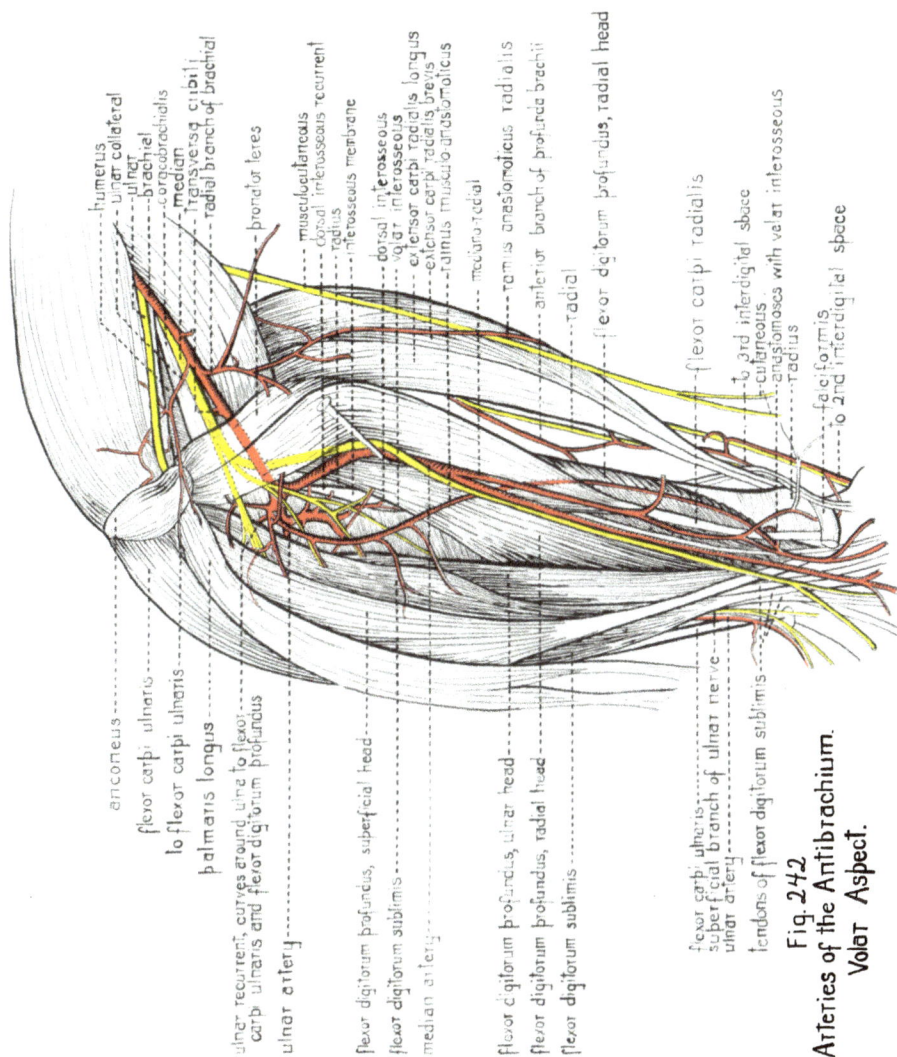

Fig. 242
Arteries of the Antibrachium.
Volar Aspect.

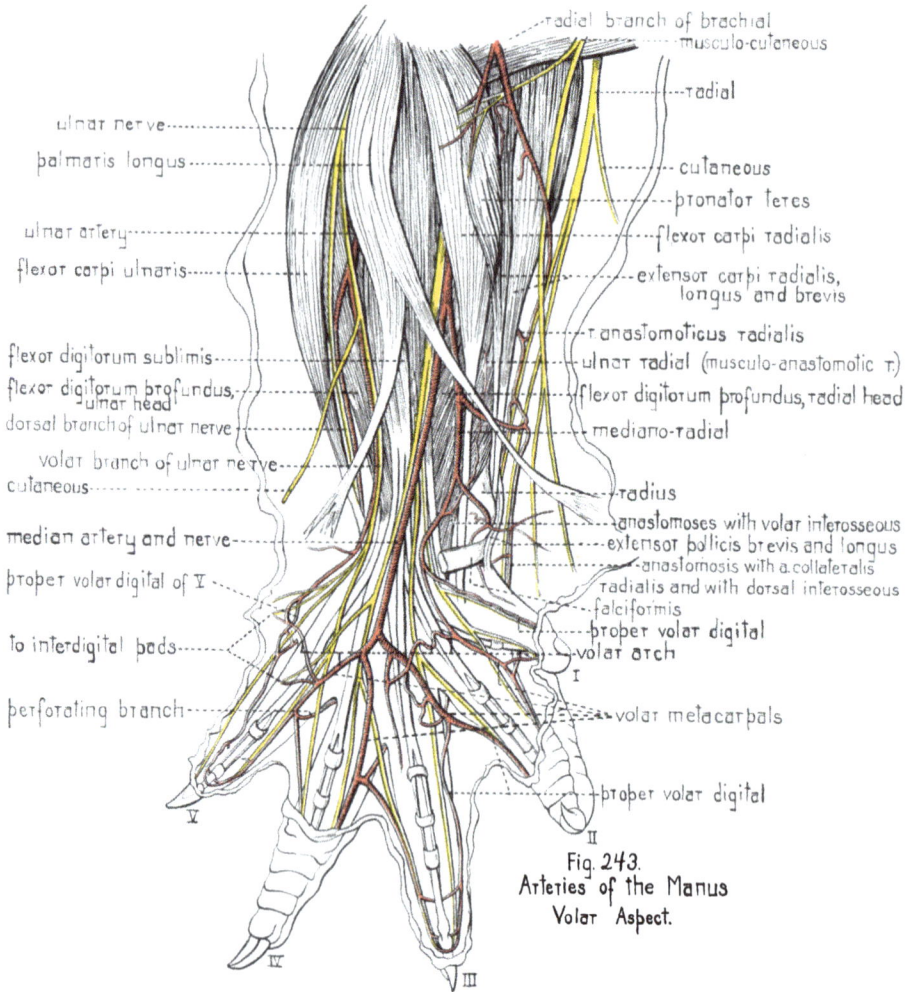

ulnar nerve

palmaris longus

ulnar artery

flexor carpi ulnaris

flexor digitorum sublimis

flexor digitorum profundus,
ulnar head

dorsal branch of ulnar nerve

volar branch of ulnar nerve

cutaneous

median artery and nerve

proper volar digital of V

to interdigital pads

perforating branch

radial branch of brachial

musculo-cutaneous

radial

cutaneous

pronator teres

flexor carpi radialis

extensor carpi radialis,
longus and brevis

r.anastomoticus radialis

ulnar radial (musculo-anastomotic r.)

flexor digitorum profundus, radial head

mediario-radial

radius

anastomoses with volar interosseous

extensor pollicis brevis and longus

anastomosis with a.collateralis

radialis and with dorsal interosseous

falciformis

proper volar digital

volar arch

I

volar metacarpals

proper volar digital

II

V

IV

III

Fig. 243.
Arteries of the Manus
Volar Aspect.

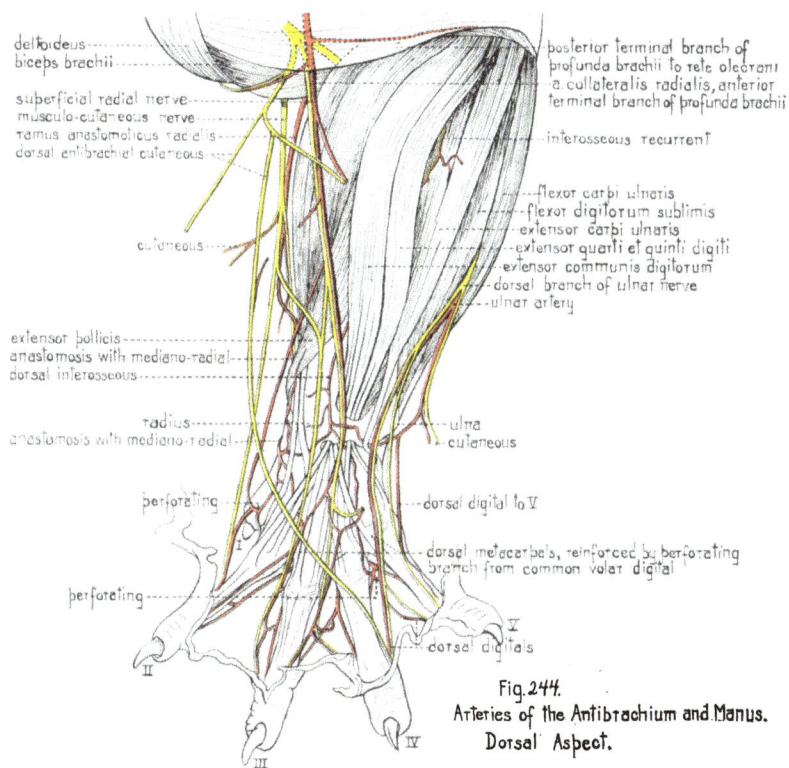

deltoideus
biceps brachii

superficial radial nerve
musculo-cutaneous nerve
ramus anastomoticus radialis
dorsal antibrachial cutaneous

cutaneous

extensor pollicis
anastomosis with mediano-radial
dorsal interosseous

radius
anastomosis with mediano-radial

perforating

perforating

posterior terminal branch of
profunda brachii to rete olecrani
a.collateralis radialis, anterior
terminal branch of profunda brachii

interosseous recurrent

flexor carpi ulnaris
flexor digitorum sublimis
extensor carpi ulnaris
extensor quarti et quinti digiti
extensor communis digitorum
dorsal branch of ulnar nerve
ulnar artery

ulna
cutaneous

dorsal digital to V

dorsal metacarpals, reinforced by perforating
branch from common volar digital

dorsal digitals

Fig. 244.
Arteries of the Antibrachium and Manus.
Dorsal Aspect.

extensor carpi radialis
extensor digitorum communis
accessory cephalic vein
cephalic vein
extensor carpi ulnaris
basilic vein
extensor pollicis
dorsal metacarpals

Fig. 245.
Superficial Veins
of the Antibrachium
Dorsal.

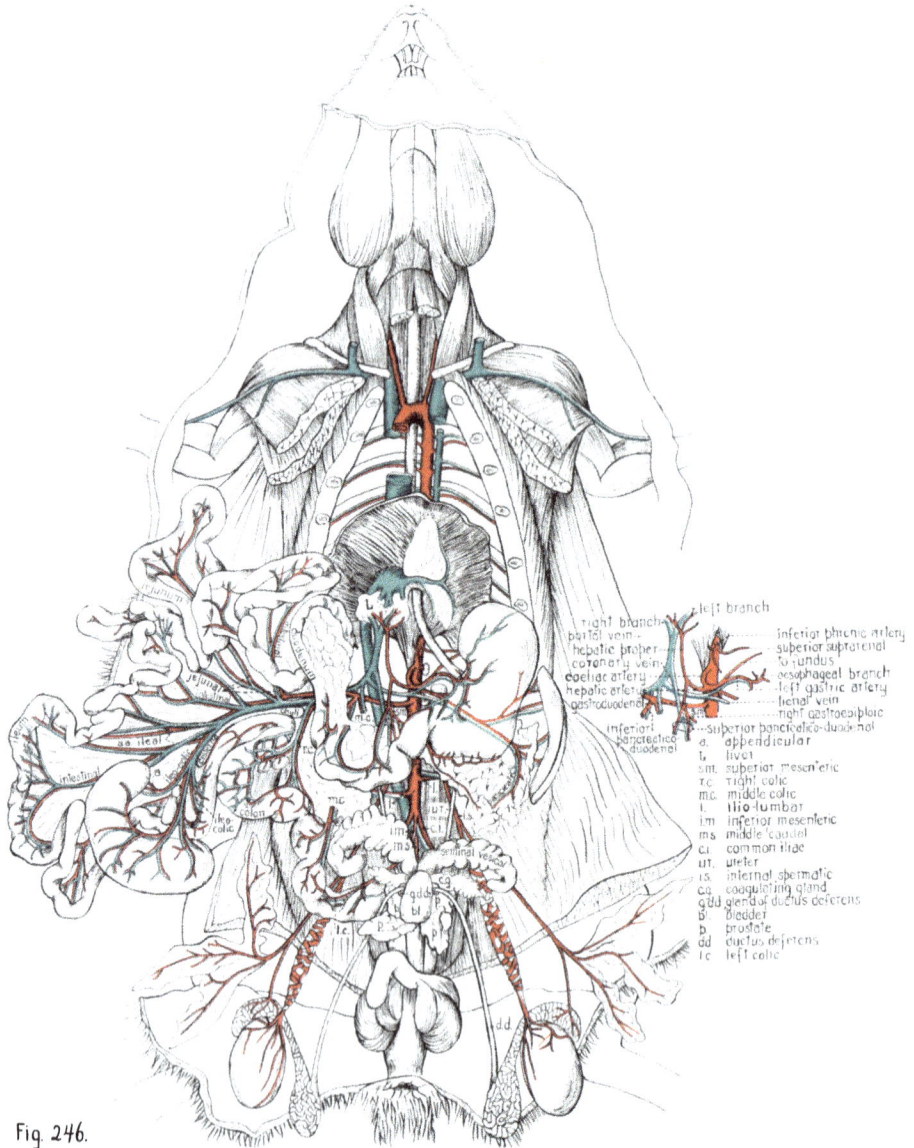

Fig. 246.

Liver removed and viscera displaced to show the portal system of veins and the abdominal branches of the aorta.

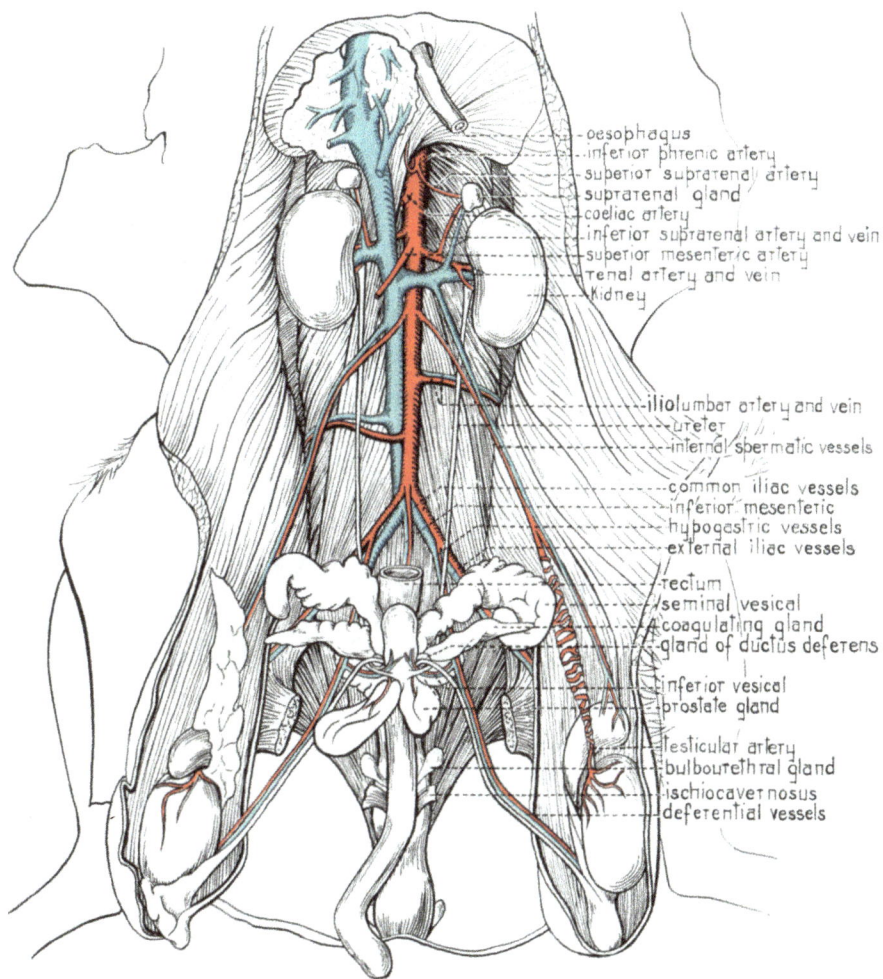

Fig. 247. Branches of the Abdominal Aorta and Vena Cava Inferior.

Labels (top to bottom, right side):
- oesophagus
- inferior phrenic artery
- superior suprarenal artery
- suprarenal gland
- coeliac artery
- inferior suprarenal artery and vein
- superior mesenteric artery
- renal artery and vein
- kidney
- iliolumbar artery and vein
- ureter
- internal spermatic vessels
- common iliac vessels
- inferior mesenteric
- hypogastric vessels
- external iliac vessels
- rectum
- seminal vesical
- coagulating gland
- gland of ductus deferens
- inferior vesical
- prostate gland
- testicular artery
- bulbourethral gland
- ischiocavernosus
- deferential vessels

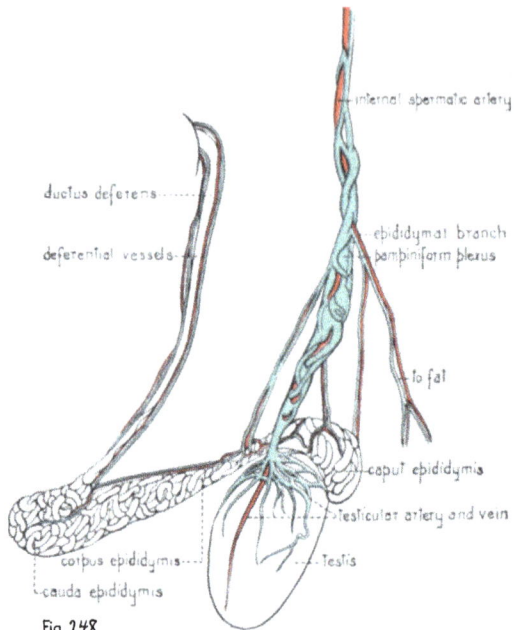

Fig. 248
Internal Spermatic Artery to the Testis and Epididymis.

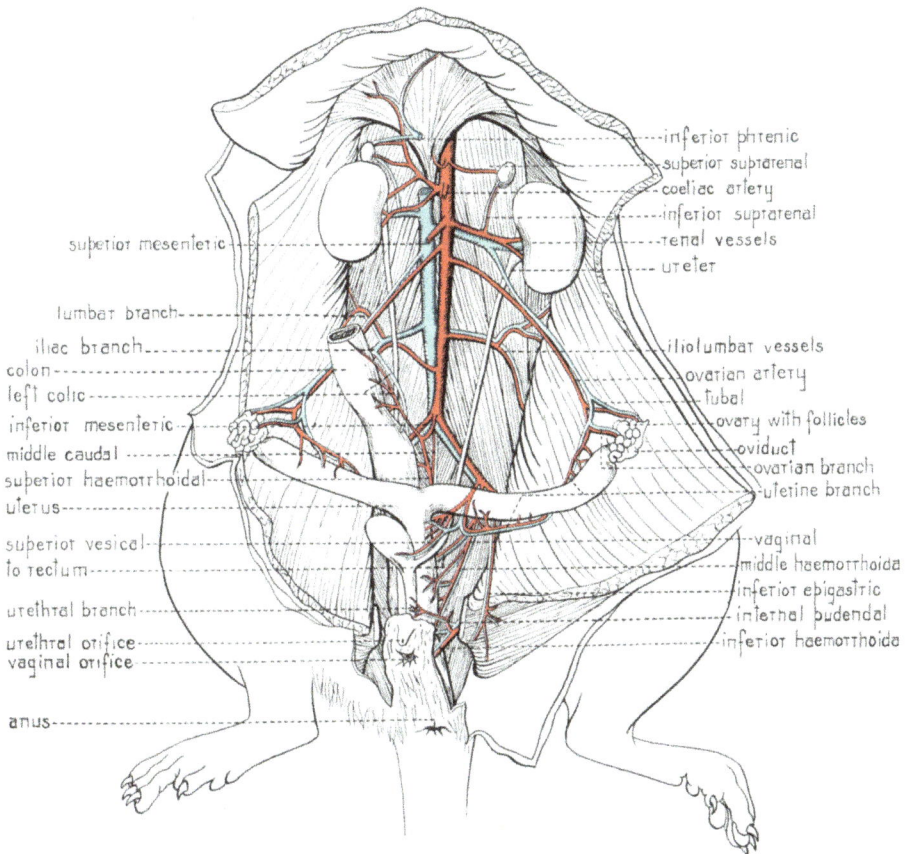

Fig. 249. Blood Vessels of the Female Urogenital System. Ovarian Artery and Vein.

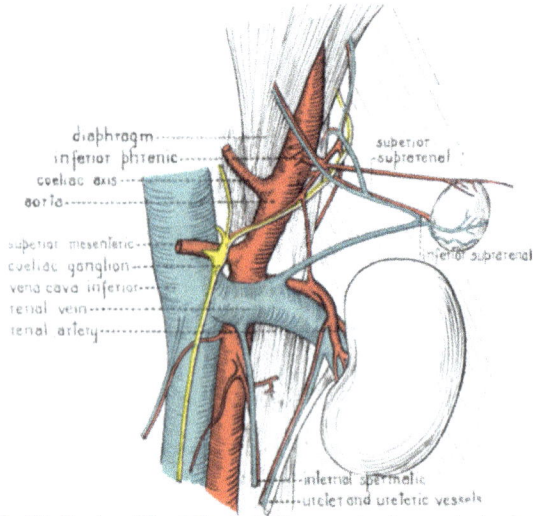

diaphragm
inferior phrenic
coeliac axis
aorta

superior mesenteric
coeliac ganglion
vena cava inferior
renal vein
renal artery

superior subrenal

inferior subrenal

internal spermatic
ureter and ureteric vessels

Fig.250. Blood supply of the left Kidney and subrenal gland

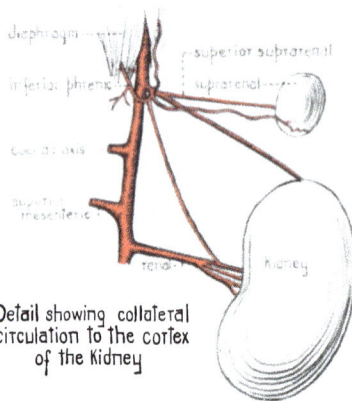

diaphragm
inferior phrenic
coeliac axis
superior mesenteric
renal

superior subrenal
subrenal

kidney

Fig.251. Detail showing collateral circulation to the cortex of the Kidney

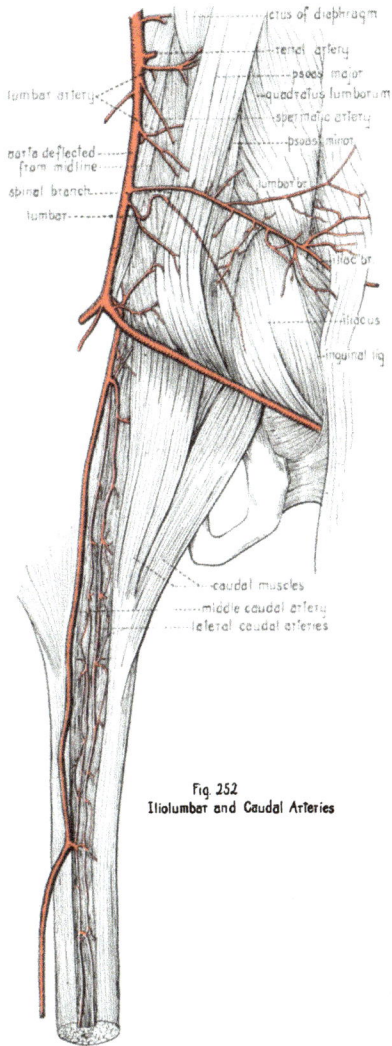

lumbar artery

aorta deflected
from midline

spinal branch

lumbar

ictus of diaphragm

renal artery

psoas major

quadratus lumborum

spermatic artery

psoas minor

lumbar br

iliac br

iliacus

inguinal lig

caudal muscles

middle caudal artery

lateral caudal arteries

Fig. 252
Iliolumbar and Caudal Arteries

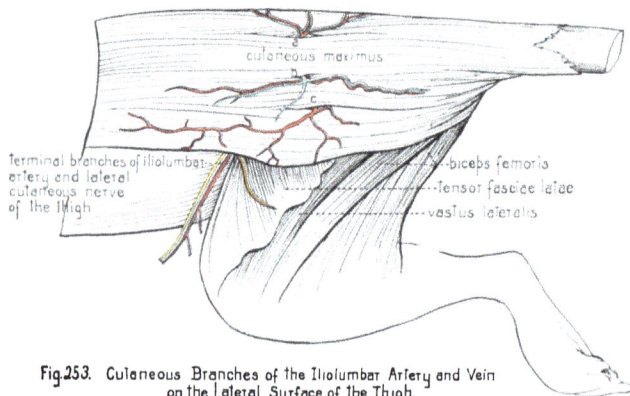

Fig. 253. Cutaneous Branches of the Iliolumbar Artery and Vein
on the Lateral Surface of the Thigh.

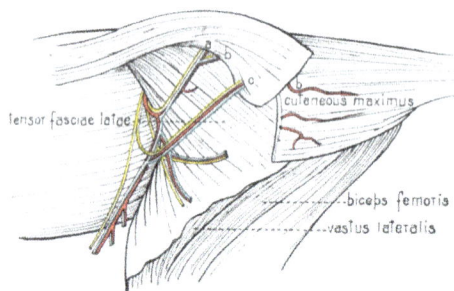

Fig. 254. Superficial Dissection of the Thigh to show
Cutaneous Branches of the Iliolumbar Artery and Vein
and Lateral Cutaneous Nerve of the Thigh.

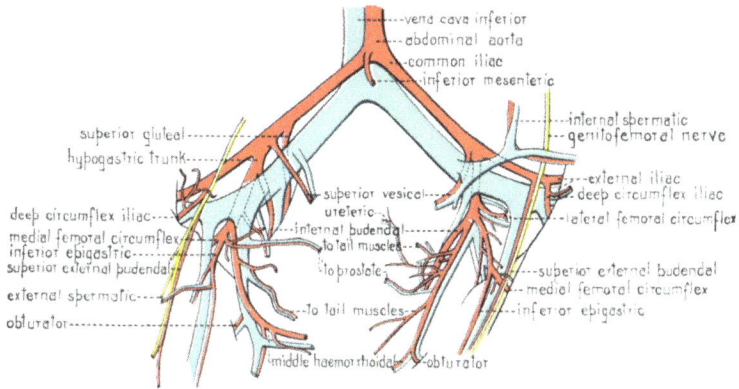

Fig. 255. Plan of Branching of the Common Iliac Arteries and Veins

Labels (Fig. 255):
vena cava inferior
abdominal aorta
common iliac
inferior mesenteric
internal spermatic
genitofemoral nerve
superior gluteal
hypogastric trunk
external iliac
deep circumflex iliac
lateral femoral circumflex
superior vesical
ureteric
deep circumflex iliac
medial femoral circumflex
internal pudendal
inferior epigastric
to tail muscles
superior external pudendal
to prostate
superior external pudendal
medial femoral circumflex
external spermatic
inferior epigastric
to tail muscles
obturator
middle haemorrhoidal
obturator

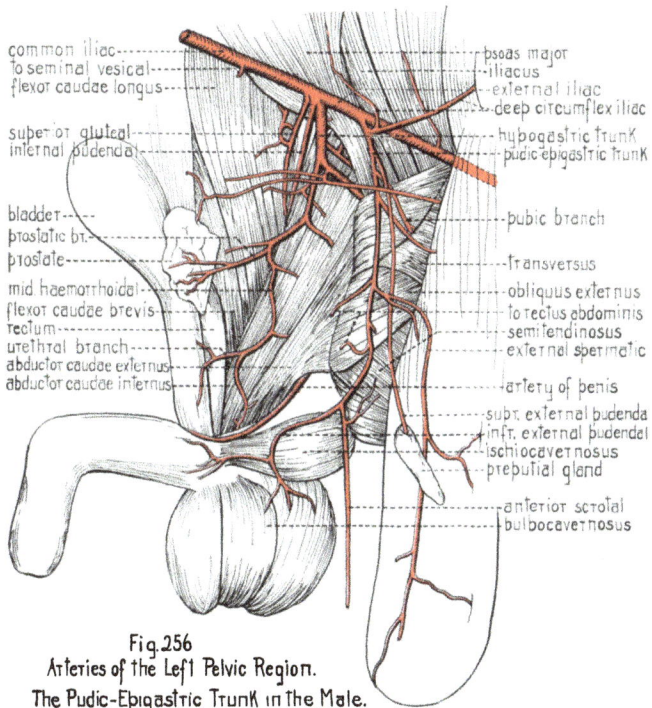

Fig. 256
Arteries of the Left Pelvic Region.
The Pudic-Epigastric Trunk in the Male.

Labels (Fig. 256):
common iliac
to seminal vesical
flexor caudae longus
psoas major
iliacus
external iliac
deep circumflex iliac
superior gluteal
internal pudendal
hypogastric trunk
pudic-epigastric trunk
bladder
prostatic br.
prostate
pubic branch
mid haemorrhoidal
flexor caudae brevis
rectum
urethral branch
abductor caudae externus
abductor caudae internus
transversus
obliquus externus
to rectus abdominis
semitendinosus
external spermatic
artery of penis
subr. external pudenda
infr. external pudendal
ischiocavernosus
preputial gland
anterior scrotal
bulbocavernosus

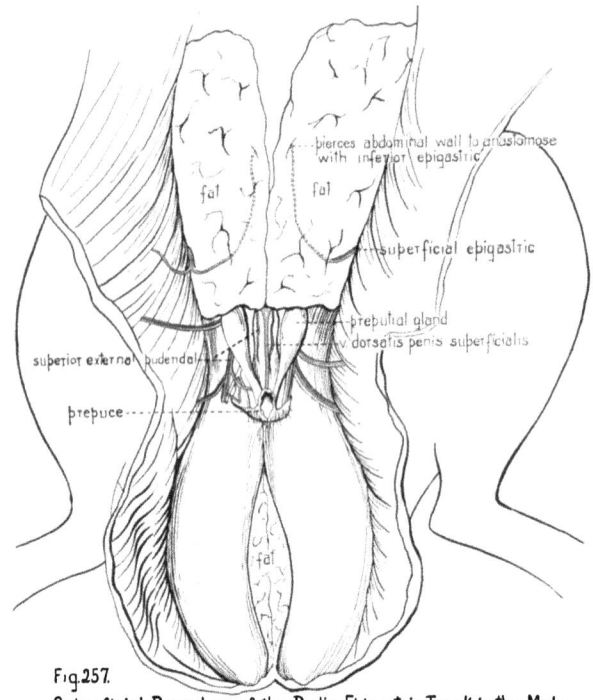

Fig.257.
Superficial Branches of the Pudic-Epigastric Trunk in the Male.

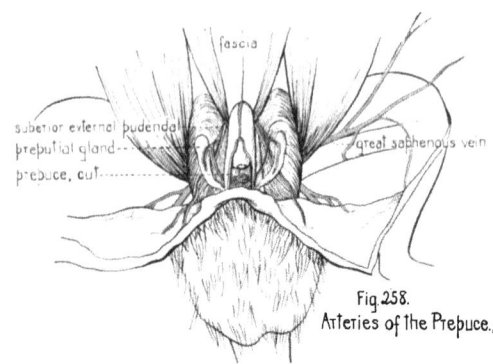

Fig.258.
Arteries of the Prepuce.

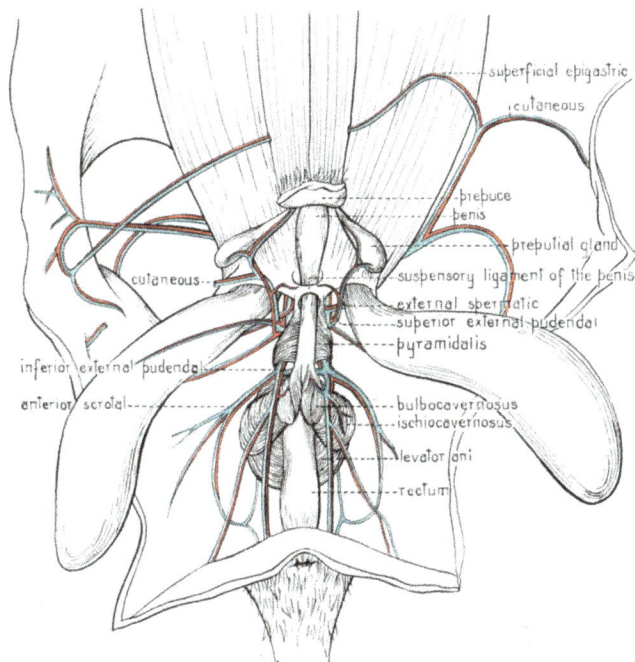

Fig. 259. Superficial Branches of Superior and Inferior External Pudendal
in the Male.

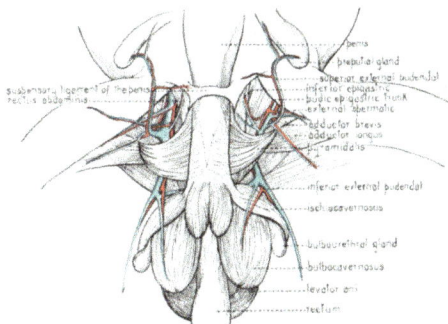

Fig. 260 Branches of the Pudic-Epigastric Trunk in Relation to Pyramidalis.

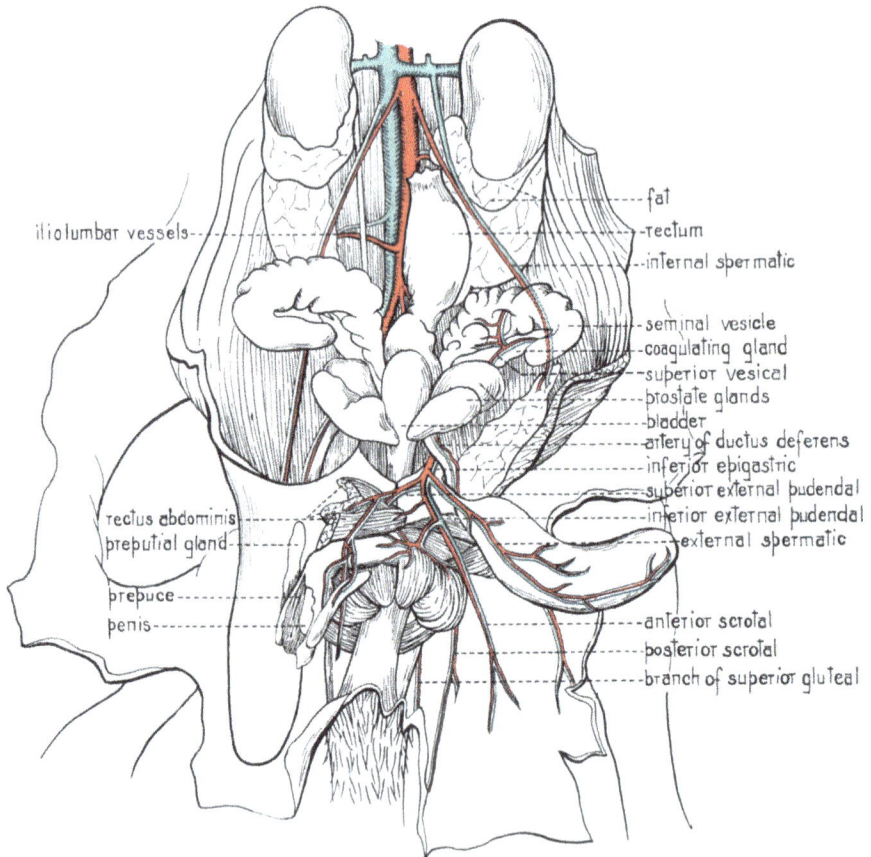

Fig. 261.    External Spermatic Branch of Inferior Epigastric.

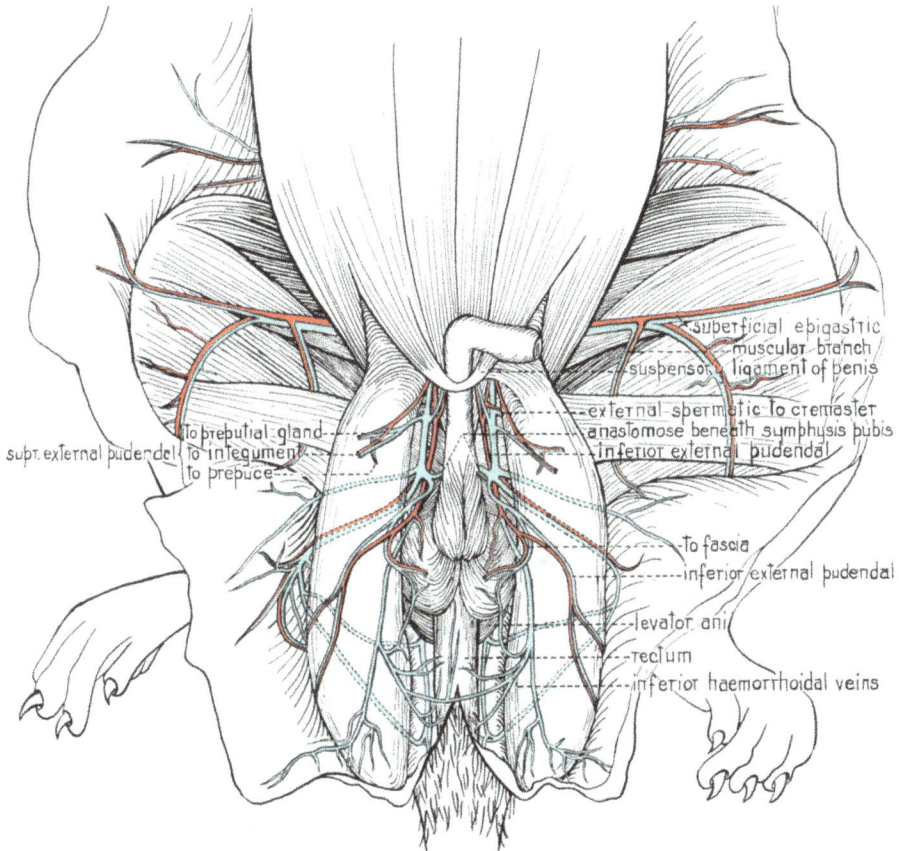

Fig. 262.   Deeper Dissection of Inferior External Pudendal Vessels.

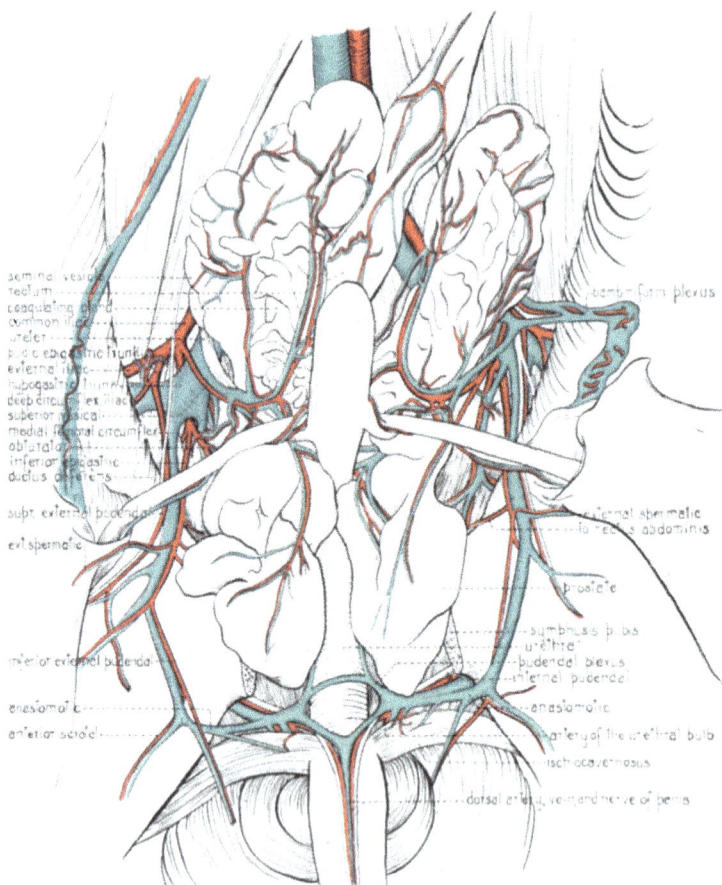

Fig. 263.  Pudendal Plexus in the Male.

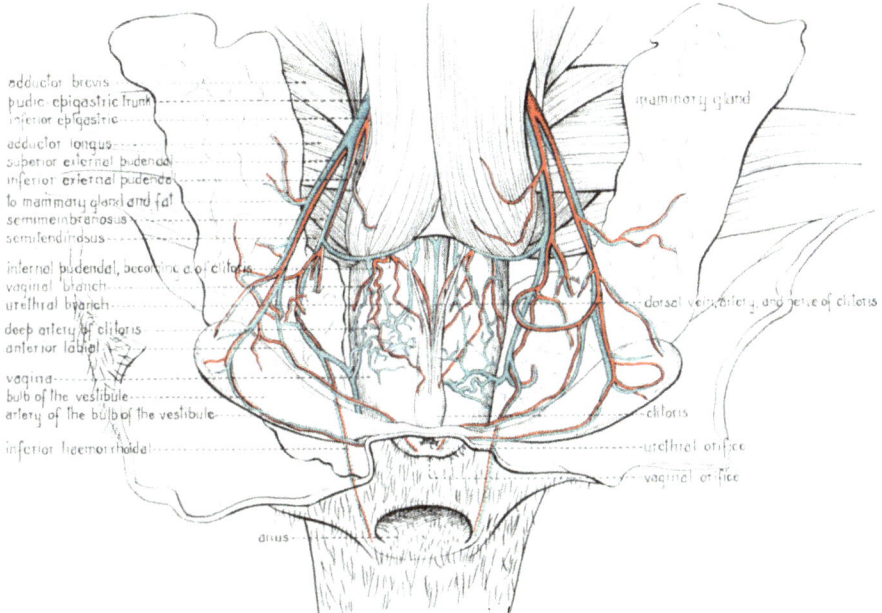

Fig.264. Branches of the Pudic-Epigastric Trunk in the Female and Vessels of The Clitoris.

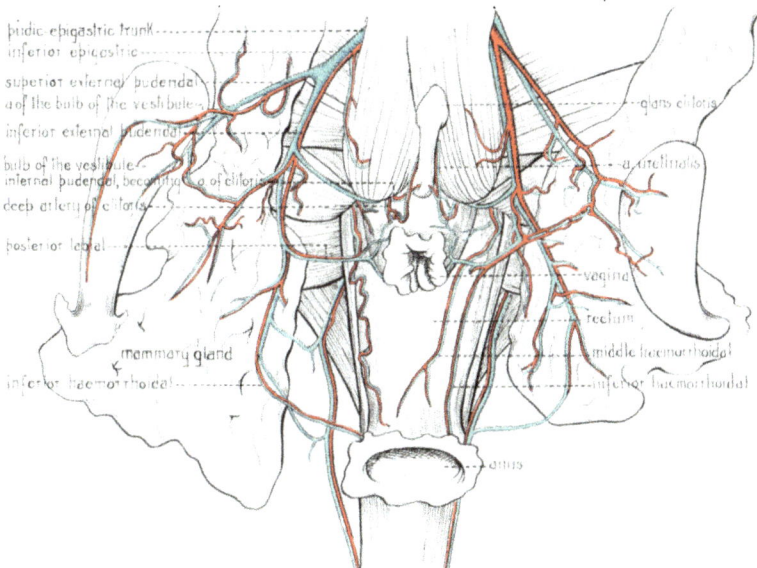

Fig.265. The Inferior External Pudendal Vessels in the Female.

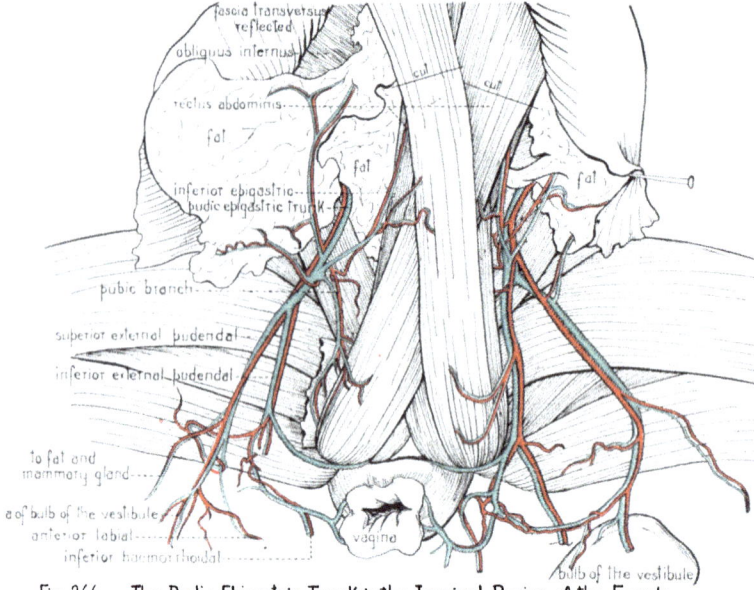

Fig. 266.    The Pudic-Epigastric Trunk in the Inguinal Region of the Female.

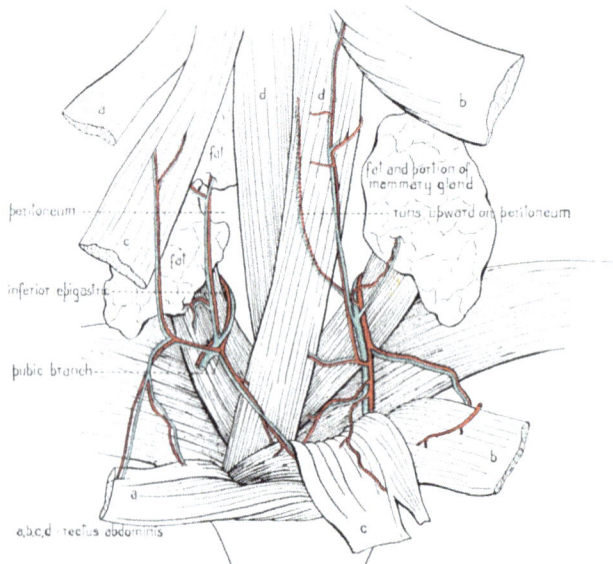

Fig. 267.    Pubic Branch of Inferior Epigastric Artery to Rectus Abdominis.

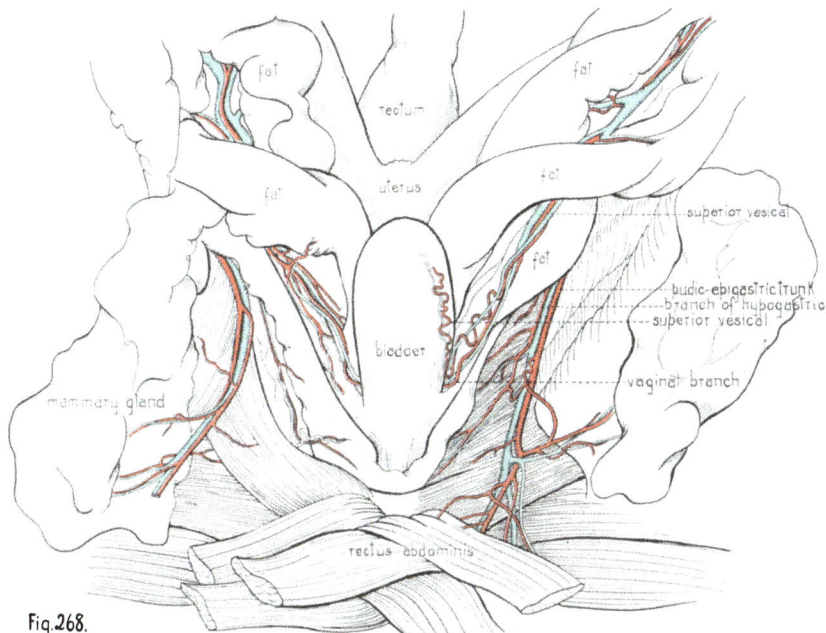

Fig.268.
Pudic-Epigastric Trunk as seen when Abdominal Cavity is first opened.

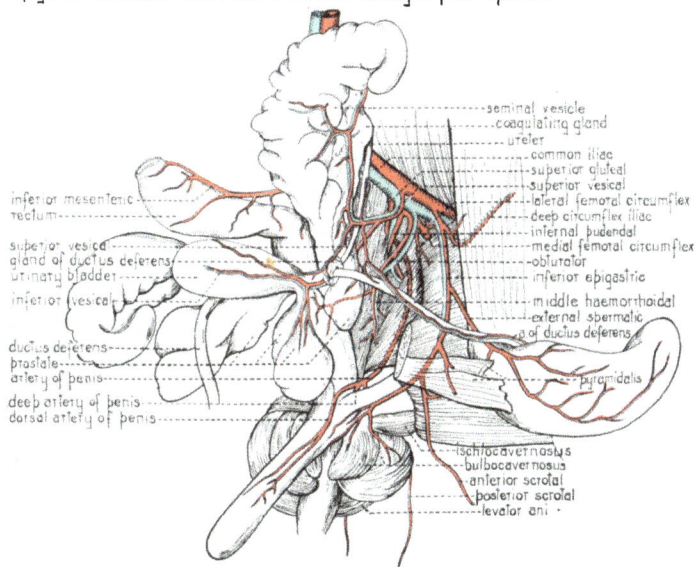

Fig.269.   Branches of Superior Vesical. Artery and Vein in the Male.

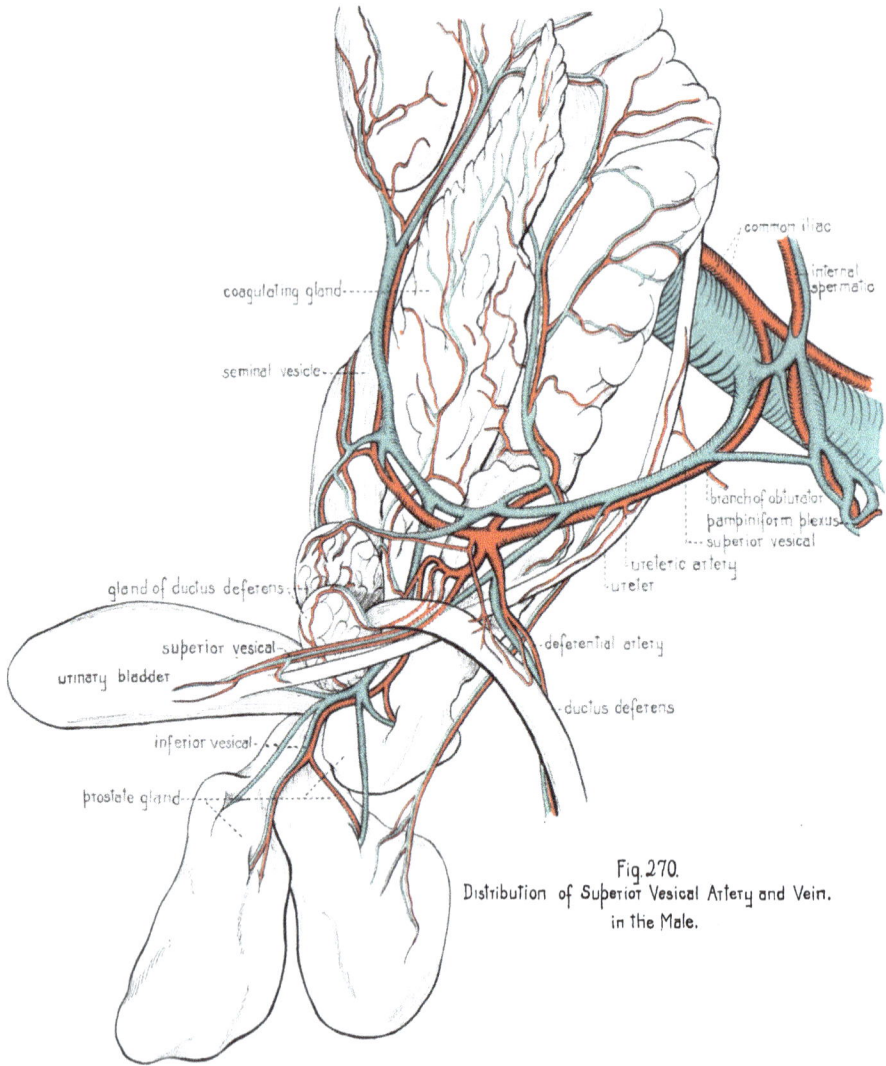

Fig. 270.
Distribution of Superior Vesical Artery and Vein.
in the Male.

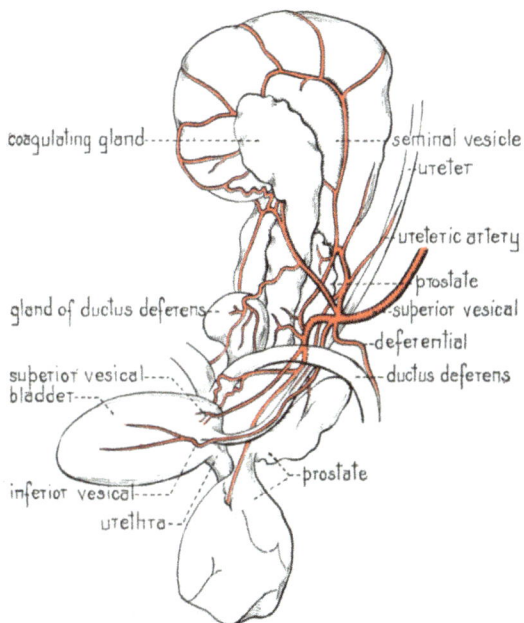

Fig. 271. Detail of arterial supply of seminal vesicle
and coagulating gland.

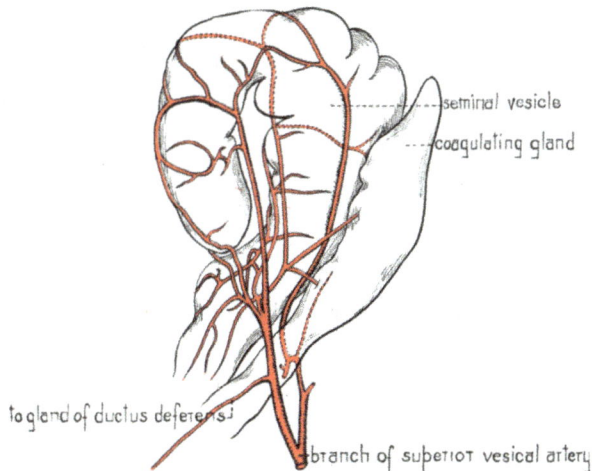

Fig 272. Detail of arterial supply of seminal vesicle
and coagulating gland, with the latter displaced.

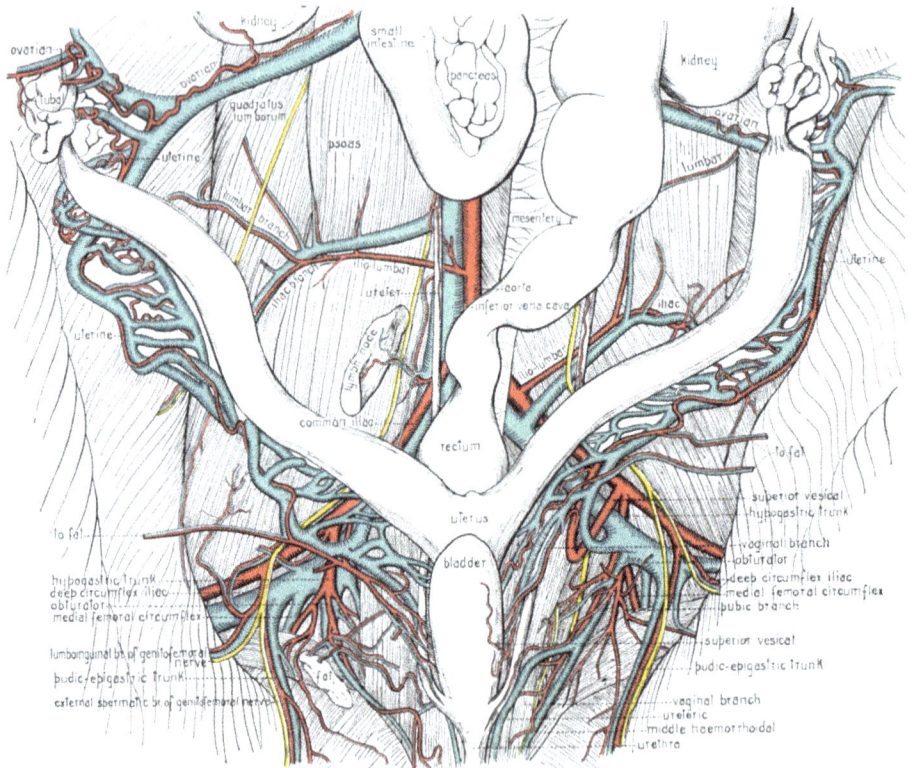

Fig. 273.   Blood Vessels of the Female Urogenital System.  Superior Vesical Artery and Vein.

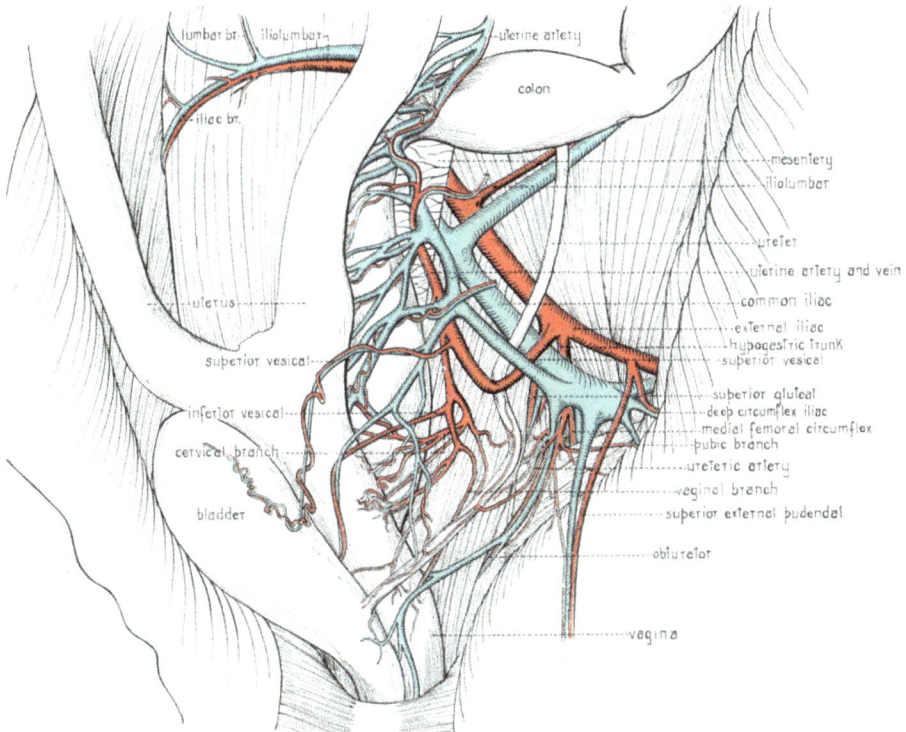

Fig. 274. Blood Vessels of the Female Urogenital System.   Left Side of Pelvis showing Superior Vesical Branches.

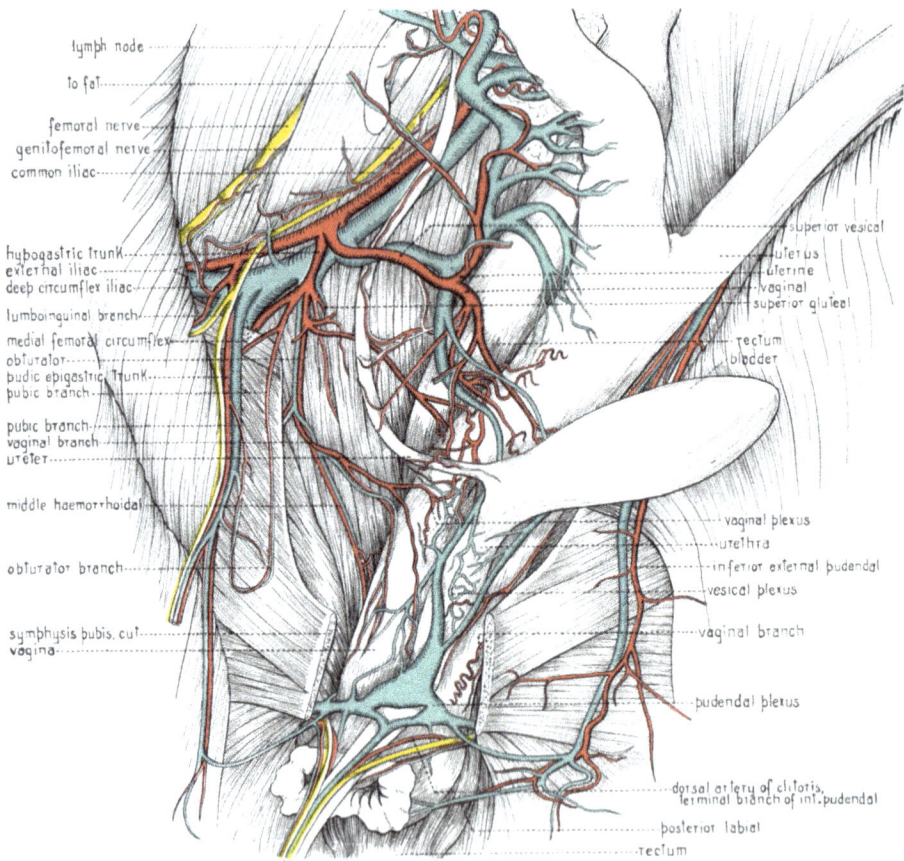

Fig. 275   Right Side of Pelvis showing Superior Vesical Artery and Vein to Female Urogenital System.

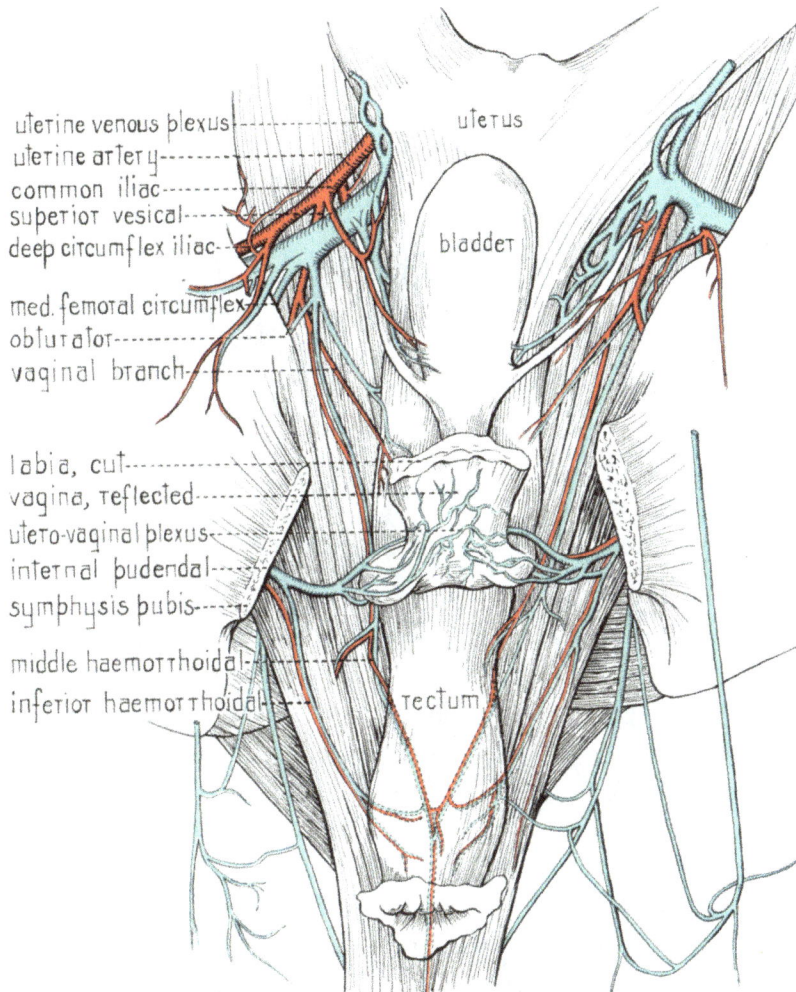

uterine venous plexus
uterine artery
common iliac
superior vesical
deep circumflex iliac

med. femoral circumflex
obturator
vaginal branch

labia, cut
vagina, reflected
utero-vaginal plexus
internal pudendal
symphysis pubis

middle haemorrhoidal
inferior haemorrhoidal

uterus
bladder
rectum

Fig.276 The Utero-vaginal Plexus and Haemorrhoidal Vessels.

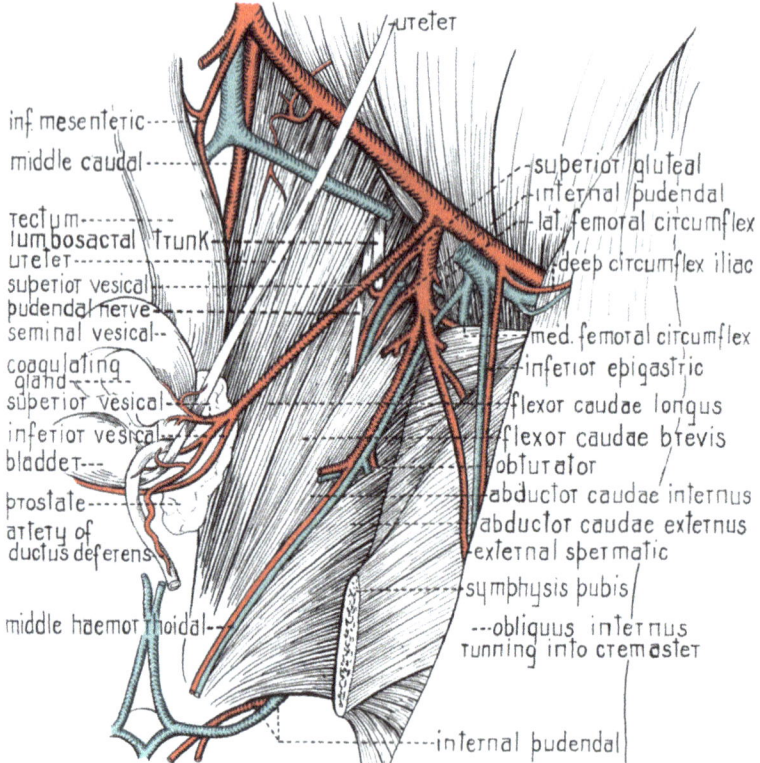

Fig. 277.
Arteries and Veins of the Left Half of the Male Pelvis.
showing relationships of Internal Pudendal.

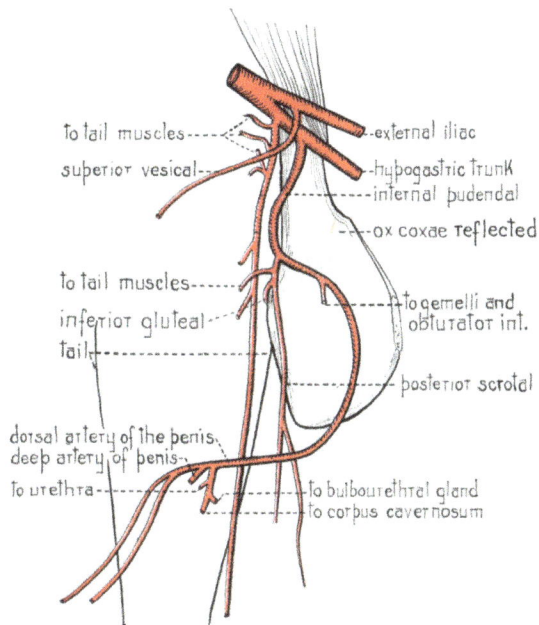

Fig. 278. Detail of Internal Pudendal Artery in the Male.

to tail muscles
superior vesical
to tail muscles
inferior gluteal
tail
dorsal artery of the penis
deep artery of penis
to urethra

external iliac
hypogastric trunk
internal pudendal
ox coxae reflected
to gemelli and
obturator int.
posterior scrotal
to bulbourethral gland
to corpus cavernosum

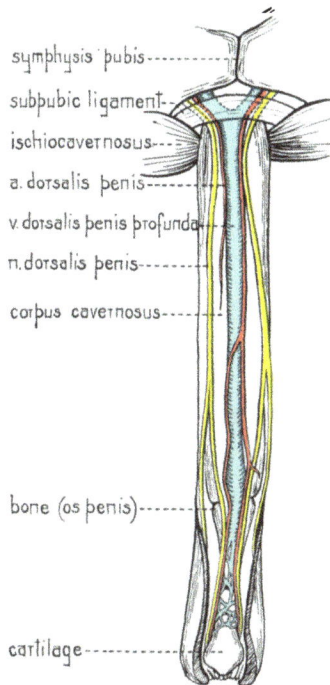

Fig. 280. Blood Vessels of the Penis.

symphysis pubis
subpubic ligament
ischiocavernosus
a. dorsalis penis
v. dorsalis penis profunda
n. dorsalis penis
corpus cavernosus
bone (os penis)
cartilage

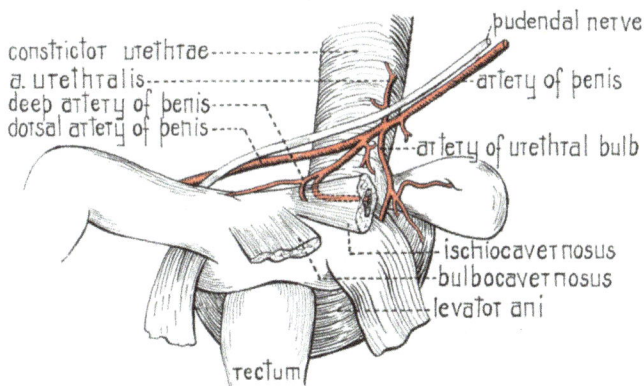

Fig. 279. Artery of the Penis.

constrictor urethrae
a. urethralis
deep artery of penis
dorsal artery of penis

pudendal nerve
artery of penis
artery of urethral bulb
ischiocavernosus
bulbocavernosus
levator ani
rectum

bladder

middle haemorrhoidal.
branch of obturator

to rectum

vesicovaginal

symphysis pubis

vaginal branch

internal pudendal nerve

vagina

superior external pudendal

plexus around urethra

clitoris

bulb of the vestibule

urethral orifice

prepuce, cut open

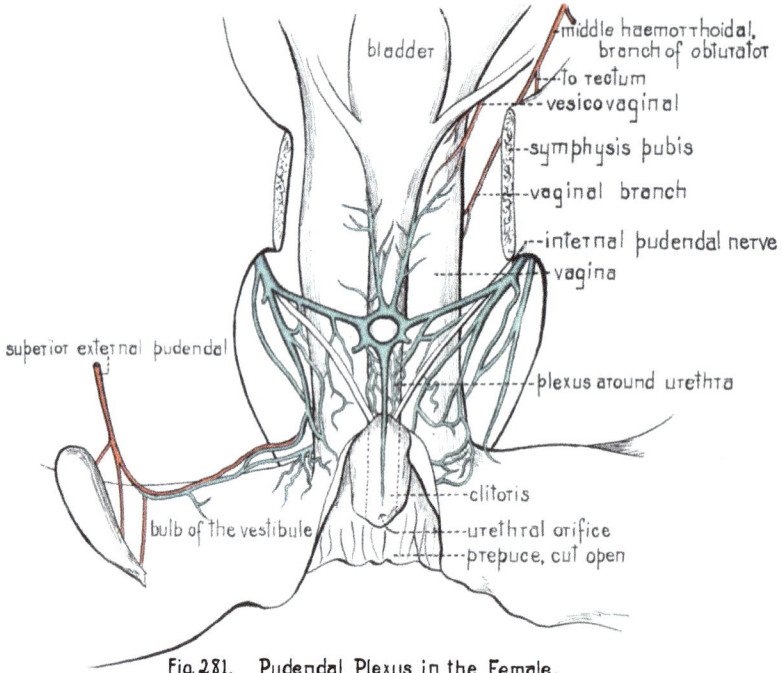

Fig. 281.   Pudendal Plexus in the Female.

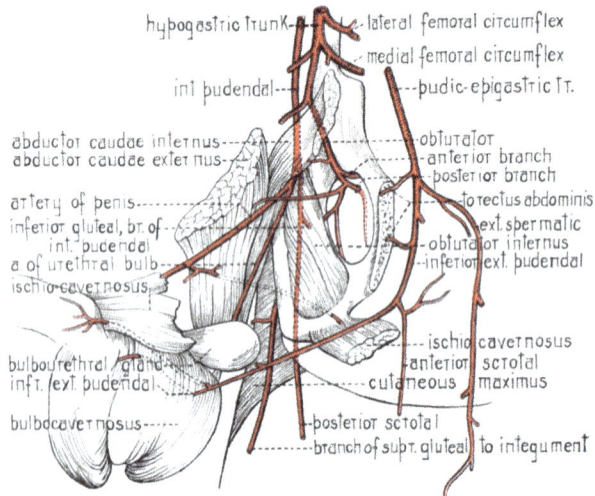

hypogastric trunk

lateral femoral circumflex

medial femoral circumflex

int. pudendal

pudic-epigastric tr.

abductor caudae internus

obturator

abductor caudae externus

anterior branch

posterior branch

artery of penis

to rectus abdominis

inferior gluteal, br. of
int. pudendal

ext. spermatic

a of urethral bulb

obturator internus

ischio-cavernosus

inferior ext. pudendal

bulbourethral gland

ischio cavernosus

inf. ext. pudendal

anterior scrotal

cutaneous   maximus

bulbocavernosus

posterior scrotal

branch of supr. gluteal. to integument

Fig. 282.   Deep Arteries of the Pelvis. Obturator Artery in the Male.

---hypogastric trunk

---medial femoral circumflex

---pubic branch of obturator

middle haemorrhoidal---

abductor caudae int---

abductor caudae ext---

---obturator artery and nerve

---anterior branch of obturator

---acetabular branch

---to gemellus inf., obturator externus
   and quadratus femoris

---to obturator externus

---posterior branch of obturator

---obturator internus

pudendal-----

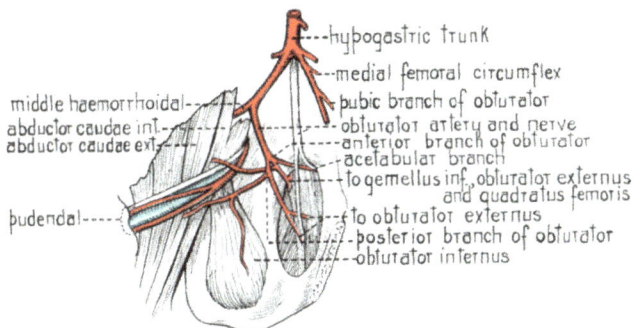

Fig.283. Detail of Obturator Artery seen from the Pelvis.

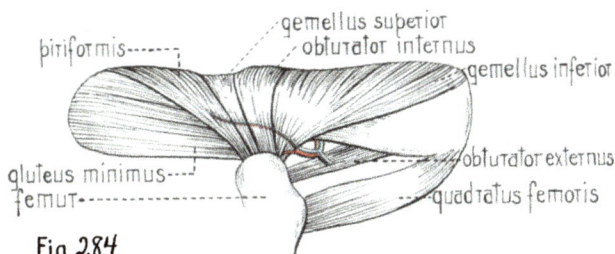

gemellus superior

obturator internus

piriformis------

gemellus inferior

gluteus minimus---

femur------

obturator externus

quadratus femoris

Fig. 284.
Anterior Branch of Obturator Artery.

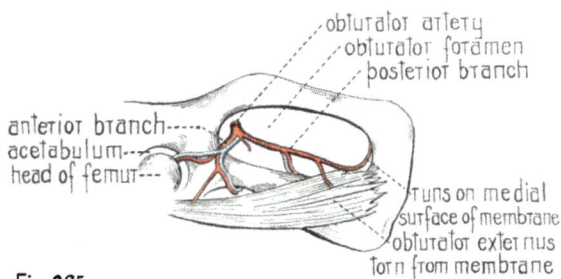

obturator artery

obturator foramen

posterior branch

anterior branch---

acetabulum---

head of femur---

runs on medial
surface of membrane

obturator externus
torn from membrane

Fig. 285.
Obturator Artery on Lateral Surface of Obturator Membrane

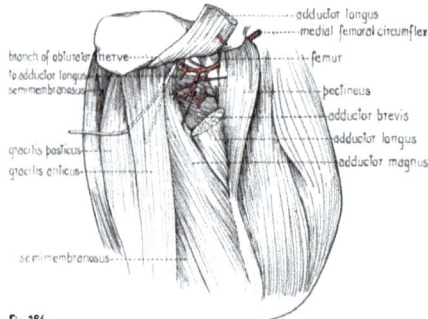

Fig.286.
Adductor Longus reflected to show Medial Femoral Circumflex Artery
entering Thigh.

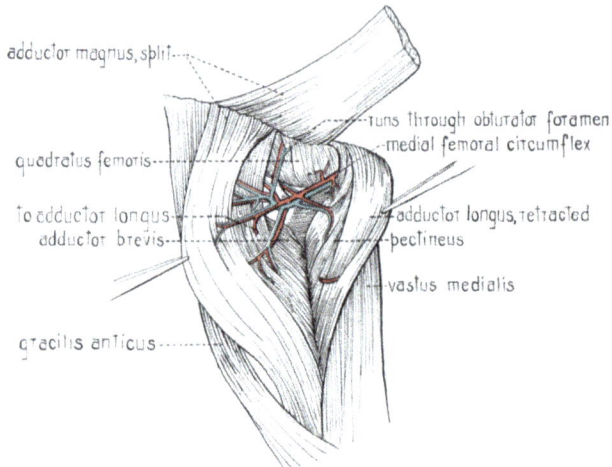

Fig.287  Medial Femoral Circumflex Artery and Vein.

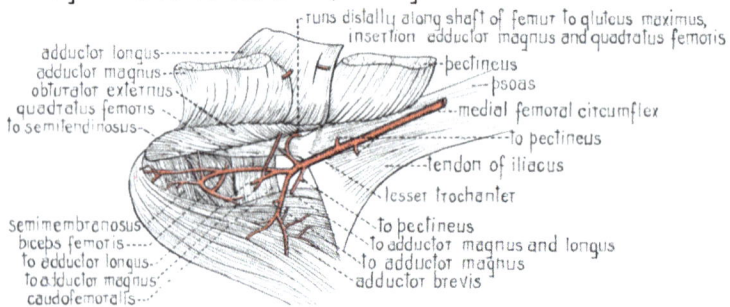

Fig.288.  Adductor Muscles reflected to show Branches of Medial Femoral Circumflex Artery.

artery of the urethral bulb
infr br. of superior gluteal
obturator internus
and gemelli

gluteus maximus
biceps femoris

posterior scrotal
semitendinosus

internal pudendal
inferior gluteal
gluteus medius

caudofemoralis
quadratus femoris
medial femoral circumflex

greater trochanter

to semimembranosus

gluteus maximus

adductor brevis
biceps femoris

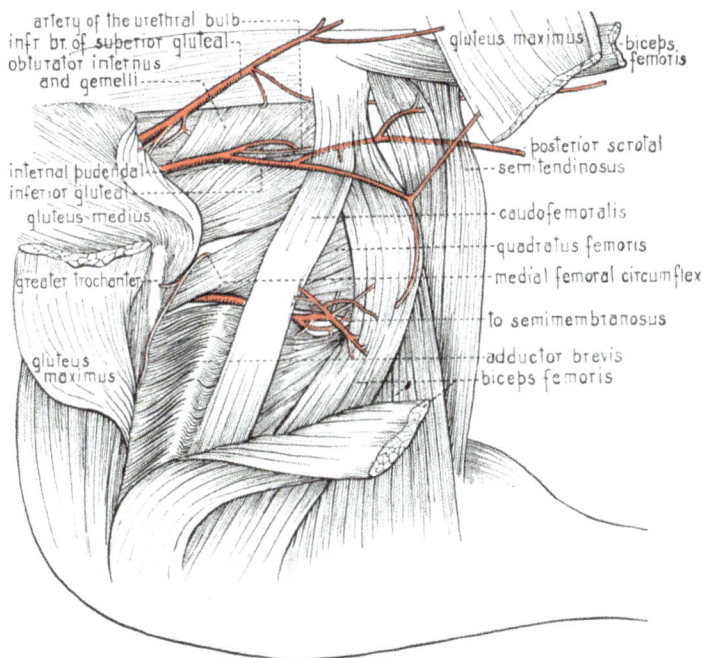

Fig.289. Lateral Aspect of the Thigh showing Medial Femoral Circumflex Artery.

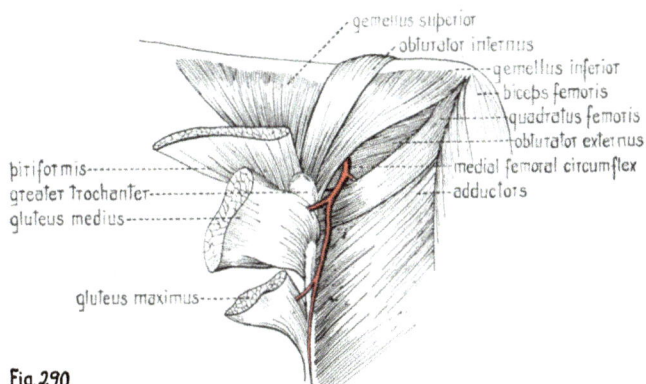

gemellus superior
obturator internus
gemellus inferior
biceps femoris
quadratus femoris
obturator externus
medial femoral circumflex

piriformis
greater trochanter
gluteus medius

adductors

gluteus maximus

Fig.290.
Detail of Medial Femoral Circumflex Artery on Lateral Surface of Thigh.

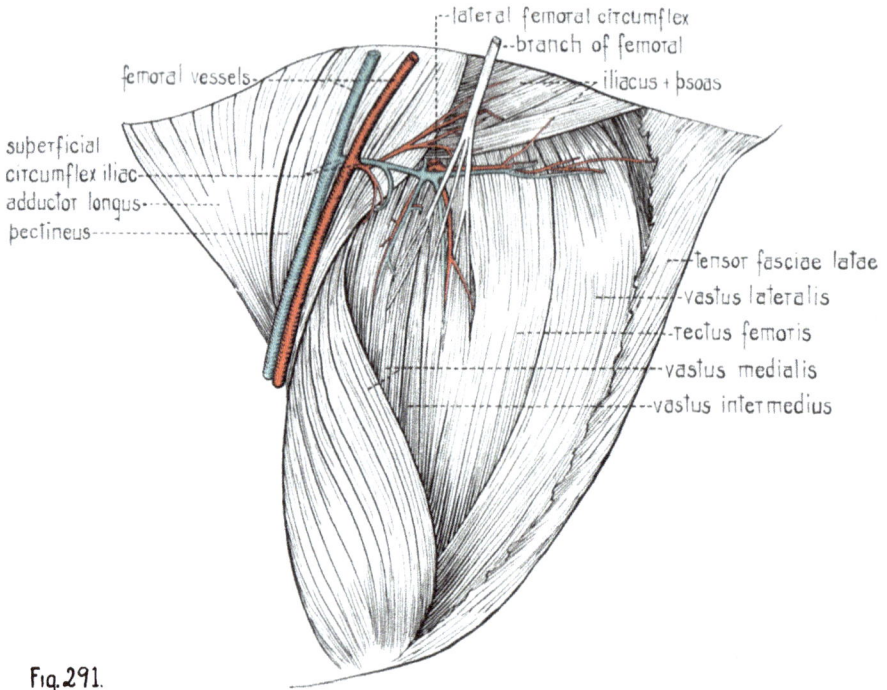

Fig. 291.
Medial Aspect of the Thigh showing Ascending Branch of Lateral Femoral Circumflex
Artery and Vein.

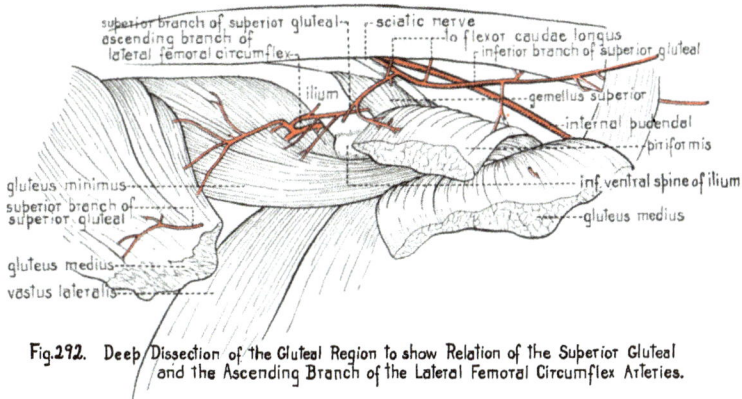

Fig. 292. Deep Dissection of the Gluteal Region to show Relation of the Superior Gluteal and the Ascending Branch of the Lateral Femoral Circumflex Arteries.

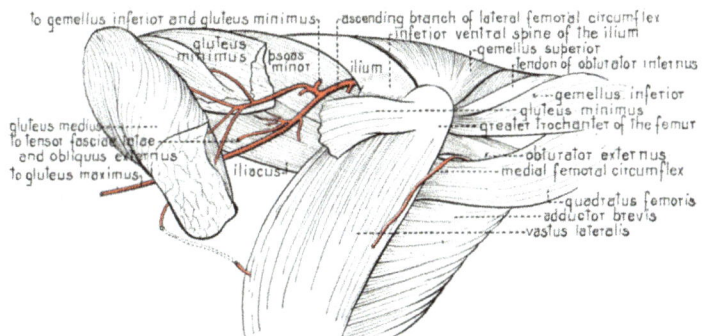

Fig. 293 Detail of the Ascending Branch of the Lateral Femoral Circumflex Artery.

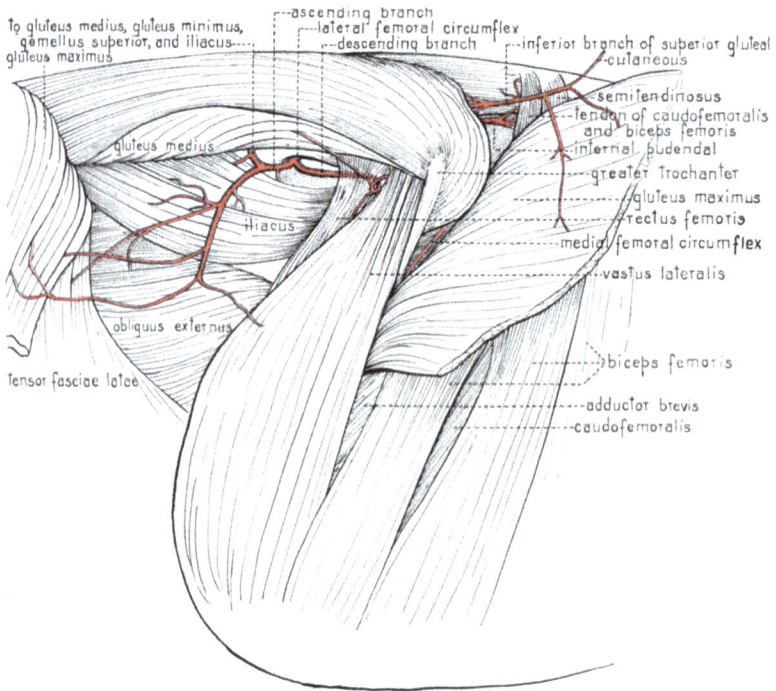

Fig.294. Dissection of the Gluteal Region Anterior to the Trochanter to show
Ascending and Descending Branches of Lateral Femoral Circumflex Artery.

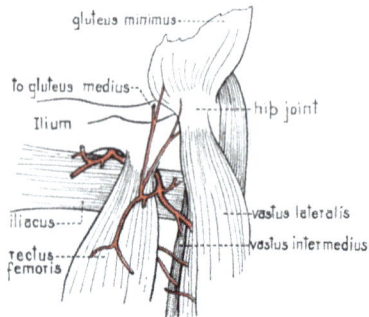

Fig.295 Detail showing Descending Branch of
Lateral Femoral Circumflex Artery.

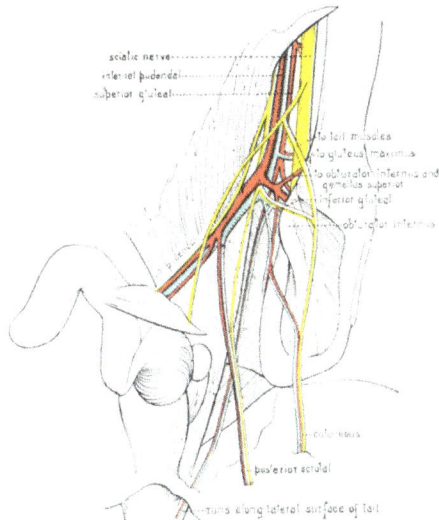

sciatic nerve
internal pudendal
superior gluteal

to tail muscles
to gluteal maximus
to obturator internus and
gemellus superior
inferior gluteal
to obturator internus

cutaneous

posterior scrotal

runs along lateral surface of tail

Fig.296. Deep Dissection of the Pelvis to show Gluteal and Internal Pudendal Arteries
and the Corresponding Veins.

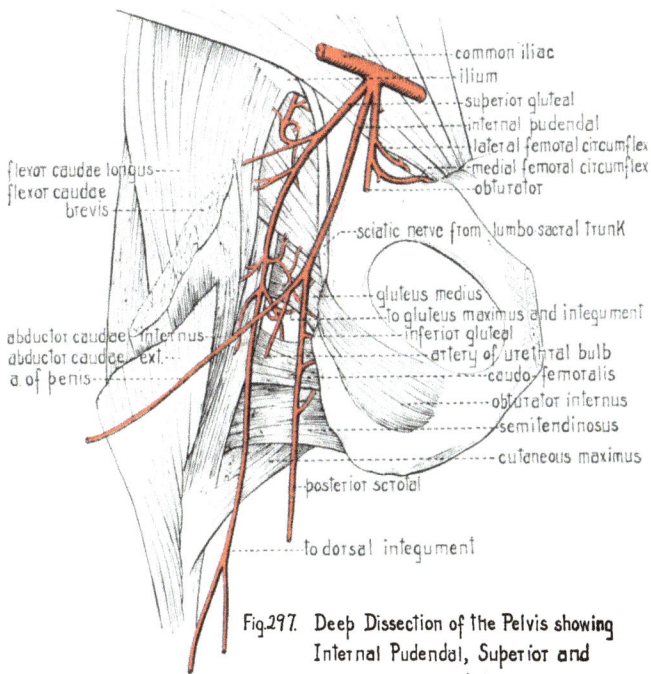

common iliac
ilium
superior gluteal
internal pudendal
lateral femoral circumflex
medial femoral circumflex
obturator

flexor caudae longus
flexor caudae
brevis

sciatic nerve from lumbo-sacral trunk

gluteus medius
to gluteus maximus and integument
inferior gluteal
artery of urethral bulb
caudo-femoralis
obturator internus
semitendinosus
cutaneous maximus

abductor caudae internus
abductor caudae ext.
a. of penis

posterior scrotal

to dorsal integument

Fig.297. Deep Dissection of the Pelvis showing
Internal Pudendal, Superior and
Inferior Gluteal Arteries.

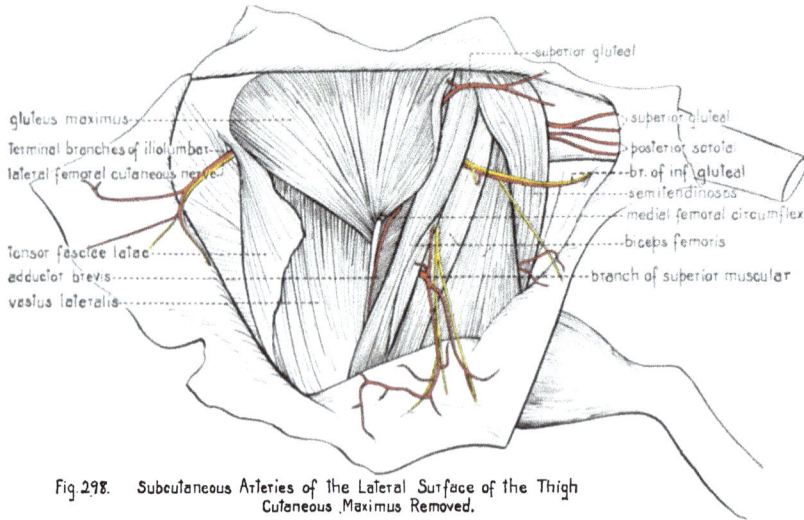

Fig. 298. Subcutaneous Arteries of the Lateral Surface of the Thigh
Cutaneous Maximus Removed.

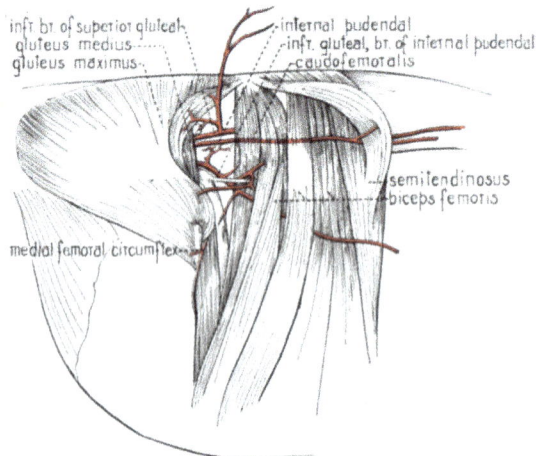

Fig. 299. Superficial Dissection of the Thigh to show branches of
Superior and Inferior Gluteal and Internal Pudendal Arteries.

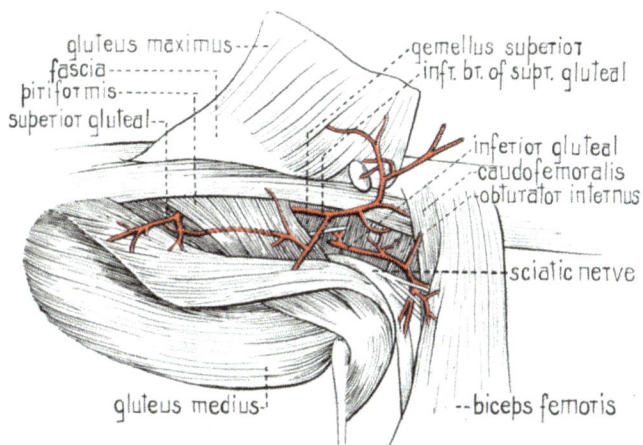

gluteus maximus
fascia
piriformis
superior gluteal

gemellus superior
inft. br. of supr. gluteal

inferior gluteal
caudofemoralis
obturator internus

sciatic nerve

gluteus medius

biceps femoris

Fig. 300.

Gluteus Maximus removed to show branches of
Superior and Inferior Gluteal Arteries.

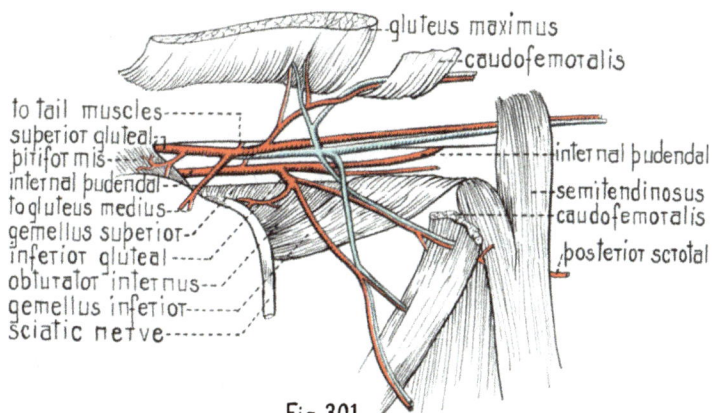

gluteus maximus
caudofemoralis

to tail muscles
superior gluteal
piriformis
internal pudendal
to gluteus medius
gemellus superior
inferior gluteal
obturator internus
gemellus inferior
sciatic nerve

internal pudendal
semitendinosus
caudofemoralis

posterior scrotal

Fig. 301.

Deep Dissection of the Hip to show Superior and
Inferior Gluteal and Internal Pudendal Arteries.

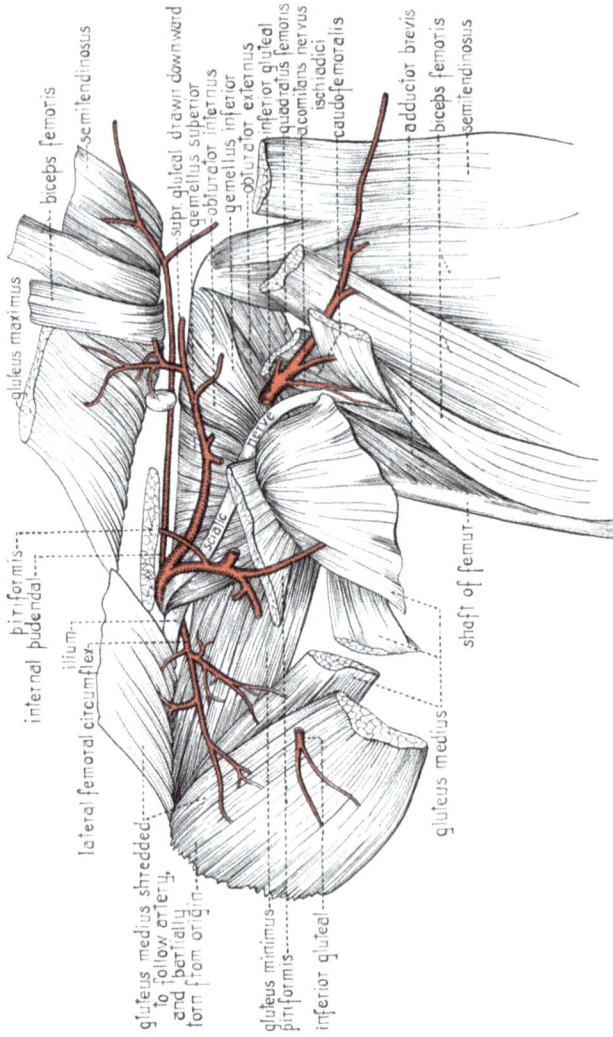

Fig.302 Deep Dissection of the Gluteal Region to show Superior and Inferior Gluteal and Internal Pudendal Arteries.

internal spermatic; caudal aorte
internal pudendal

middle haemorrhoidal
obturator psoas
external pudendal common iliac
gracilis anticus external iliac
quadratus femoris deep circumflex iliac
gracilis posticus
adductor magnus femoral artery
pectineus iliacus
adductor longus lat fem circumflex
semitendinosus superficial
caudofemoralis circumflex
muscular branch iliac
tensor fasciae latae
rectus femoris
vastus medialis
musculo-articular
br. of genu suprema
popliteal artery saphenous branch

superficial epigastric

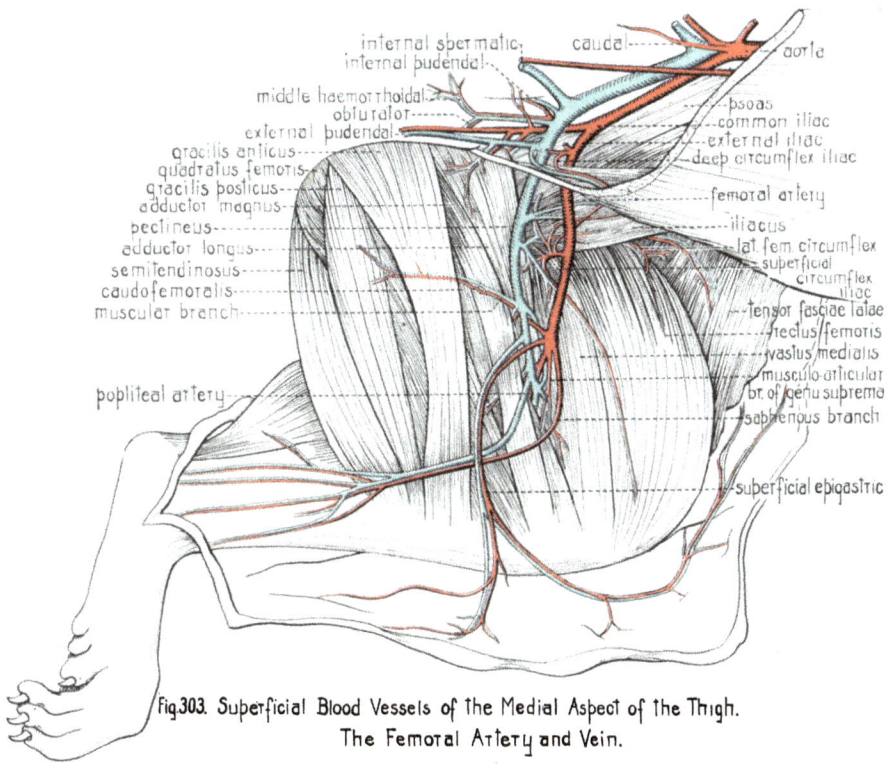

Fig. 303. Superficial Blood Vessels of the Medial Aspect of the Thigh.
The Femoral Artery and Vein.

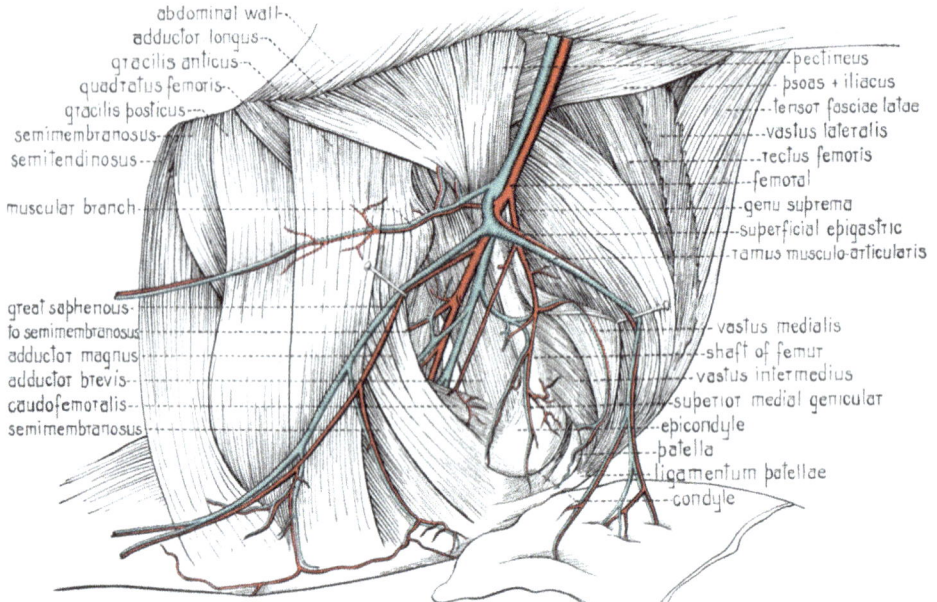

**Fig.304.** Medial Asbect of the Thigh Dissected to show Branches of the Femoral Artery and Vein.

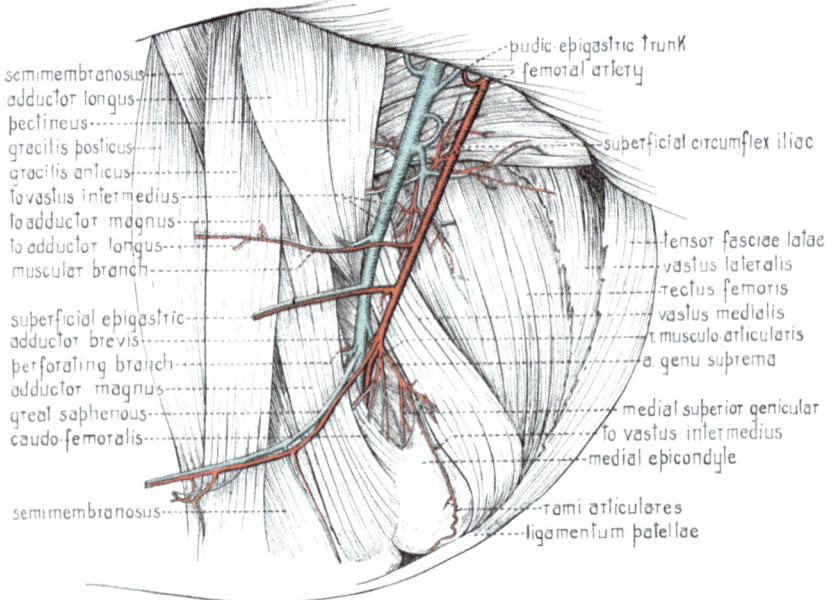

**Fig.305.** Muscular Branches of the Femoral Artery and Vein in the Inguinal Region.

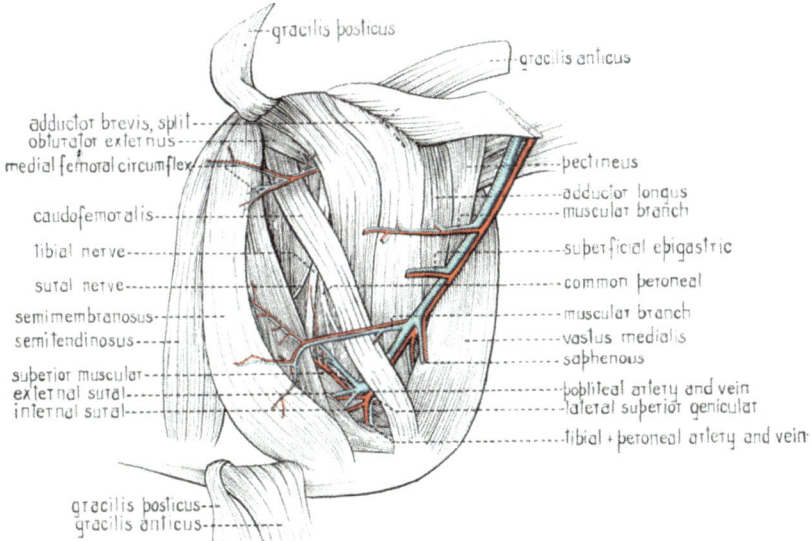

Fig.306. Muscular Branch of the Femoral Artery to Semimembranosus.

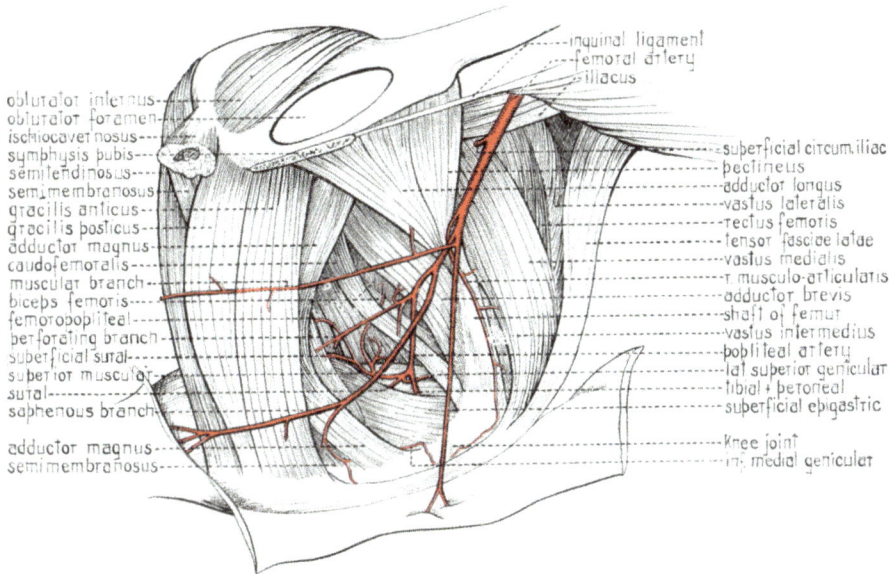

Fig.307. Case in which Popliteal Artery runs medial to Caudofemoralis Muscle. Medial Aspect.

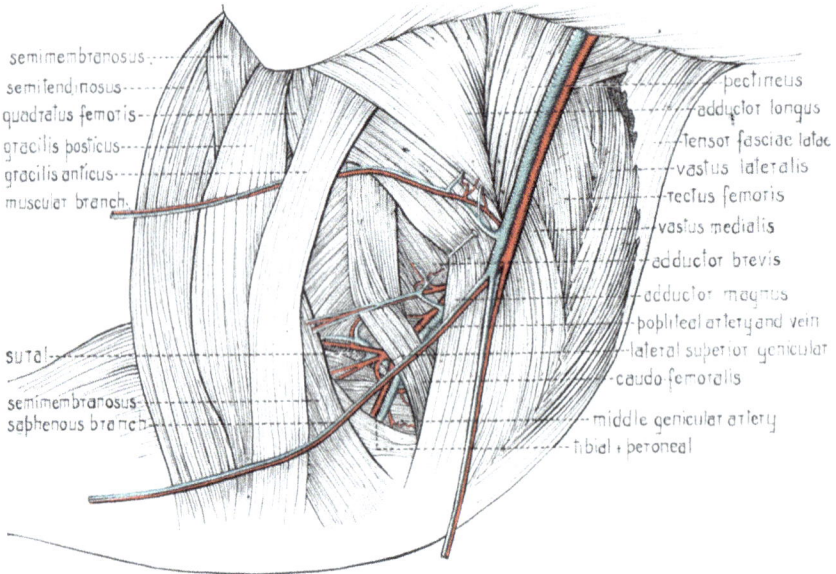

semimembranosus
semitendinosus
quadratus femoris
gracilis posticus
gracilis anticus
muscular branch

sural

semimembranosus
saphenous branch

pectineus
adductor longus
tensor fasciae latae
vastus lateralis
rectus femoris
vastus medialis
adductor brevis
adductor magnus
popliteal artery and vein
lateral superior genicular
caudo-femoralis
middle genicular artery
tibial + peroneal

Fig. 308. Arteries and Veins of the Popliteal Space. Medial Aspect.

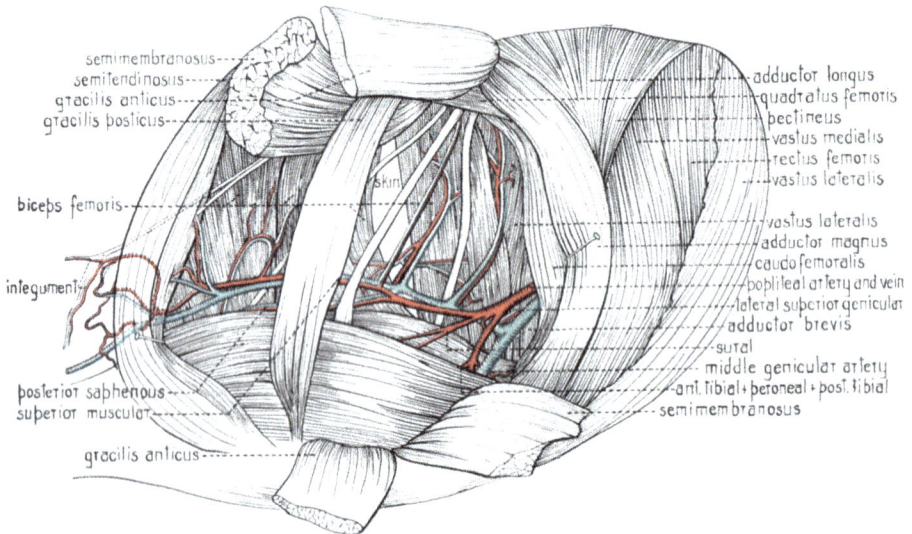

semimembranosus
semitendinosus
gracilis anticus
gracilis posticus

biceps femoris

integument

posterior saphenous
superior muscular

gracilis anticus

skin

adductor longus
quadratus femoris
pectineus
vastus medialis
rectus femoris
vastus lateralis
vastus lateralis
adductor magnus
caudo-femoralis
popliteal artery and vein
lateral superior genicular
adductor brevis
sural
middle genicular artery
ant. tibial + peroneal + post. tibial
semimembranosus

Fig. 309. Branches of the Popliteal Artery and Vein. Medial Aspect.

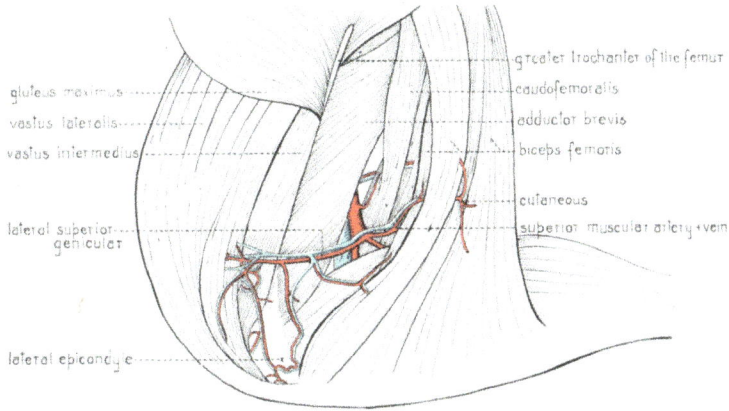

Fig. 310.  Superficial  Dissection to show the Lateral Superior Geniculat Artery and Vein.

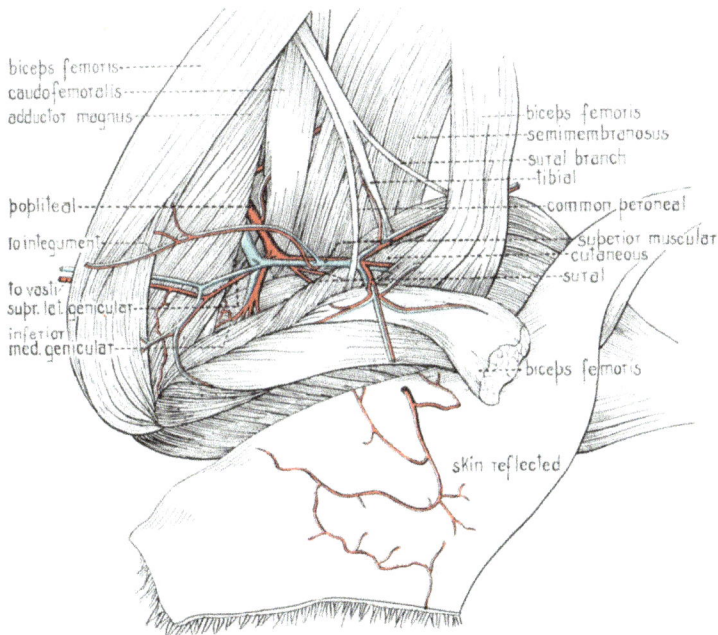

Fig. 311.  Arteries and Veins of the Popliteal Space.  Lateral Aspect.

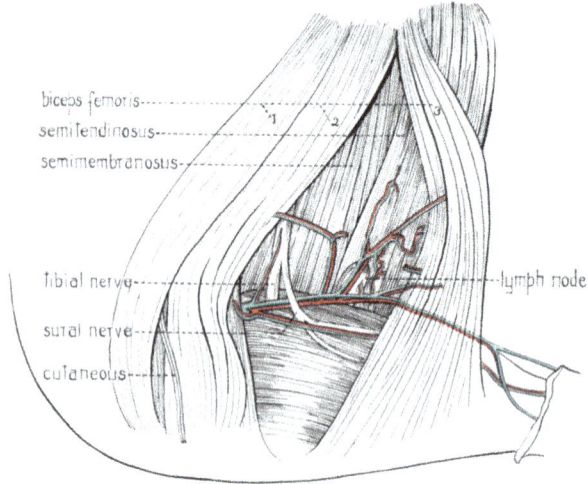

Fig. 312. Distribution of the Superior Muscular Artery and Vein
to Hamstring Muscles. Lateral Aspect.

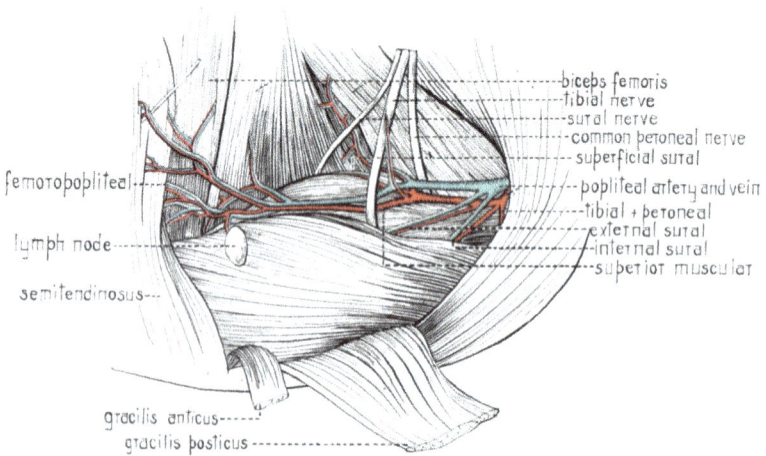

Fig. 313. Superior Muscular Artery and Vein. Medial Aspect.

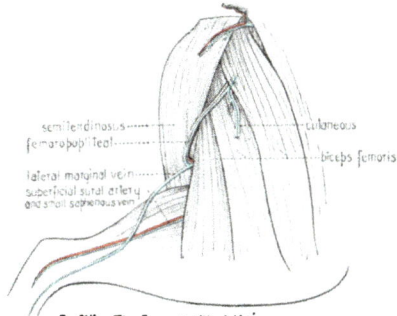

Fig. 314. The Femoropopliteal Vein.

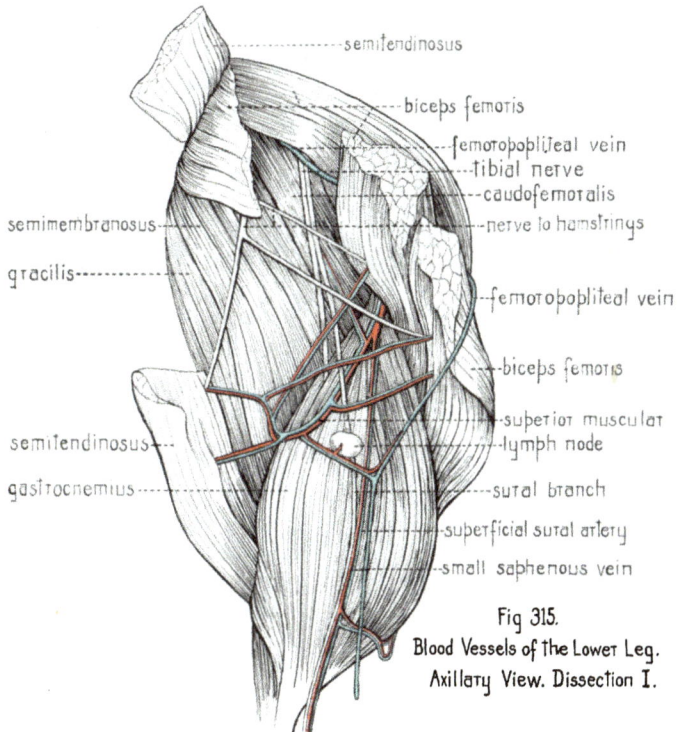

Fig 315.
Blood Vessels of the Lower Leg.
Axillary View. Dissection I.

common peroneal nerve------                    ------semimembranosus
tibial nerve------                             --caudofemoralis
biceps femoris------                           ----adductor brevis
                                               --popliteal art.+ vein
lateral superior genicular------               ----semitendinosus
lateral sural nerve------                       ----anterior tibial+peroneal
superior muscular------
biceps femoris------                           ---internal sural
femoropopliteal------
                                               ----external sural
                                               --sural nerve
                                               --posterior tibial
                                               superficial sural artery
                                               and small saphenous vein
lateral marginal------                          ---gastrocnemius
                                               --plantaris

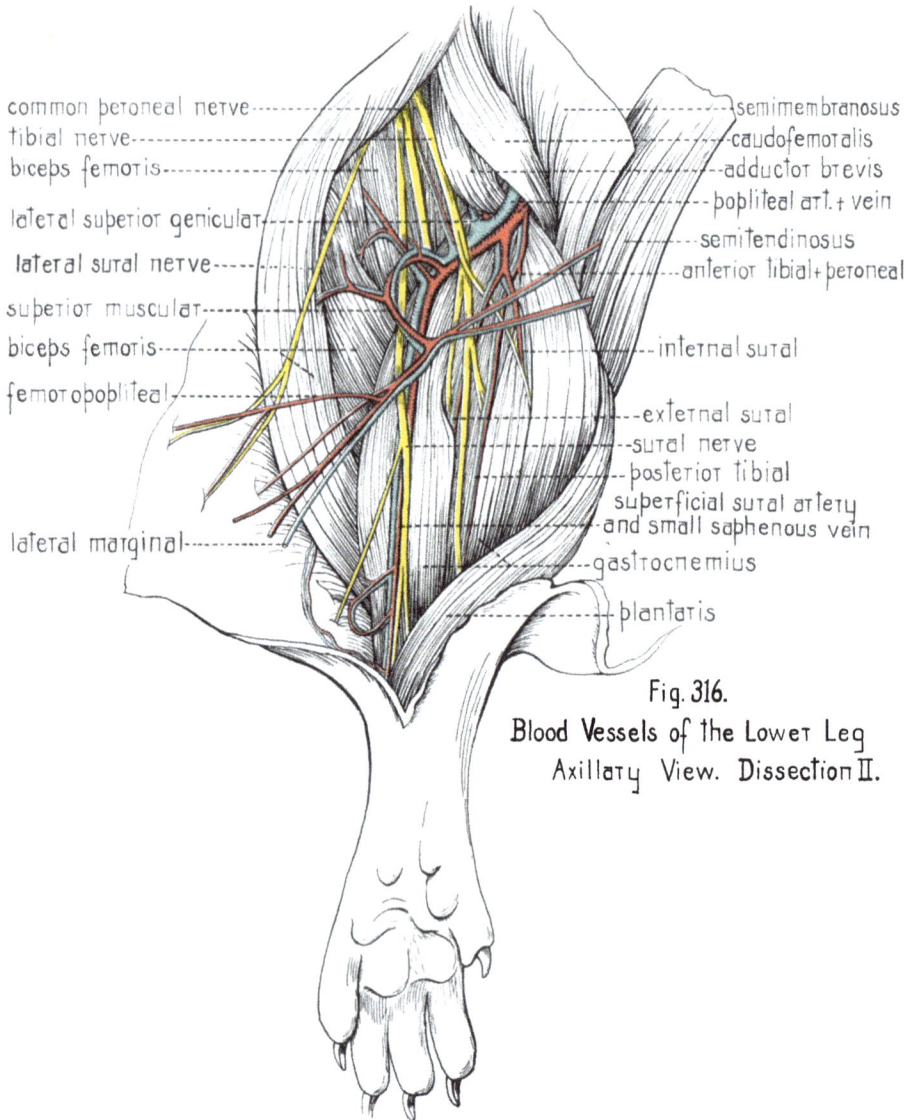

Fig. 316.
Blood Vessels of the Lower Leg
Axillary View. Dissection II.

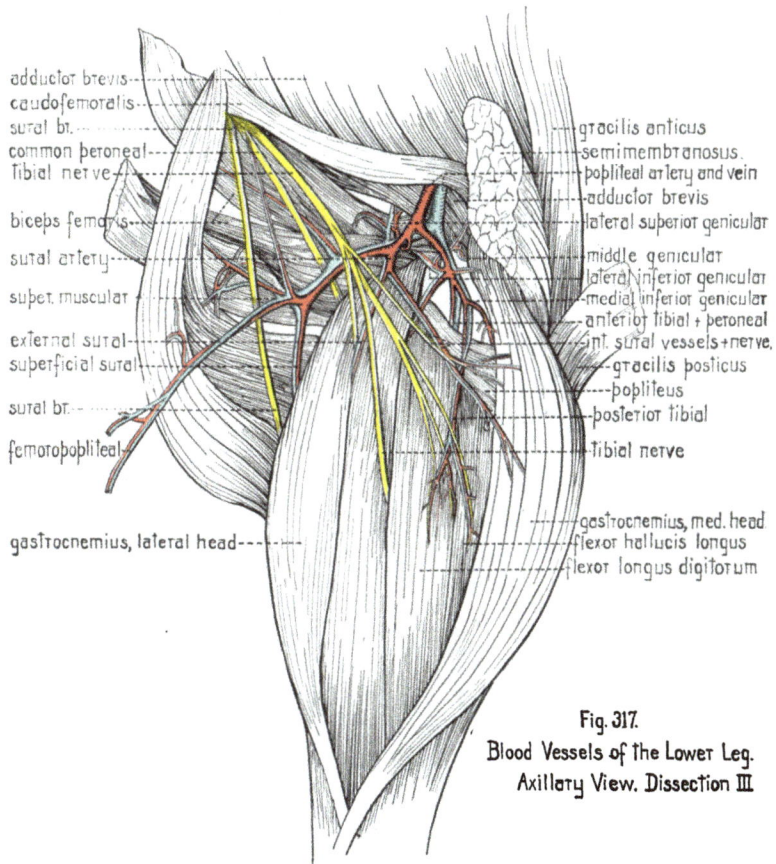

adductor brevis
caudofemoralis
sural br.
common peroneal
tibial nerve

biceps femoris

sural artery

super. muscular

external sural
superficial sural

sural br.

femoropopliteal

gastrocnemius, lateral head

gracilis anticus
semimembranosus
popliteal artery and vein
adductor brevis
lateral superior genicular

middle genicular
lateral inferior genicular
medial inferior genicular
anterior tibial + peroneal
int. sural vessels + nerve
gracilis posticus
popliteus
posterior tibial

tibial nerve

gastrocnemius, med. head
flexor hallucis longus
flexor longus digitorum

Fig. 317.
Blood Vessels of the Lower Leg.
Axillary View. Dissection III

vastus lateralis----

lat. superior genicular----
biceps femoris----

external sural----
superficial sural----
superior muscular----
sural nerve----

femoropopliteal----

small saphenous vein----

lateral marginal vein----

--tibial nerve
--adductor brevis
--semimembranosus
--popliteal artery + vein

--caudofemoralis
--med. epicondyle
--semitendinosus
--tibial nerve
--anterior tibial
--posterior tibial
--internal sural

--lymph node

--gastrocnemius, med. head

--gastrocnemius, lateral head

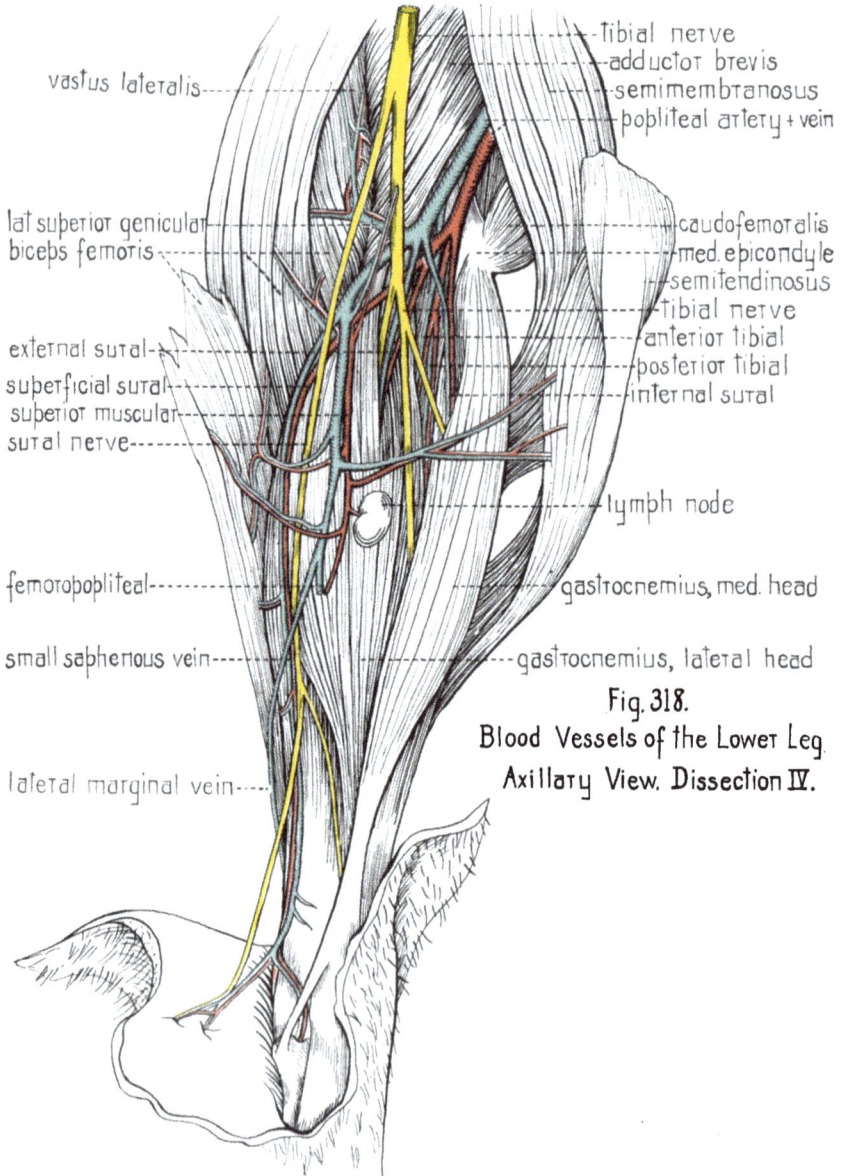

Fig. 318.
Blood Vessels of the Lower Leg.
Axillary View. Dissection IV.

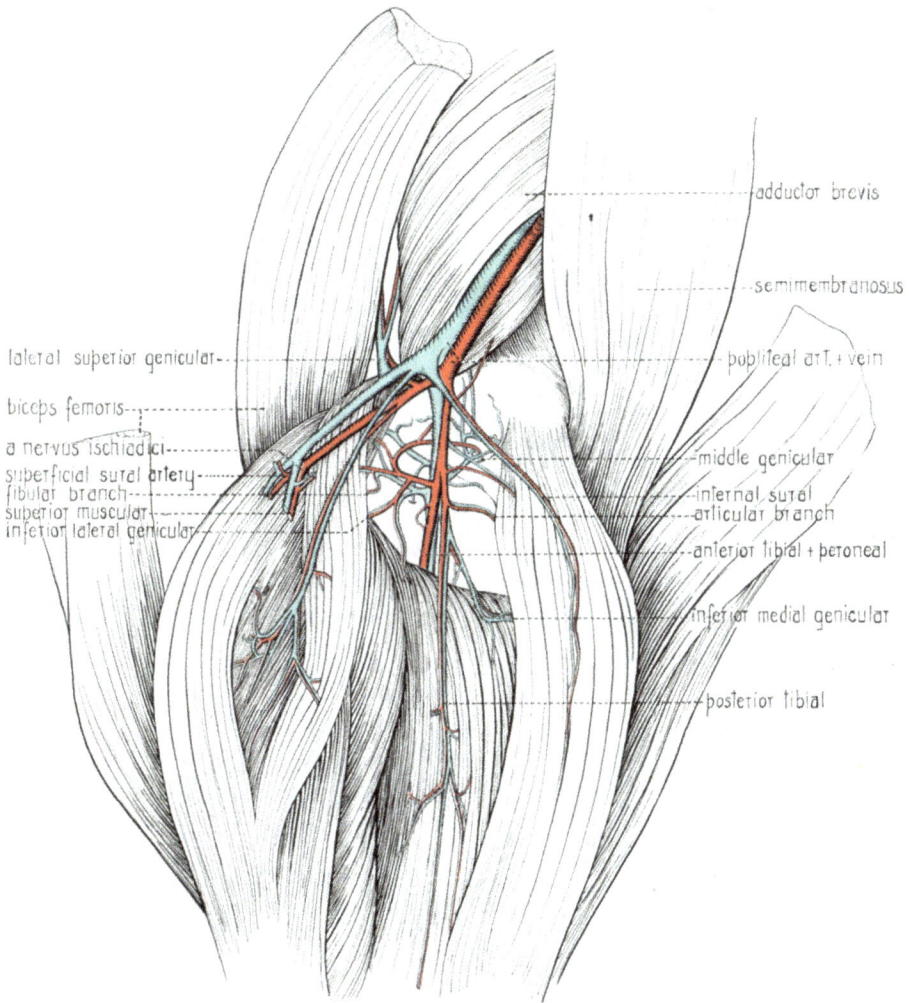

adductor brevis

semimembranosus

poplifeal art. + vein

lateral superior genicular

biceps femoris

a nervus ischiadici

superficial sural artery

fibular branch

superior muscular

inferior lateral genicular

middle genicular

internal sural

articular branch

anterior tibial + peroneal

inferior medial genicular

posterior tibial

Fig.319. Deep Dissection of the Popliteal Space. Axillary View.

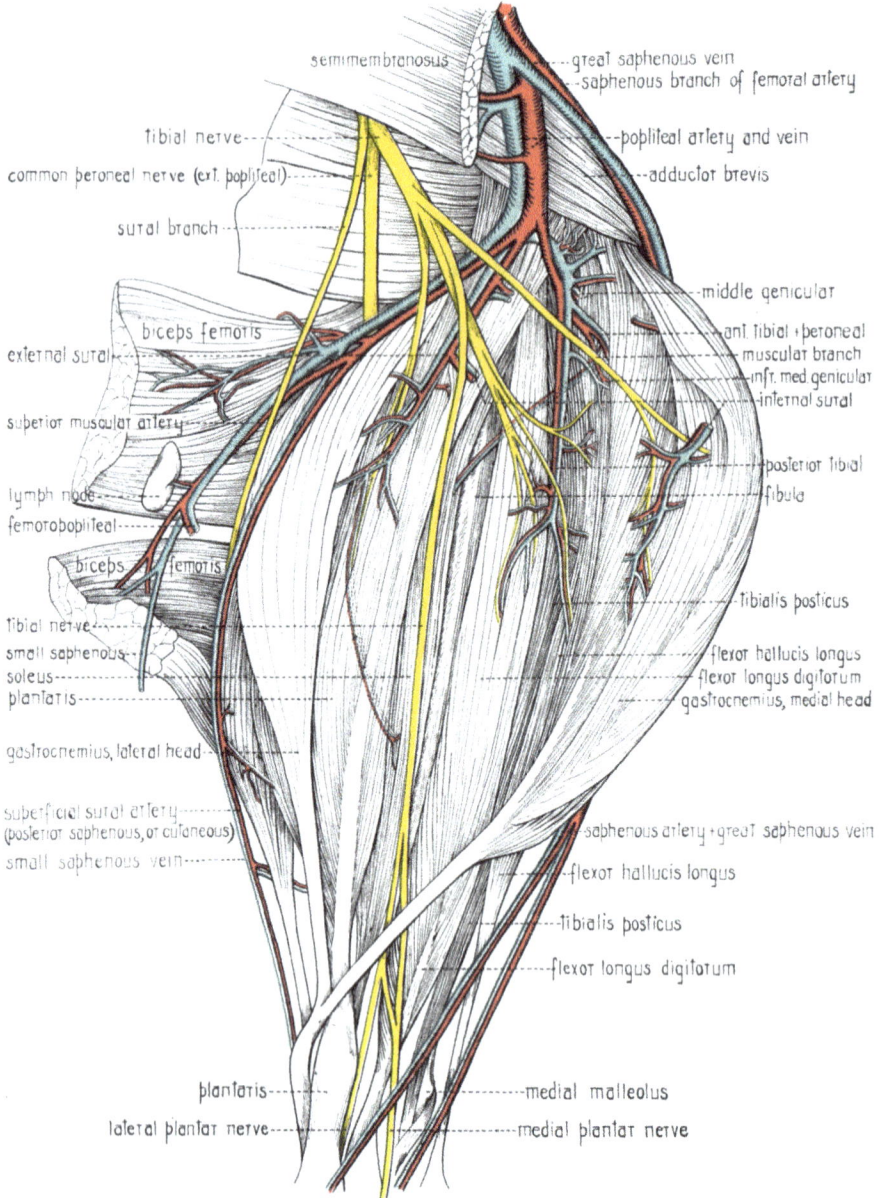

semimembranosus

great saphenous vein
saphenous branch of femoral artery

tibial nerve

poplifeal artery and vein

common peroneal nerve (ext. poplifeal)

adductor brevis

sural branch

middle genicular

biceps femoris

ant. tibial + peroneal
muscular branch
infr. med genicular
internal sural

external sural

superior muscular artery

posterior tibial
fibula

lymph node
femorobopliteal

biceps femoris

tibialis posticus

tibial nerve
small saphenous
soleus
plantaris

flexor hallucis longus
flexor longus digitorum
gastrocnemius, medial head

gastrocnemius, lateral head

superficial sural artery
(posterior saphenous, or cutaneous)

sabhenous artery + great saphenous vein

small saphenous vein

flexor hallucis longus

tibialis posticus

flexor longus digitorum

plantaris

medial malleolus

lateral plantar nerve

medial plantar nerve

Fig.320. Blood Vessels of the Lower Leg. Axillary View

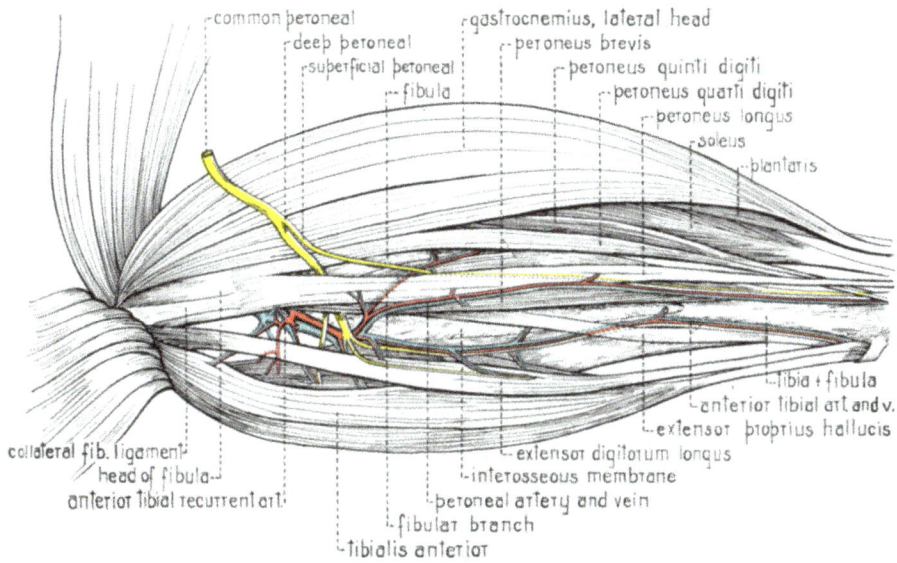

common peroneal
deep peroneal
superficial peroneal
fibula

gastrocnemius, lateral head
peroneus brevis
peroneus quinti digiti
peroneus quarti digiti
peroneus longus
soleus
plantaris

tibia + fibula
anterior tibial art. and v.
extensor proprius hallucis
extensor digitorum longus
interosseous membrane
peroneal artery and vein
fibular branch
tibialis anterior

collateral fib. ligament
head of fibula
anterior tibial recurrent art.

Fig. 321. Blood Vessels of the Lower Leg. Lateral Aspect.

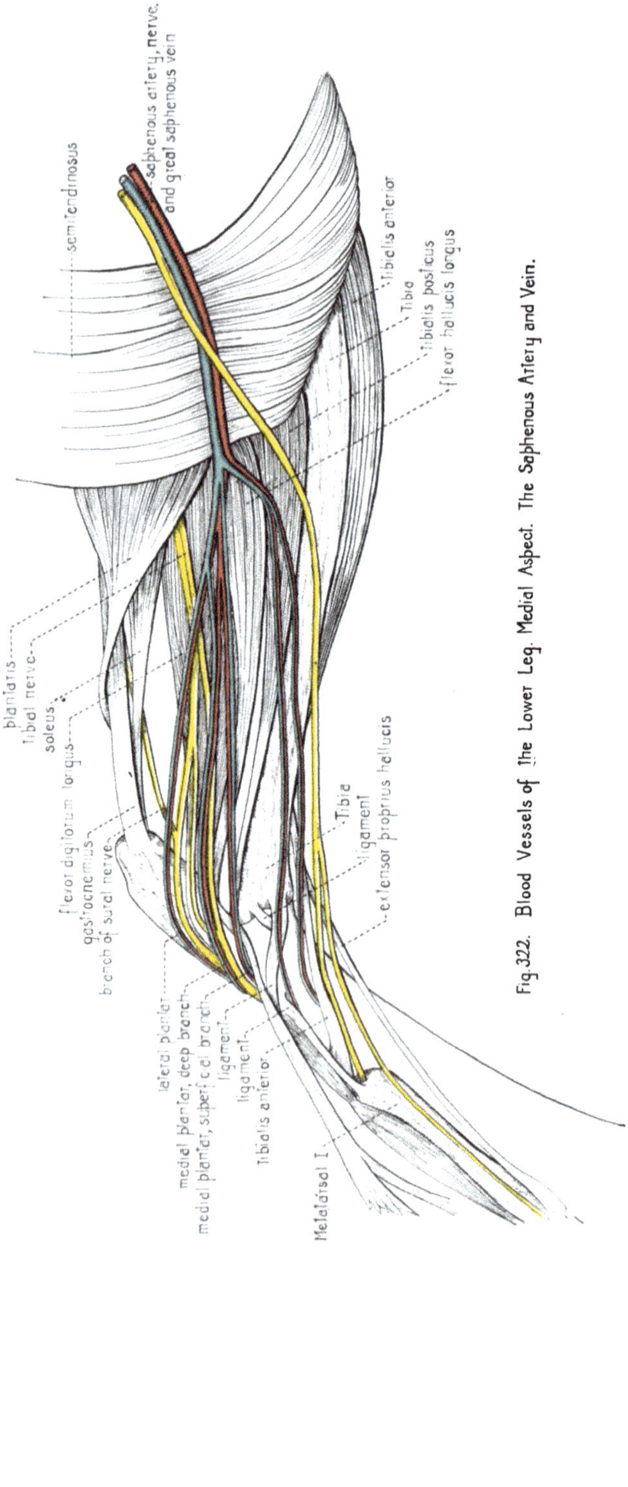

Fig. 322. Blood Vessels of the Lower Leg. Medial Aspect. The Saphenous Artery and Vein.

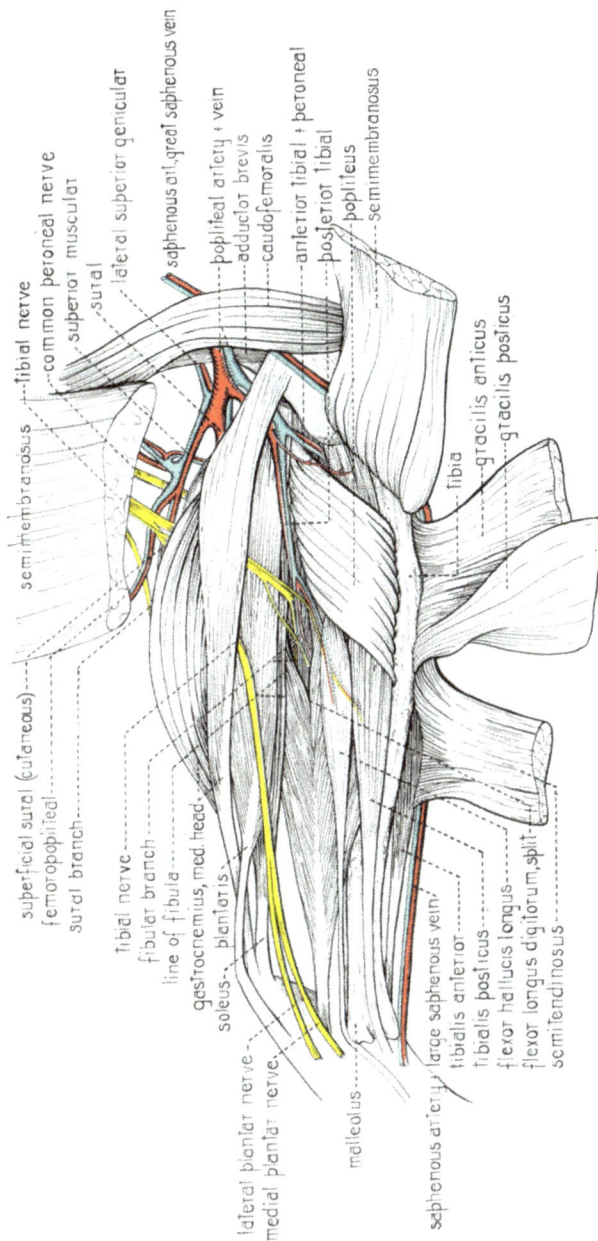

Fig. 323 · Blood Vessels of the Lower Leg. Medial Aspect. Posterior Tibial Artery and Vein.

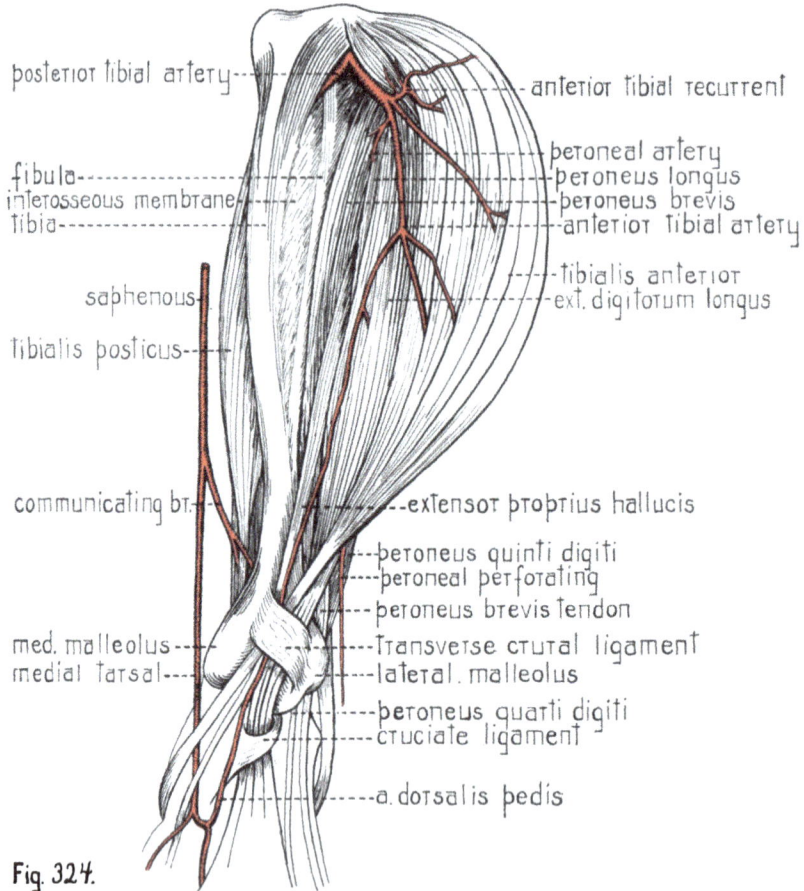

posterior tibial artery---

fibula------
interosseous membrane
tibia------

saphenous-

tibialis posticus--

communicating br.

med. malleolus---
medial tarsal---

anterior tibial recurrent

peroneal artery
peroneus longus
peroneus brevis
anterior tibial artery

tibialis anterior
ext. digitorum longus

extensor proprius hallucis

peroneus quinti digiti
peroneal perforating
peroneus brevis tendon

Transverse crural ligament
lateral. malleolus

peroneus quarti digiti
cruciate ligament

a. dorsalis pedis

Fig. 324.

Blood Vessels of the Lower Leg.   The Anterior Tibial Artery.

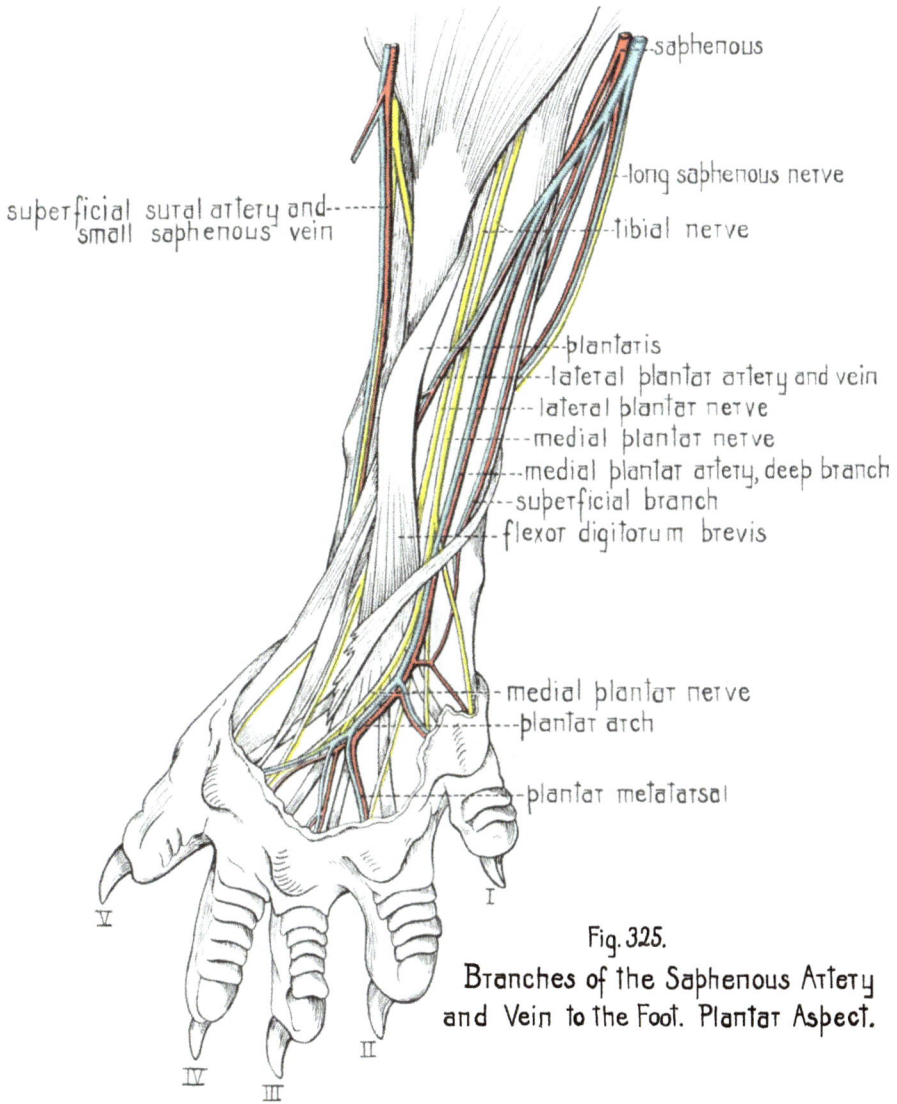

superficial sural artery and
small saphenous vein

saphenous

long saphenous nerve

tibial nerve

plantaris
lateral plantar artery and vein
lateral plantar nerve
medial plantar nerve
medial plantar artery, deep branch
superficial branch
flexor digitorum brevis

medial plantar nerve
plantar arch

plantar metatarsal

V

I

IV

III

II

Fig. 325.
Branches of the Saphenous Artery
and Vein to the Foot. Plantar Aspect.

sural branch
common peroneal nerve (ext poplitea)
tibial nerve
a nervus ischiadici
adductor brevis
biceps femoris

superior muscular branch

gastrocnemius, lateral head

plantaris

superficial sural artery and
small saphenous vein

semimembranosus
lat. superior genicular
popliteal art. + vein

ant. tibial + peroneal
post. tibial

internal sural

external sural
gastrocnemius,
medial head

saphenous artery
great saphenous vein

cutaneous

medial plantar nerve, vein + artery
lateral plantar nerve, vein + artery
medial tarsal, joins a dorsalis pedis
superficial branch of medial plantar

plantaris tendon

communicating branch

abductor quinti digiti
lateral plantar, (superficial branch
deep branch)

flexor digitorum brevis
flexor quinti digiti brevis
flexor digitorum longus
plantar arch
interdigital bed
dorsal
interosseous
ant perforating
plantar
metatarsal
to interdigital pad
anterior perforating
fdl

flexor hallucis brevis
plantar digital
interdigital pad

dorsal digital artery
fdl
lumbricales

interdigital bed
fdb - flexor digitorum brevis
fdl - flexor digitorum longus

plantar digital
arteries

fdb
fdl

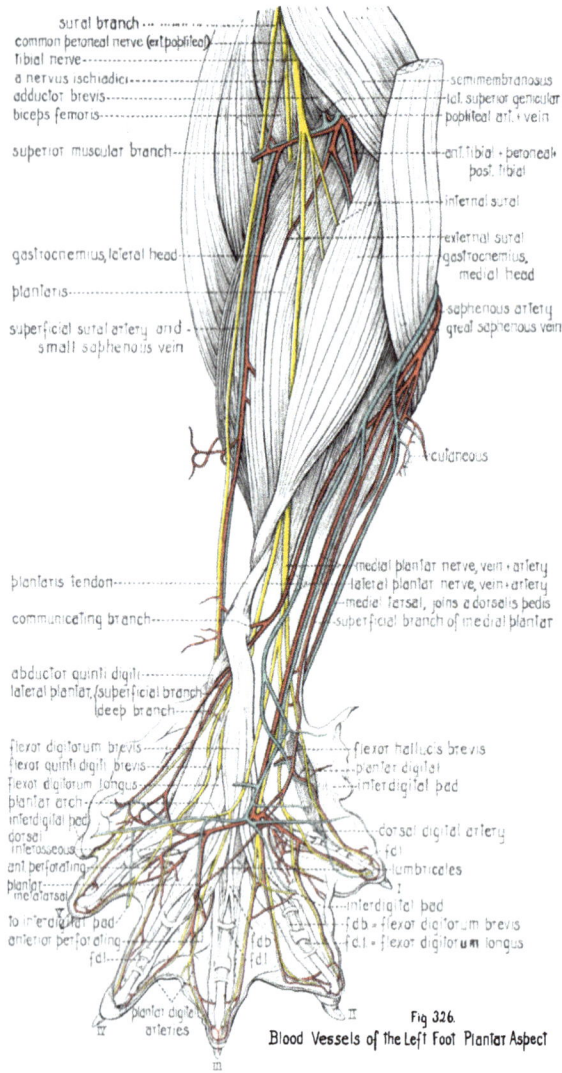

Fig 326.
Blood Vessels of the Left Foot Plantar Aspect

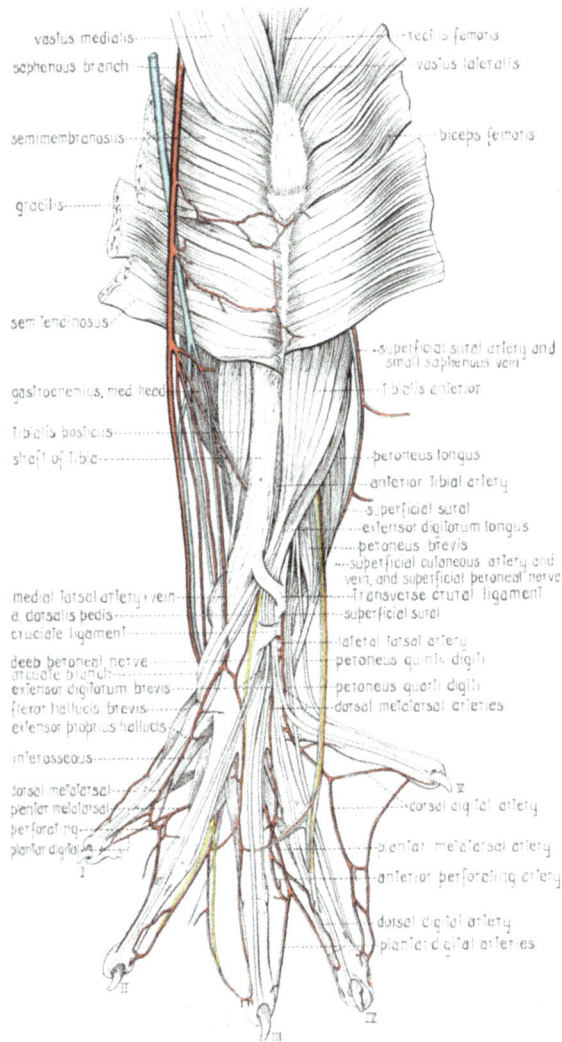

vastus medialis
saphenous branch
semimembranosus
gracilis
semitendinosus
gastrocnemius, med. head
tibialis posticus
shaft of tibia

medial tarsal artery + vein
a. dorsalis pedis
cruciate ligament
deep peroneal nerve
arcuate branch
extensor digitorum brevis
flexor hallucis brevis
extensor proprius hallucis
interosseous
dorsal metatarsal
plantar metatarsal
perforating
plantar digital

rectus femoris
vastus lateralis
biceps femoris

superficial sural artery and
small saphenous vein
tibialis anterior
peroneus longus
anterior tibial artery
superficial sural
extensor digitorum longus
peroneus brevis
superficial cutaneous artery and
vein, and superficial peroneal nerve
transverse crural ligament
superficial sural
lateral tarsal artery
peroneus quinti digiti
peroneus quarti digiti
dorsal metatarsal arteries

dorsal digital artery
plantar metatarsal artery
anterior perforating artery
dorsal digital artery
plantar digital arteries

Fig. 327. Blood Vessels of the Left Foot. Dorsal Aspect.

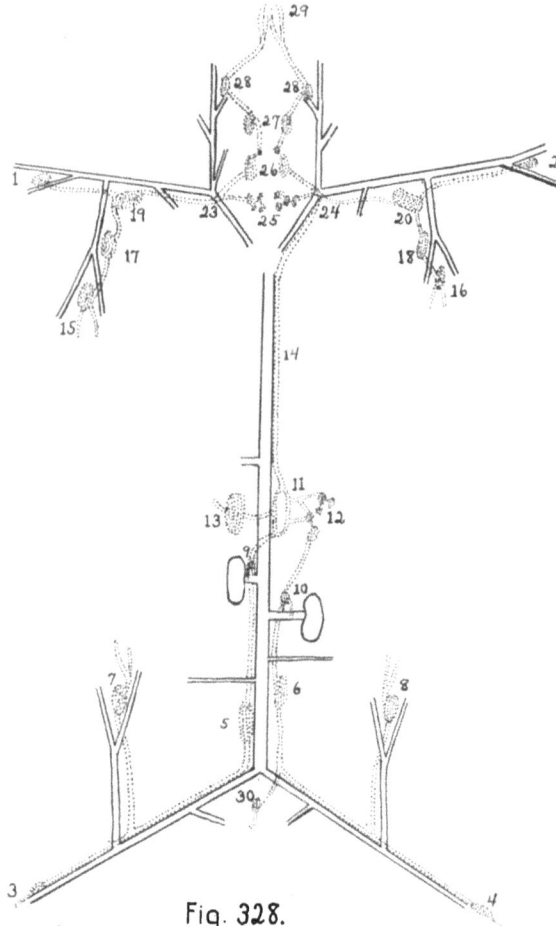

Fig. 328.
Diagram of Lymphatic System.
Venous system solid lines, lymphatic stippled.

1-2, right and left elbow nodes
3-4, right and left knee nodes
5-6, lumbar nodes
7-8, inguinal nodes
9-10, renal nodes
11, cisterna chyli
12, cisternal group of nodes
13, intestinal node
14, thoracic duct
15-20, axial nodes
23-24, jugulo-subclavian taps
25, thoracic group
26, posterior cervical nodes
27, anterior cervical nodes
28, submaxillary nodes
29, Tongue and lip plexus
30, caudal lymph node
Job, Thesle T., Anat. Rec. Vol. 9. No 6.

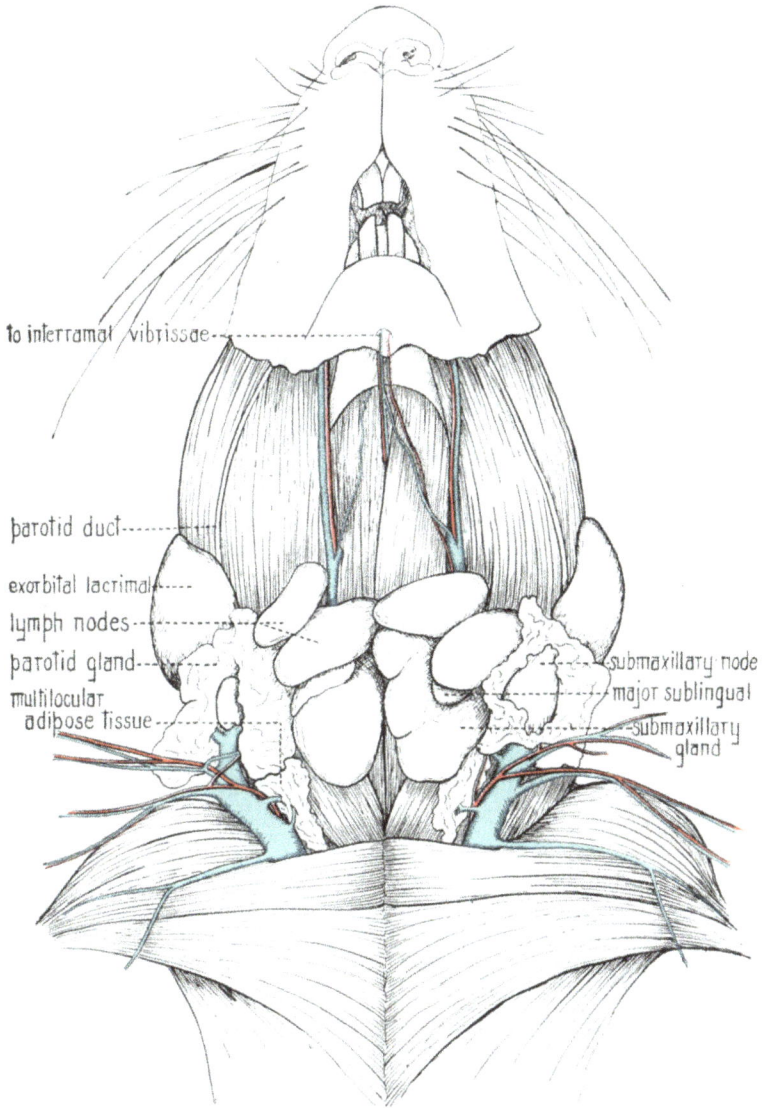

Fig.329. Superficial Dissection of Neck Viscera

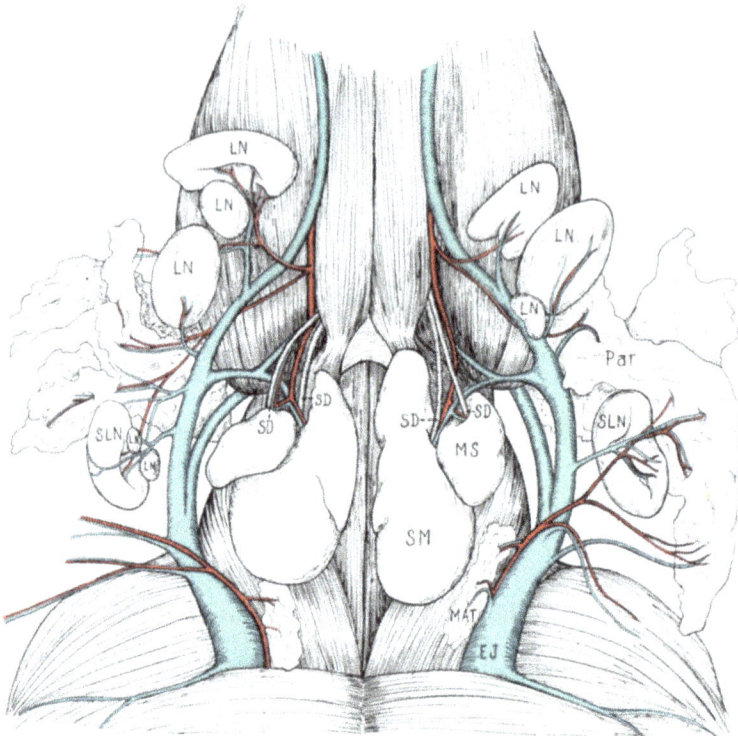

Fig. 330. Superficial glands of neck reflected to show relation to underlying structures

LN- lymph node                    MAT- multilocular adipose tissue
Par- parotid gland                SD - salivary duct
MS- major sublingual gland        EJ - external jugular vein
SM- submaxillary gland
SLN- submaxillary lymph node

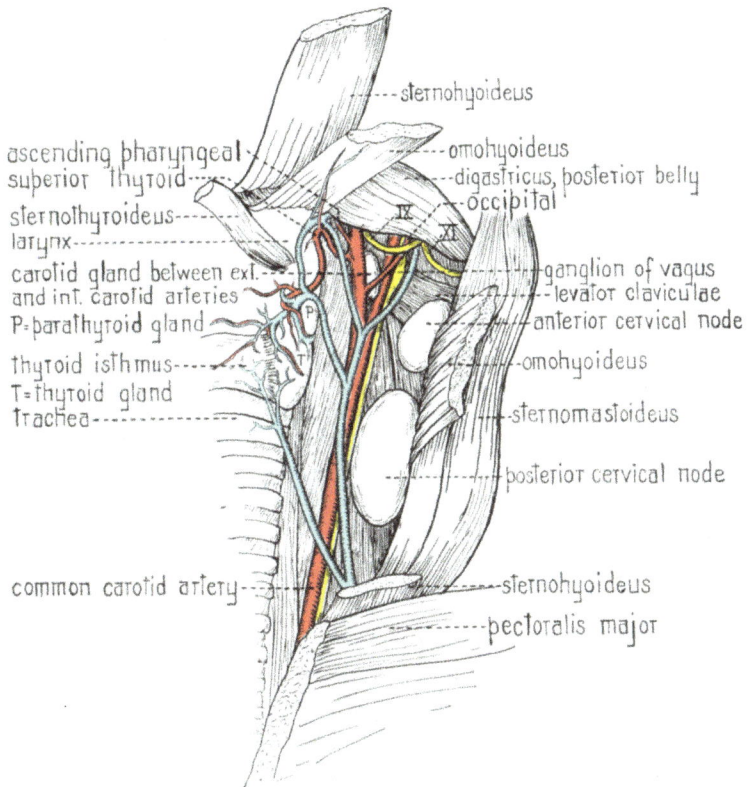

Fig. 331. Deep dissection of left side of neck to show carotid gland, cervical lymph nodes, thyroid and parathyroid glands.

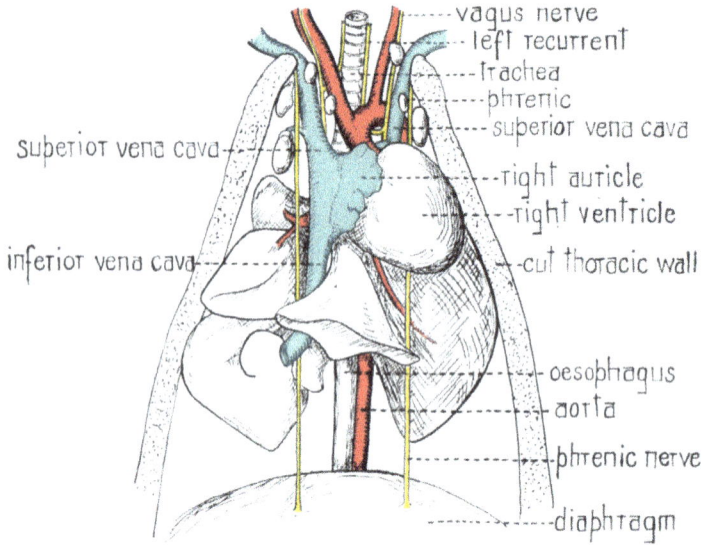

Fig. 332. Ventral view of thoracic cavity showing thoracic lymph nodes and vagus and phrenic nerves.

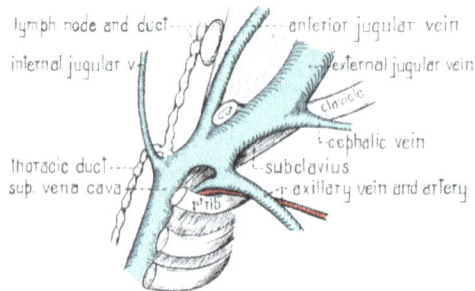

Fig. 333. Detail of lymph duct entering jugular vein. In this specimen the subclavius muscle perforated the vein.

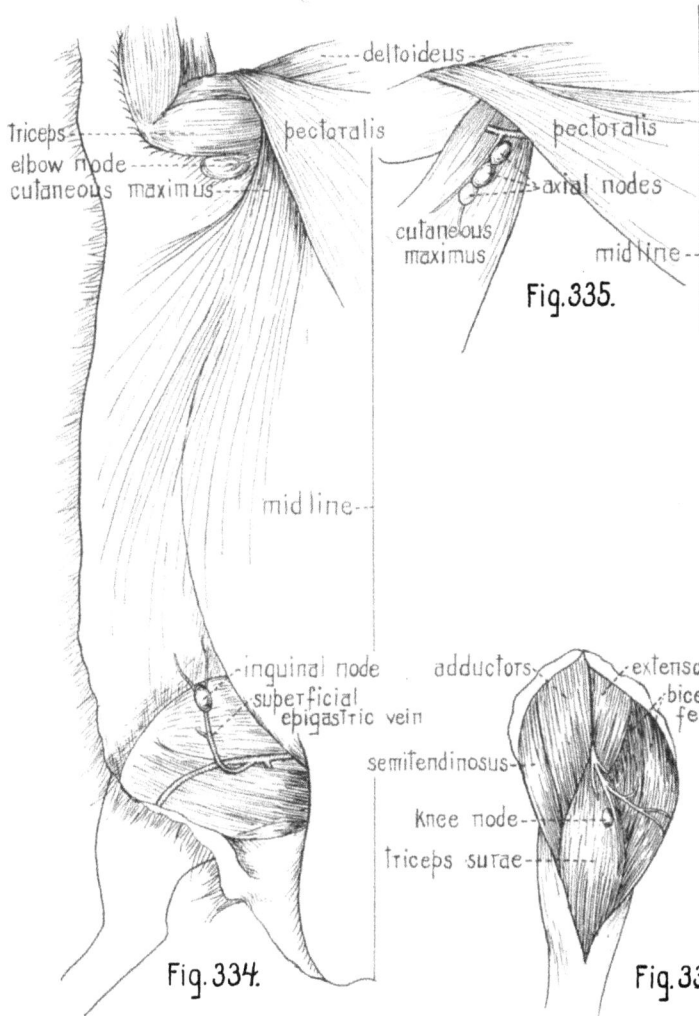

Fig. 335.

Fig. 334.

Fig. 336.

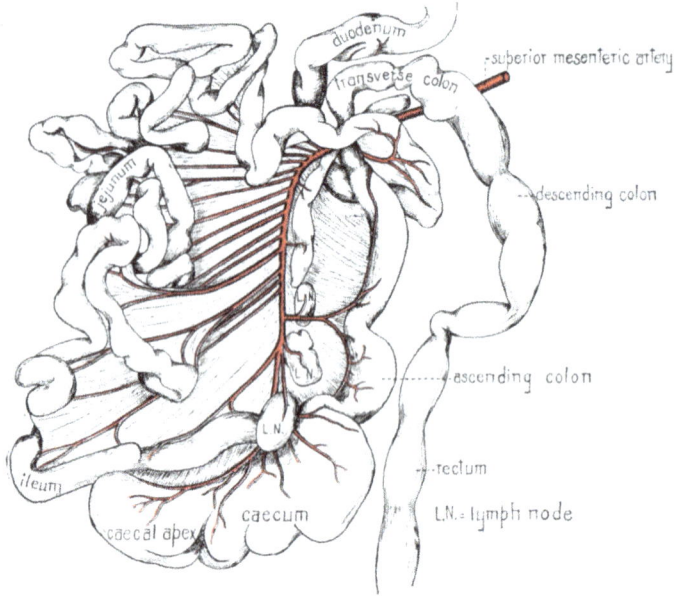

Fig. 337. Intestinal group of lymph nodes.

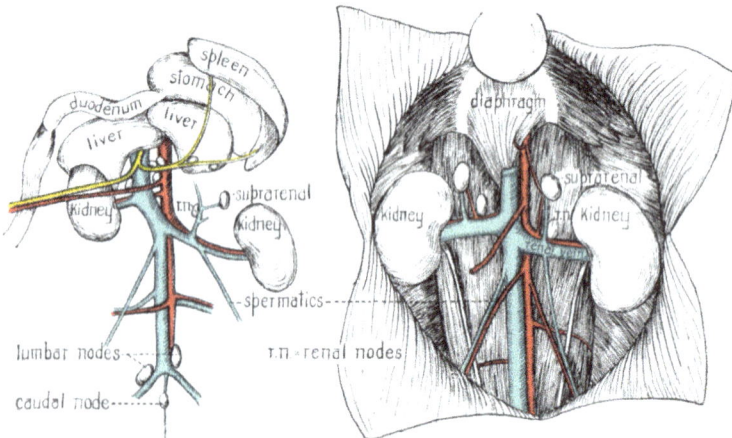

Fig. 338. Cisternal, lumbar, and caudal nodes.

Fig. 339. Renal Lymph Nodes.

# BIBLIOGRAPHY

The following Textbooks and Atlases of human anatomy were used extensively: Cunningham—Robinson; Gray—Lewis; Morris—McMurrich; Piersol; Poirier et Charpy; Toldt; Spalteholtz—Barker.

ADACHI, BUNTARO. 1928. Das Arteriensystem der Japaner. Bd. I und II., Kaiserlich-Japanischen Universität zu Kyoto.

———. 1933. Das Venensystem der Japaner. I. Lieferung, Kaiserlich-Japanischen Universität zu Kyoto.

ADDISON, WM. H. F., and HOW, H. W. 1921. The development of the eyelids of the albino rat, until the completion of disjunction. *Am. J. Anat.*, v. 29, pp. 1–31.

ADDISON, WM. H. F., and FRASER, DORIS A. 1932. Variability of pigmentation in the hypophysis and parathyroids of the gray rat (Mus norvegicus). *J. Comp. Neur.*, v. 55, no. 2.

ALEZAIS, H. 1898–1903. Etude anatomique du cobaye (Cavia cobaya). *J. Anat. Phys.*, Ann. 34, pp. 735–756. Ann. 35, pp. 333–381, 18 figg. Ann. 36, pp. 635–648, 4 figg. Ann. 37, pp. 102–126, 270–290, 20 figg. Ann. 38, pp. 259–275, 624–646, 16 figg. Apart, Paris 8° 172 pp. 58 figg.

———. 1902. Les adducteurs de la cuisse chez les rongeurs. *J. de l'Anat. et de la Physiol. normales et pathologiques de l'homme et des animaux.*, v. 38, pp. 1–13.

ANGLAS, J. 1910. Les animaux de laboratoire La Souris (Anatomie et dissection). Vigot Frères, Paris.

BARDELEBEN, VON K. 1894. On the bones and muscles of the mammalian hand and foot. *Proc. Zool. Soc.*, London, pp. 354–375.

BAYER, L. 1893. Beitrag zur vergleichenden Anatomie der Oberarmarterien. *Gegenbauer's Morph. Jahrb.*, Bd. XIX, pp. 1–41 Tafel. I.

BEILING, KARL. 1906. Beiträge zur makroskopischen und mikroskopischen Anatomie der Vagina und des Uterus der Säugetiere. *Archiv. f. mikr. Anat.*, Bd. 67, S. 573–637. Mus decumanus S. 588.

BERRY. R. J. A. 1901. The true caecal apex, or the vermiform appendix: its minute and comparative anatomy. *J. Anat. and Phys.*, v. 35, p. 92.

BIGNOTTI, G. 1900. Sul tarso del Mus decumanus. *Monit. Zoöl. ital.*, v. 11, suppl., pp. 17–19.

BLUNTSCHLI, H. 1906. Die Arteria femoralis und ihre Äste bei den nierderen catarrhinen Affen. Eine vergleichend-anatomische Untersuchung. *Morph. Jahrb.*, Bd. XXXVI, 85 Fig. im Text.

———. 1907. Varietäten der Arteria profunda femoris und der Arteria circumflexa femoris medialis des Menschen. *Morph. Jahrb.*, Bd. XXXVII, 5 Fig. im Text.

BRADLEY, O. CHARNOCK. 1927. Topographical anatomy of the dog. 2nd ed., Macmillan & Co., New York.

BRESSLAU, ERNST. 1920. The mammary apparatus of the Mammalia. Methuen & Co., Ltd., London.

BUTCHER, EARL ORLO. 1934. The hair cycles in the albino rat. *Anat. Rec.*, v. 61, no. 1.

BUTLER, E. G. 1927. The relative rôle played by the embryonic veins in the development of the mammalian vena cava posterior. *Am. J. Anat.*, v. 39, p. 267.

CHAUVEAU, A. 1888. The comparative anatomy of domesticated animals. Revised by S. Arloing and translated and edited by George Fleming. D. Appleton & Company, New York.

CLASEN, F. 1897. Die Muskeln und Nerven des proximalen Absnittes des vorderen Extremität des Kaninchens. Halle.

CUNNINGHAM, D. J. 1879. Intrinsic muscles of the mammalian foot. *J. Anat. and Phys.*, v. XIII, pp. 1–16.

DANDY, W. E., and GOETSCH, E. 1911. The blood supply of the pituitary body of the dog. *Am. J. Anat.*, v. 11, pp. 137–150.

DECASTRO, F. 1926. Sur la Structure et l'Innervation de la Glande Intercarotidienne (Glomus caroticum) de l'Homme et des Mammifères, et sur un nouveau Système d'Innervation Autonome du Nerf Glossopharyngien. Etudes anatomiques et expèrimentales Trauvaux du Laboratoire de Recherches Biologiques de L'Université de Madrid, Tome XXIV, Fascicule 4 eme, pp. 365-432.

DISSELHORST, RUDOLPH. 1904. Ausführapparat und Anhangsdrüsen der Männlichen Geschlechtsorgane. In Oppel A, "Lehrbuch der Vergleichenden Mikroskopischen Anatomie der Wirbelthiere." Vierter Teil. Gustav Fischer, Jena. Rodentia. Mus decumanus, pp. 263-282.

DOBSON, G. E. 1883. On the homologies of the long flexor muscles of the feet of Mammalia, with remarks on the value of their leading modifications in classification. *J. Anat. and Phys.*, v. XVII, pp. 142-179.

———. 1884. On the myology and visceral anatomy of Capromys melanurus. *Proc. Zool. Soc. of London*, No. 16.

DONALDSON, H. H. 1924. The Rat: data and reference tables. Second edition, revised and enlarged. (Memoir of the Wistar Institute of Anatomy and Biology—Number 6.) Philadelphia: The Wistar Institute of Anatomy and Biology.

EISLER, PAUL. 1912. Die Muskeln des Stammes Bardeleben: Handbuch der Anatomie des Menschen. Jena.

ELLENBERGER, WILHELM, und GUENTHER, G. 1908. Grundriss der vergleichenden Histologie der Haussaugetiere. Berl. Parey 3rd ed., rev. and enlarged.

ELLIOTT, F. R., and TUCKETT, I. 1906. Cortex and medulla in the suprarenal glands. *J. Phys.*, v. 34, pp. 332-369.

EVANS, C. L. 1930. Recent advances in physiology. 4th ed., P. Blakiston's Son & Co., Phila.

FISCHEL, ALFRED. 1914. Zur normalen Anatomie und Physiologie der weiblichen Geschlechtsorgane von Mus decumanus sowie über die experimentelle Erzeugung von Hydro- und Pyosalpinx. *Arch. f. Entwicklungsmechn. d. Organ.*, v. 39, pp. 578-616.

FLOWER, W. H. 1872. Lectures on the comparative anatomy of the Mammalia. *Med. Times & Gaz.*, v. 1 and 2.

FRASER, DORIS A. 1931. The winter pelage of the adult albino rat. *Am. J. Anat.*, v. 47, pp. 55-87.

FRETS, G. P., 1909. Über den Plexus lumbo-sacralis,

sein Verbreitungs-gebiet und die Beziehungen zwischen Plexus und Wirbelsäule bei den Monotremen nebst vergleichend-myologischen Bemerkungen. *Morph. Jahrb.*, Bd. 40, Erstes Heft. pp. 1-104.

GADOW, H. 1882. Beiträge zur Myologie der hinteren Extremität der Reptilien. *Morph. Jahrb.*, VII, pp. 382-466.

GOEPPERT, E. 1909. Ueber die Entwicklung von Varietäten im Arteriensystem. Untersuchungen an der Vorder-gliedmasse der weissen Maus. *Gegenbauer's Morph. Jahrb.*, Bd. XL.

GRAY, ALBERT A. 1907 and 1908. The labyrinth of animals, including mammals, birds, reptiles and amphibians. 2 vol., 450 pp. J. & H. Churchill.

GREENMAN, M. J. 1913. Studies on the regeneration of the peroneal nerve of the albino rat: number and sectional areas of fibers: area relation of axis to sheath. *J. Comp. Neur.*, v. 23, pp. 479-513.

GREENMAN, M. J., and DUHRING, F. L. 1923. Breeding and care of the albino rat for research purposes. Philadelphia: The Wistar Institute of Anatomy and Biology.

GREGORY, W. K., and CAMP, C. L. 1918. Studies in comparative myology and osteology. No. III, *Bull. Am. Mus. Nat. Hist.*, XXXVIII, pp. 447-563.

HALDEMAN, K. O. 1921. The method of opening of the vagina in the rat. *Proc. of Am. Assoc. Anat.*, in *Anat. Rec.*, v. 21, p. 60.

HARMON, N. BISHOP. 1898. Caudal limit of lumbar visceral efferent nerves in man. *J. Anat. and Phys.*, v. 32, pp. 403-421.

———. 1899. The pelvic splanchnic nerves: an examination into their range and character. *J. Anat. and Phys.*, v. 33, pp. 386-399.

HENNEBERG, BRUNNO. 1900. Die erste Entwickelung der Mammarorgane bei der Ratte. *Anat. Hefte.*, v. 13.

———. 1905. Beitrag zur Kenntnis der lateralen Schilddrüsenanlage. *Anat. Hefte.*, v. 28, pp. 287-302.

———. 1914. Beitrag zur Entwickelung der äusseren Genitalorgane beim Säuger. Erster Teil. *Anat. Hefte.*, v. 50, pp. 425-498.

HEYMANS, C. 1929. Le sinus carotidien. Louvain and Paris. London: H. K. Lewis.

HIGGINS, GEORGE M. 1925. On the lymphatic system of the newborn rat. *Anat. Rec.*, v. 30, no. 4.

HILZHEIMER, M. 1913. Handbuch der Biologie der Wirbeltiere. Unter Mitwirkung von O. Haempel. Ferdinand Enke, Stuttgart.

HOAG, L. A. 1918. Histology of the sensory root of the trigeminal nerve of the rat (Mus norvegicus). *Anat. Rec.*, v. 14, pp. 165-182.

HOFFMAN, C. K., und WEYENBERGH, H. 1870. Osteologie und Myologie von Sciurus vulgaris.

HOLLISTER, N. 1916a. The generic names Epimys and Rattus. *Proc. Biol. Soc. of Wash.*, v. 29, pp. 125-218.

———. 1916b. The type species of Rattus. *Proc. Biol. Soc. of Wash.*, v. 29, pp. 205-208.

HOSKINS, M. M., and CHANDLER, S. B. 1925. Accessory parathyroids in the rat. *Anat. Rec.*, v. 30, No. 2, p. 95.

HOWELL, A. B. 1926. Anatomy of the wood rat. Comparative anatomy of the subgenera of the American wood rat (Genus Neotoma), No. 1, Monographs of the Amer. Soc. of Mammalogists. Williams & Wilkins Co., Baltimore.

HOWELL, W. H. 1888. Dissection of a dog as a basis for the study of physiology. Henry Holt & Company, New York.

HUBER, G. CARL. 1915a. The development of the albino rat from the end of the first to the tenth day after insemination. *Anat. Rec.*, v. 9, pp. 84-88.

———. 1915b. The development of the albino rat (Mus norvegicus albinus). Part I. From the pronuclear stage to the stage of mesoderm anlage; end of the first to the end of the ninth day. *J. Morph.*, v. 26, pp. 247-358.

———. 1915c. The development of the albino rat (Mus norvegicus albinus). Part II. Abnormal ova. End of the first to the end of the ninth day. *J. Morph.*, v. 26, pp. 359-386.

HUMPHREY, G. M. 1872. Observations in myology. Macmillan and Company, Cambridge and London.

HUNT, H. R. 1924. A laboratory manual of the anatomy of the rat. The Macmillan Co., N. Y.

HUNTINGTON, GEO. S., and SCHULTE, H. VON W., 1912. Studies in cancer and allied subjects. v. IV. Contributions to the anatomy and development of the salivary glands in the Mammalia. Conducted under the George Crocker Special Research Fund at Columbia University. Columbia Univ. Press, New York, 1913.

JACKSON, C. M. 1913. Postnatal growth and variability of the body and of the various organs in the albino rat. *Am. Anat.*, v. 15, No. I, July, pp. 1-68.

JOB, THESLE T. 1915. The adult anatomy of the lymphatic system in the common rat (Epimys norvegicus). *Anat. Rec.*, v. 9, No. 6, pp. 447-459.

———. 1918. Lymphatico-venous communications in the common rat and their significance. *Am. J. Anat.*, v. 24, pp. 467-491.

———. 1922. Studies on lymph nodes. I. Structure. Introductory paper. *Am. J. Anat.*, v. 31, pp. 125-137.

KINGSLEY, J. S. 1899. Vertebrate Zoology, Henry Holt and Company, New York.

———. 1912. Comparative Anatomy of Vertebrates. P. Blakiston's Son & Co., Phila.

KOLSTER, RUD. 1901. Vergleichend anatomische Studien über der m. pronator teres der Säugetiere. *Anat. Hefte.*, Bd. 17, S. 673-834. Mus Rattus, p. 714.

LEITCH, MARY SANFORD. 1934. (Unpublished manuscript.) Embryology of the principal arterial vessels in the rat.

LEWIS, W. B. 1881. On the comparative structure of the brain in rodents. *Phil. Trans.*, 1882, pp. 699-749.

LEYDIG, F. 1850. Sur Anatomie der männlichen Geschlechtsorgane und Analdrüsen der Säugetiere. *Ztschr. f. wiss. Zool.*, Bd. 2, S. 1-57.

———. 1854. Kleinere Mitteilungen zur tierischen Gewebelehre. *Archiv. f. Anat. u. Entwickelungsgeschichte*, S. 296-348.

LOEWENTHAL, N. 1892. Beitrag zur Kenntnis der Harder'schen Drüse bei den Säugetieren. *Anat. Anz.*, v. 7, nr. 16, 17, pp. 546-556.

———. 1894. Zur Kenntnis der Glandula submaxillaris einiger Säugetiere. *Anat. Anz.*, v. 9, pp. 223-229.

———. 1895. Zur Kenntnis der Glandula infraorbitalis einiger Säugetiere. *Anat. Anz.*, v. 10. pp. 123-130.

———. 1897. Note sur le structure fine des glandes de Cowper du rat blanc. *Bibliogr. Anat.*, v. 4, pp. 168-170.

———. 1900. Drüsenstudiern II. Die Gl. infraorbitalis und eine besondere der parotis angliegende Drüse bei der weissen Ratte. *Arch. f. mikr. Anat.*, v. 56, pp. 535–552.

——. 1908. Drüsenstudien III. Die Unterkieferdrüse des Igels und der weissen Ratte. *Arch. f. mikr. Anat.*, Bd. 71, S. 588–666.

LONG, J. A. 1922. On the structure and development of a fat body or gland in the rat. *Anat. Rec.*, v. 23, p. 107.

MACALISTER, ALEXANDER. 1872. The myology of the Cheiroptera. Phil. Trans. 1872, pp. 125–172.

MAISONNEUVE, PAUL. 1878. Osteologie et Myologie du Vespertilio murinis. Paris.

MANN, F.C. 1920. A comparative study of the anatomy of the sphincter at the duodenal end of the common bile duct, with special reference to species of animals without a gall-bladder. *Anat. Rec.*, v. 18, pp. 355–360.

MARTIN, H. N., and MOALE, W. A. 1884. Handbook of vertebrate dissection, Part III. How to dissect a rodent. Macmillan and Company, New York.

MCALPINE, D. 1884. Zoölogical Atlas (including comparative anatomy). Century Co., New York.

MCMASTER, P. D. 1922. Do species lacking a gallbladder possess its functional equivalent? *J. Exp. Med.*, v. 35, pp. 127–140.

METCALF, ZENO PAYNE. 1932. An introduction to zoology through the study of the vertebrates, with special reference to the rat and man. Charles C. Thomas, Springfield.

MILLER, G. S., JR. 1895. On the introitus vaginae of certain Muridae. *Proc. Boston Soc. Nat. Hist.*, v. 26, pp. 459–468.

MIVART, ST. GEORGE J. 1881. The Cat. Chas. Scribner's Sons, New York.

MORRELL, G. HERBERT. 1872. Comparative anatomy and dissection of the Mammalia. Part I, pp. 180–208. Longman & Co., London.

——. 1872. Supplement to the anatomy of the Mammalia, containing dissections of sheep's heart and brain, rat, sheep's head, and ox's eye. Longman & Co. London.

MURIE, JAMES, and MIVART, ST. GEORGE J. 1866. On the anatomy of the Lemuroidea. *Proc. Zool. Soc. London.*

MYERS, J. A. 1916. Studies of the Mammary Gland. *Am. J. Anat.*, v. 19, No. 3, pp. 353-390.

MYERS, J. A., and MYERS, F. J. 1921a. Studies on the mammary gland. VII. The distribution of the subcutaneous fat and its relation to the developing mammary glands in male and female albino rats from birth to ten weeks of age. *Anat. Rec.*, v. 22, pp. 353–362.

——. 1921b. Studies on the mammary gland. VIII. Gross changes in the mammary gland in the female albino rat during the period of involution. *Proc. Am. Assoc. Anat. in Anat. Rec.*, v. 21, p. 4.

NOBLE, G. K. 1922. The Phylogeny of the Salientia. I. The Osteology and the thigh musculature; their bearing on classification and phylogeny. *Bull. Am. Mus. Nat. Hist.*, v. XLVI, Art. 1, pp. 1–87, New York.

OPPEL, ALBERT. 1896–1914. Lehrbuch der vergleichenden mikroscopischen Anatomie der Wirbeltiere. 8 vols. Gustav Fischer. Jena.

OUDEMANS, J. TH. 1892. Die accessorischen Geschlechtsdrüsen der Säugetiere. Haarlem 96, p. 16 pl. 4ᵉ.

OWEN, RICHARD. 1868. On the anatomy of vertebrates, London 1866–8., v. 3, Mammals.

PARKER, T. J. 1893. A course of instruction in zootomy. Macmillan and Co., London.

PARSONS, F. G. 1894a. On the morphology of the tendo-Achilles. *J. Anat. & Phys.*, v. XXVIII, pp. 414–418.

——. 1894b. On the myology of the Sciuromorphine and Hystricomorphine rodents. pp. 251–296. *Proc. Zool. Soc. London* 1894.

——. 1896. Myology of rodents. Part II. An account of the myology of the Myomorpha together with a comparison of the muscles of various suborders of rodents. *Proc. Zool. Soc. London.* 1896, pp. 159–192.

PATERSON, A. M. 1887a. The morphology of the sacral plexus in man. *J. Anat. & Phys.*, v. 21, pp. 407–413.

——. 1887b. The limb plexuses of mammals. *J. Anat. and Phys.*, v. 21, pp. 611–634.

——. 1890. Development of the sympathetic nervous system in mammals. *Philos. Trans. Royal Soc. of London* (B), pp. 159–186.

——. 1891. The pectineus muscle and its nerve supply. *J. Anat. and Phys.*, v. 26, pp. 43–48.

——. 1894. The origin and distribution of the nerves to the lower limb. *J. Anat. and Phys.*, v. 28, pp. 84–96.

——. 1896. A discussion of some points in the

distribution of the spinal nerves. *J. Anat. and Phys.*, v. 30, pp. 530–538.

——. 1901. The sternum; its early development and ossification in man and mammals. *J. Anat. and Phys.*, v. 35, pp. 21–32.

PETERS, ALBERT. 1890. Beitrag zur Kenntniss der Harder'schen Drüse. *Arch. f. mikr. Anat.*, Bd. 36, S. 192–203.

POCOCK, R. I. 1914. On the facial vibrissae of mammalia. *Proc. Zool. Soc. London.* 1914, pp. 889–912. Rodentia p. 901.

RAMSTRÖM, M. 1905. Undersuchungen und Studien über die Innervation des Peritoneum der vorderen Bauchwand. *Anat. Hefte.*, Bd. 29, S. 351–443. Mus decumanus, p. 372.

RASMUSSEN, A. T. 1916. Theories of Hibernation. *Am. Natur.*, v. 50, No. 598, pp. 609–625.

——. 1922. The glandular status of brown multilocular adipose tissue. *Endocrinology*, v. 6, pp. 760–770.

RAUTHER, MAX. 1903. Ueber den Genitalapparat einiger Nager und Insektivoren, insbesondere die accessorischen Genitaldrüsen derselben. *Jenaische Ztsch. f. Naturw.*, Bd. 38, S. 377–472, 3 pl. Neue. Folge Bd. 31.

REYNOLDS, S. H. 1897. The vertebrate skeleton. Cambridge Univ. Press.

ROBINSON, ARTHUR. 1887. On the position and peritoneal relations of the mammalian ovary. *J. Anat. & Phys.*, v. XXI, pp. 169–179.

ROLLESTON, GEORGE A. 1870. Forms of animal life; being outlines of zoological classification based upon anatomical investigation and illustrated by descriptions of specimens and figures. Oxford, Clarendon Press.

ROMER, A. S. 1922. The locomotor apparatus of certain primitive and mammal-like reptiles. *Bull. Am. Mus. of Nat. Hist.*, v. XLVI, Art. X, pp. 517–606.

RÖMER, F. 1896. Studien über das Integument der Säugetiere. I. Entwickel. d. Schuppen u. Haare am Schwanze u. an d. Füssen v. Mus decumanus und einigen anderen Muriden. *Jenaische Ztsch. f. Naturw.*, v. 30, pp. 603–622.

ROSENFELD, CARL. 1899. Zur vergleichenden Anatomie des musculus tibialis posticus. *Anat. Hefte.*, v. 11, pp. 361–388. Mus rattus, p. 364.

RUGE, G. 1878a. Untersuchung über die Extensorengruppe am Unterschenkel und Fuss der Säugetiere. *Morph. Jahrb.*, Bd. IV, p. 592.

——. 1878b. Vergleichende Anatomie der Muskeln Fussohle. *Morph. Jahrb.*, Bd. IV, p. 644.

SALTER, H. H. 1859. Pancreas (Article in R. B. Todd's "The Cyclopaedia of anatomy and physiology"). v. 5 (suppl. vol.), pp. 81–114. London.

SCAMMON, R. E. 1916. On the development of the biliary system in animals lacking a gall-bladder in post-natal life. *Anat. Rec.*, v. 10, pp. 543–558, Rat. p. 553.

SCHAPIRO, B. 1913. Das Verhältnis der Gattung Dipus zu den Myomorphen; Mus Rattus and Meriones auf grund vergleichend-anatomischer Untersuchung der Muskeln der hinteren Extremitäten. *Morph. Jahrb.*, v. 46, p. 209.

SHELLSHEAR, J. L. 1920. The basal arteries of the forebrain and their functional significance. *Brit. J. Anat.*, v. 55, p. 27.

——. 1927. The arteries of the brain of the orangutan. *Brit. J. Anat.*, v. 61, p. 167.

SISSON, S. 1911. Veterinary anatomy. W. B. Saunders Co., Philadelphia.

SMITH, CHRISTIANA. 1924. The origin and development of the carotid body. *Am. J. Anat.*, v. 34, pp. 87–133.

STECHE, OTTO. 1919. Grundriss der Zoologie. Eine Einführung in die Lehre vom Bau und von den Lebenserscheinungen der Tiere für Studierende der Naturwissenschaften und der Medizin. Verlag. von Veit & Comp. Leipzig. 1919.

STENDELL, W. 1913. Zur vergleichenden Anatomie und Histologie der Hypophysis cerebri. *Arch. f. mikr. Anat.*, v. 82, pp. 289–332.

STIRLING, WM. 1883. A simple method of demonstrating the nerves of the epiglottis. *J. Anat. & Phys.*, v. 17, p. 203. Rats included in mammals observed.

TANDLER, J. 1899. Zur vergleichenden Anatomie der Kopfarterien bei den Mammalia (1898). Denkschr. der kais. *Akad. der Wissensch. in Wien.* Bd. 67, S. 729; Mus rattus—albino among those examined.

——. 1902. Zur Entwickelungsgeschichte der Kopfarterien bei den Mammalia. *Morph. Jahrb.*, Bd. 30, S. 275–373.

VORIS, HAROLD C. 1928. The arterial supply of the brain and spinal cord of the Virginian opossum (Didelphis virginiana). *J. Comp. Neur.*, v. 44, pp. 403–424.

WALKER, GEORGE. 1899. Beitrag zur Kenntniss der

Anatomie und Physiologie der Prostata, etc. Arch. für Anatomie und Entwickl. Anat. Abthl., pp. 313–352.

———. 1910. The nature of the secretion of the vesiculae seminales and of an adjacent glandular structure in the rat and guinea pig, with special reference to the occurrence of histone in the former. Johns Hopkins Hosp. Bull., vol. 21, pp. 185–192.

———. 1910a. A special function discovered in a glandular structure hitherto supposed to form a part of the prostate gland in rats and guinea pigs. Johns Hopkins Hosp. Bull., vol. 21, pp. 182–185.

WARREN, JOHN, 1915. On the early development of the inguinal region in Mammalia. Anat. Rec., v. 9, pp. 131–133.

WIEDERSHEIM, ROBERT. 1897. Comparative anatomy of the vertebrates. Parker's translation, 2nd edition, London.

WILDER, H. H. 1909. History of the human body. Henry Holt and Co., New York.

———. 1912. The appendicular muscles of Necturus maculosus. Zool. Jahrb., Suppl. 15. Jena.

WINDLE, B. C. A., and PARSONS, F. G. 1899. On the myology of the Edentata. Proc. Zool. Soc. London.

WOOD, J. 1867. On human muscular variations and their relation to comparative anatomy. J. Anat. & Phys., v. I, p. 44.

———. 1870. On a group of varieties of the muscles of the human neck, shoulder and chest, with their transitional forms and homologies in the Mammalia. Phil. Trans., clx, pt. 1, pp. 83–116.

YUDKIN, ARTHUR M. 1922. Ocular manifestations of the rat which result from deficiency of vitamin A in the diet. J. Am. Med. Assoc., v. 79, pp. 2206–2208.

YUDKIN, A. M., and LAMBERT, R. A. 1922a. Location of the earliest changes in experimental xerophthalmia of rats. Proc. Soc. Exper. Biol. and Med., v. 19, p. 375.

———. 1922b. Lesions in the lacrimal glands of rats in experimental xerophthalmia. Proc. Soc. Exper. Biol. and Med., v. 19, pp. 376–377.

———. 1923. Pathogenesis of the ocular lesions produced by a deficiency of vitamine A. J. Exper. Med., v. 38, pp. 17–24.

ZUCKERKANDL, E. 1894. Zur anatomie und Entwicklungsgeschichte der Arterien des Vorderarmes. I. Teil. Anat. Hefte., Heft. XI, Bd. IV, 1894, S. 1–95.

———. 1895. Zur Anatomie und Entwicklungsgeschichte der Arterien des Vorderarmes. II. Teil. Anat. Hefte., Heft. XV, Bd. V, 1895.

———. 1903. Die Entwickelung der Schilddrüse und der Thymus bei der Ratte. Anat. Hefte., Bd. 21, S. 3–28.

# INDEX

The numbers for the more important figures are printed in bold face type.

343

www.ingramcontent.com/pod-product-compliance
Lightning Source LLC
Chambersburg PA
CBHW080933240326
41458CB00144B/5972